Chemical Demonstrations

Volume 4

Volume 4 Collaborators and Contributors

JERRY A. BELL, PH.D.
Professor of Chemistry, Simmons College; Director, Institute for Chemical Education, University of Wisconsin–Madison, 1986–1989

GLEN E. DIRREEN, PH.D.
Associate Director, Institute for Chemical Education, University of Wisconsin–Madison

RONALD I. PERKINS, M.S.T.
Senior Teacher, Greenwich High School, Greenwich, Connecticut; Assistant Director, Institute for Chemical Education, University of Wisconsin–Madison

LARRY E. JUDGE, M.S.
Science Teacher, Webb High School, Reedsburg, Wisconsin; Project Assistant, University of Wisconsin–Madison, 1984–1985

DORIS KOLB, PH.D.
Professor of Chemistry, Bradley University, Peoria, Illinois; 1991 Chair, Division of Chemical Education, American Chemical Society

RODNEY SCHREINER, PH.D.
Associate Scientist, University of Wisconsin–Madison

EARLE S. SCOTT, PH.D.
Emeritus Professor of Chemistry, Ripon College; Visiting Professor, University of Wisconsin–Madison, June–December 1980, Summer 1981, Summer 1982, Summer 1984, Summer 1987, Summer 1988

MARY ELLEN TESTEN, M.S.
Environmental Technical Services Supervisor, Department of Public Health, City of Madison; Project Assistant, University of Wisconsin–Madison, 1980–1982

Chemical Demonstrations

A Handbook for Teachers of Chemistry

Bassam Z. Shakhashiri

VOLUME **4**

THE UNIVERSITY OF WISCONSIN PRESS

The University of Wisconsin Press
114 North Murray Street
Madison, Wisconsin 53715

3 Henrietta Street
London WC2E 8LU, England

5 4 3 2 1

Printed in the United States of America

For LC CIP information see the colophon

ISBN 0-299-12860-1

To
Don Herbert,
Television's Mr. Wizard,
who has perfected the art
of communicating science
to kids of all ages.

B. Z. S.

The demonstrations and other descriptions of use of chemicals and equipment contained in this book have been compiled from sources believed to be reliable and to represent the best opinion on the subject as of 1991. However, no warranty, guarantee, or representation is made by the authors or the University of Wisconsin Press as to the correctness or sufficiency of any information herein. Neither the authors nor the publisher assumes any responsibility or liability for the use of the information herein, nor can it be assumed that all necessary warnings and precautionary measures are contained in this publication. Other or additional information or measures may be required or desirable because of particular or exceptional conditions or circumstances, or because of new or changed legislation. Teachers and demonstrators must develop and follow procedures for the safe handling, use, and disposal of chemicals in accordance with local regulations and requirements.

Contents

Preface

This series of volumes on chemical demonstrations is part of a major effort to enhance the quality of science education. The effort aims both at increasing the flow of talent into careers in science and science teaching and at achieving literacy in science among those who elect to pursue other endeavors. Those of us in the science-rich sector of society must become inventive and creative in communicating science to the rest of society. Our attitudes and values must be conveyed effectively to *all* students and not only to those who are college-bound or those who plan to go to graduate school. We must take special steps to nurture the talent of the best and brightest, *and* we must take other steps to develop the talent of the rest of the student population. Those of us in research and education believe we can discover and transmit knowledge, help formulate opinions, develop skills, and influence attitudes. We are mindful of the real purpose of education, namely, to enable individuals to fulfill their human potential. It is through education that we can nurture and develop talent. Chemical demonstrations are a significant vehicle to help us convey our convictions about chemistry and our attitudes about chemicals, their importance in our daily lives, their benefits, their potential hazards, and their safe handling.

The present volume is the fourth in this series aimed at providing teachers of science at all educational levels with detailed instructions and background information for using chemical demonstrations in the classroom and in public lectures. Volumes 1, 2, and 3 included demonstrations in the areas of thermochemistry, chemiluminescence, polymers, metal ion precipitates and complexes, physical behavior of gases, chemical behavior of gases, oscillating chemical reactions, acids and bases, and liquids, solutions and colloids. The demonstrations in this volume deal with clock reactions and with electrochemistry. Additional volumes, now in preparation, will include demonstrations on organic chemistry, spectroscopy and color, photochemistry, solid state and materials chemistry, cryogenics, and other topics.

The introduction to each chapter in this volume includes material aimed at reinforcing and expanding the knowledge base of the user. We believe firmly that whenever demonstrations are presented, the phenomena should be discussed and explained at a level suitable to the audience. A number of demonstrations included in this volume involve quite complex chemical concepts. The intricate details of how clock reactions work are not very well understood, and partly for that reason the phenomena are fascinating. On the other hand, electrochemical behavior is well understood; even so, many of the phenomena described here are still fascinating. Teachers in elementary and secondary schools, as well as teachers in colleges and universities, are urged to make use of the demonstrations in this volume, for many of them are designed to serve as an introduction to intriguing chemical behavior.

A new feature in this volume is an index to Volumes 1 through 4. Topical entries and key words should help in quickly locating information about particular demonstrations and in finding related demonstrations in the various volumes. We are pleased to add this feature in response to many requests, especially from secondary and elementary teachers. This index is based on the combined suggestions of several individuals and our own judgment. We invite additional comments and suggestions as we prepare the index for Volumes 1 through 5.

The reception accorded to Volumes 1, 2, and 3 continues to be gratifying. Teachers of chemistry and other sciences have commented most favorably about this series and its usefulness. We are pleased with the unequivocal statements about the quality of our work, and have undertaken to continue our commitment of time and effort to maintain the same standards of excellence. We reaffirm the importance of reliable and effective source materials and the exercise of good judgment in the use of demonstrations in communicating chemistry.

I am thankful for the blessing of having expert coauthors. Rod Schreiner, as principal coauthor of these chapters, deserves major credit for making this volume possible. His collaboration on this series, and on many other projects, has truly enhanced the quality of our work and has enriched the scope of my work immensely. Earle Scott, an excellent teacher, contributed greatly to both chapters in this book. Jerry Bell and Ron Perkins, both master teachers and outstanding communicators of science, made significant contributions toward the completion of this volume. Mary Ellen Testen and Larry Judge received their master's degrees in chemistry, in part for their contributions to the chapters in this volume.

Rod Schreiner and I are grateful for the suggestions and comments made by numerous colleagues around the country. In particular, we thank Roland Stout of Drake University, L. K. Brice of Virginia Polytechnic University, and John J. Fortman of Wright State University, for their contributions to Chapter 10; Joe Conrad of the University of Wisconsin–River Falls, Ken Watkins of Colorado State University, Bob Olson, formerly an undergraduate student in our laboratory, Ronald R. Esman of Abilene (Texas) High School, Jerry Bell of Simmons College, and Floyd Sturtevant and Kenneth Hartman of Ames (Iowa) High School for their contributions to Chapter 11; Doris Kolb of Bradley University for her essay, "The Joy of Teaching Chemistry"; Gil Haight, emeritus professor at the University of Illinois, Tim Watt of Montgomery College (Rockville, Maryland), Richard Noyes of the University of Oregon, Worth Vaughan of the University of Wisconsin–Madison, Richard W. Ramette of Carleton College, and Lois Nicholson of West Springfield High School (Springfield, Virginia) and the National Science Foundation, for reviewing Chapter 10; Lois Nicholson, Ethel Schultz of Marblehead High School (Massachusetts) and the National Science Foundation, Lee Marek of Naperville North High School (Illinois), Doris Kolb, and Fred Juergens of the University of Wisconsin–Madison for reviewing Chapter 11. Their advice and contributions have been very helpful. The preparation of the index benefitted from discussions with Ron Perkins, Fred Juergens, and others interested in topical organization of demonstrations.

Others at the University of Wisconsin–Madison provided invaluable assistance: Glen Dirreen, David Shaw, and Patti Puccio in proofreading and in other matters which helped to complete this volume, copy editor Robin Whitaker for her suggestions to improve the style of the text, and Gardner Wills of the University of Wisconsin Press in expediting production of the volume.

I wish to express special gratitude to Elizabeth A. Steinberg, Assistant Director of the University of Wisconsin Press, for all her sustained efforts, professionalism, and kindness.

Madison, Wisconsin
October 1991

Bassam Z. Shakhashiri
Professor of Chemistry
University of Wisconsin–Madison

The Joy of Teaching Chemistry

Doris Kolb

SCIENCE IS FUN! That is the message that Bassam Shakhashiri personifies as he presents one of his programs on chemical demonstrations. Not only is that the slogan on his familiar T-shirt, but it is also the feeling conveyed by the smile on his face, the enthusiasm in his voice, and the laughter in his eyes. It is obvious that everyone in the audience is having fun, too.

Reading about science is interesting, but seeing it in action is fun. And *CHEMISTRY* is the science that is the most fun of all. It has color, light, bubbles, fire, explosions, and many other kinds of excitement [*1*]. Chemistry offers a varied menu of interesting topics, most of them relevant to everyday life. It deals with all the matter in the universe, from giant stars to tiny particles inside the atom. It covers foods, fabrics, metals, plastics, plants, animals, and just about everything else. It includes all the chemical elements and their millions of compounds, along with their many kinds of interactions, and the instruments and equipment used to study them. Everything we do involves chemistry, from taking a shower or driving a car to just breathing or thinking. Life itself is chemistry in action. Surely there is no subject in the world more fascinating!

CHEMICAL DEMONSTRATIONS

One of the joys of teaching chemistry is the opportunity to do chemical demonstrations. When I get ready to do a demonstration, the class becomes visibly more attentive. No matter how simple the demonstration might be, students enjoy watching things happen. When chemical principles are illustrated with demonstrations, not only are they more interesting to students, but they are also more fun for the instructor.

I suppose it is possible for someone to be a great chemistry teacher without ever doing a demonstration. But why would anyone teaching chemistry want to do that? We have such a large repertoire of chemical demonstrations that we could make an appropriate presentation in every single class if we wanted to [*2–4*]. But if we chose to do only two or three demonstrations during the semester, I suspect that those would be the memorable moments of the course for many of our students.

There are some chemistry teachers who refrain from doing classroom demonstrations because they lack the time for elaborate equipment set-up. It is true that some chemical demonstrations are fairly complicated and time-consuming to prepare. Since most chemistry teachers have to set up their own demonstrations, preparation time is an important factor to consider when deciding whether or not to use a particular demonstration.

Fortunately, there are many demonstrations that are quite simple to prepare and carry out. It takes very little time and effort to demonstrate the formation of a precipitate, or to add acid to a carbonate to produce bubbles, or to show the color change of an acid-base indicator. Such demonstrations may sound too simple to be worthwhile, but students are interested in watching them. And when the demonstrations are done at appropriate times during the course, they can add a visual dimension to theoretical discussions, making the topics seem more real. In discussing the effect of surface area on reaction rate, for example, lighting a wooden splint from a candle flame and watching it burn, then tossing a little lycopodium powder into the flame to create a sudden fireball, makes the point much more vividly than can be done with words alone.

There are hundreds of simple chemical demonstrations that take only minutes to prepare and even less time to carry out in the classroom. Still, they can add a touch of excitement to a chemistry lecture. On the other hand, there is no need to omit the more spectacular but difficult demonstrations. Most of them can now be obtained on film or on video tape or disc, and they can be shown wherever and whenever one wishes to use them.

Overhead Projector Demonstrations

For the many chemistry teachers who are short on time and resources, there is an especially simple way to do chemical demonstrations. When carried out on the stage of an overhead projector, many demonstrations can be done very quickly and easily. They are usually inexpensive because such small amounts of chemicals are needed, and preparation time and clean up are both minimal. But the best thing about overhead projector demonstrations is that students can really see them, even the students way back in the last row. The bright light and large magnification afforded by the overhead projector allow a kind of visibility that cannot be matched on the demonstration desktop, even when a thousand times as much material is used [*5*].

Acid-base color changes can be shown on the overhead projector using only drops of indicator solution and a few milliliters of acid or base. The shifting of equilibrium can be illustrated with a solution of potassium chromate by adding a few drops of acid and then of base, the color shifting from yellow to orange and then back again. The formation of a coordination compound can be shown by adding ammonia to a solution of copper sulfate, with the initial production of an opaque precipitate of copper hydroxide and the eventual formation of the deep royal blue tetraammine complex. Chelation can be demonstrated by adding a little dimethylglyoxime reagent to a solution containing nickel ions to produce the cherry-red complex. An entire chemical demonstration can often be carried to class in one hand, or slipped into a pocket [*6*].

Demonstrations requiring the use of devices such as pH meters or voltmeters can be done using an instrument with an oversize dial or lighted digital readout placed on the benchtop, but there are also transparent versions of these meters that can be set directly on the projector screen. Not all kinds of demonstrations can be done on the overhead projector, of course. Opaque substances all show up as black spots on the screen, regardless of their actual colors, and demonstrations involving fire or explosions are obviously unsuitable for this medium.

I still recall the very first time I ever saw a chemical demonstration on the overhead projector. It was more than 25 years ago, and the demonstrator was Clark Bricker from the University of Kansas. He added a few drops of ammonia to a solution containing ferric ions. The beauty, simplicity, and clear visibility of that demonstration impressed me so much that I have been doing demonstrations on the overhead projector ever since.

The Tilted Stage

One problem with demonstrations on the overhead projector is that reactions are all viewed from the top looking down, rather than from the side, as we might normally look at a flask or a test tube. The difference is noticed especially in the case of reactions involving bubbles. With the overhead projector, bubbles often appear on the screen simply as vigorous activity. One misses the fact that bubbles rise in a liquid.

Hubert Alyea, Princeton's famous professor of chemistry, has been one of the strongest proponents of using projectors to show chemical demonstrations. But he long ago decided that he wanted students to see the bubbles rise when a gas was generated in a chemical reaction. He therefore favored vertical projection of demonstrations, and for that purpose he designed a special TOPS projector, as well as an adaptor for use with an ordinary overhead projector [*7*]. Using the TOPS equipment was a bit cumbersome, but it did provide a more realistic view of many chemical reactions, especially those involving gas generation.

Then, in 1988 he made a startling discovery. If a thin flat cell is placed on the stage of an overhead projector so that it is inclined at an angle of about 20 degrees, the container appears on the screen to be standing in a vertical position [*8*]. Objects such as bottles and test tubes can be viewed as if in a normal standing position, and bubbles can be seen to rise during chemical reactions. A simple tilted support placed on the stage of an ordinary overhead projector, along with some thin, flat reaction "cells," creates the illusion that the chemical reactions are being carried out in standing vertical containers. The technique is so simple and it works so well that one cannot help wondering why it had never been tried before!

HISTORY OF CHEMISTRY

Another reason why chemistry is such fun to teach is that it has such a rich and exciting history. Although many chemistry teachers feel they must ignore the background material in order to have enough time to discuss all the principles and theories, I think that the history is important. More than that, I think it is interesting, and I find that students do, too. The subject of atomic structure is much more impressive when you know something about how scientists were able to figure out what these things called atoms are like.

There are many fascinating stories about important chemical discoveries: the shiny little globules of potassium metal that appeared when Sir Humphry Davy sent an electric current through molten potash, causing him to dance about the room in a state of ecstatic delight; the "aniline purple" dye that Sir William Perkin made accidentally when he was trying to make quinine; the whole new family of invisible elements (the noble gases) that resulted from the work of Lord Rayleigh and Sir William Ramsay; the tons of pitchblende from which Marie Curie and her husband painstakingly extracted one-tenth of a gram of the radioactive new element, radium; and the list goes on and on.

Aaron Ihde of the University of Wisconsin, whose specialty is the history of chemistry, feels that the main reason for including history in a chemistry course is to give students an appreciation for the fact that chemistry is a human enterprise [*9*]. Those famous chemists of past centuries were real people, with human faults and frailties. They made some important discoveries, but they could also make mistakes. For example, Wilhelm Ostwald, winner of the Nobel Prize in 1909, was possibly the last chemist of his day to accept the atomic theory. In the early years of the twentieth century he was still referring to "those mythical particles called atoms." Relating anecdotes or describing personalities of famous chemists not only catches the interest of students but can help them realize that chemistry is a dynamic field: a theory that is popular today could be replaced a few years from now.

In learning about famous chemists, one realizes what a versatile group they are. The fact that many chemists started out as physicians or pharmacists is not surprising because those fields are related to chemistry. But Joseph Priestley was a Unitarian minister, and John Dalton was a Quaker schoolteacher. Georges Urbain, who did research on the rare earths and discovered the element lutetium, was a sculptor, a painter, and a musician [*10*]. Many chemists (like other scientists) started out with musical careers in mind. The most famous example of a chemist-musician was Alexander Borodin. Although best known as the composer of the opera *Prince Igor* and other famous musical scores, Borodin was both a physician and a professor of organic chemistry. Sir William Ramsay, who isolated the noble gases, was also interested in music from an early age; throughout his life he loved to sing and often accompanied himself on the piano.

A number of chemists have become involved in the law and politics. One of the best examples is Chaim Weizmann, who discovered a method for making acetone and butanol by fermentation; he became the first president of Israel. More recently there was an industrial chemist named Margaret Thatcher who became the first female prime minister of Great Britain.

HUMOR

Chemistry is a serious subject, but now and then it can have its lighter moments. A cartoon laid on the overhead projector a few minutes before the hour can start a class off with a smile. One of my favorite cartoons is the one by Sidney Harris about the elements. It shows two robed men from ancient times, one of them pointing at a stone wall on which are inscribed the words: AIR, WATER, FIRE, and EARTH. The caption is simply: "The Periodic Table" [*11*].

Once in a while, a joke comes along that is germane to chemistry. For example, there is the one about the chemist who went into a drugstore and asked for a bottle of the acetyl ester of salicylic acid. The druggist scratched his head and asked: "Do you mean aspirin?" "That's it!" exclaimed the chemist. "I never can think of that name!"

Most people can tell funny stories based on their own experiences. I recall a Halloween many years ago when one of my sons was in the second grade. I had agreed to provide entertainment for his class party at school that day, so I decided to dress up as a witch and do some "magic" chemical demonstrations. I breezed into the room on a broomstick, wearing a long black dress and cape, and a tall black pointed hat. I also wore a mask so the children would not recognize me. I began pulling bottles out of a big old briefcase I had brought along, saying, "Let's make yellow!" "Let's make blue!" or "Let's make red!" Each time the mixture produced the specific color. Then I used the "Old Nassau" mixture (see Demonstration 10.3). When nothing happened, the children yelled: "You forgot to tell it what color to turn!" So I asked them if they knew what the Halloween colors were. As they started shouting "orange and black," the liquid obeyed by turning orange and then black. By now they were really under my spell! That evening I was taking my four-year-old son "trick-or-treating," and we happened to walk past a man with his daughter. After they had passed us, I heard the little girl say: "Daddy! Do you know who that was? That was Ronnie Kolb's mother, and she's a real witch!"

A FINAL WORD

Of course, it might be objected, students enjoy watching chemical demonstrations and listening to stories, but in a chemistry course, students are supposed to learn *chemistry.* Certainly that is true, but if they happen to find the course interesting and fun, does that mean they will learn less? On the subject of teaching introductory chemistry George Kistiakowsky of Harvard University once said: "It is far more important to be interesting than to be thorough or erudite, for if we have the interest of a beginning student, we can easily lead him to read more on his own or to take further courses that will be rigorous and complete" [*12*]. I heartily agree. Furthermore, I think a class that is more interesting for the students is also more fun for the teacher.

Many chemistry teachers seem to have a genuine love for what they do. Consider Jean Baptiste Dumas, who was born back in 1800. Although he was foremost a chemist, he also held many high government positions, including Senator, Minister of Agriculture, and Master of the French Mint. In his later years Dumas said: "I have seen many phases of life; I have moved in imperial circles, and I have been a minister of state; but if I had to live my life again, I would always remain in my laboratory, for the greatest joy of my life has been to accomplish original scientific work, and, next to that, to lecture to a set of intelligent students" [*13*].

Having met hundreds of chemistry teachers from all over the United States and from many other parts of the world, I am impressed by the genial enthusiasm that most of them share. Chemistry is hardly a popular subject among students; in fact, they often rate it as "most difficult." After all, much of the time in a chemistry course is spent trying to solve a wide assortment of challenging problems. Teaching chemistry is not easy, but it can be fun!

I think that Harry Gray of the California Institute of Technology, in accepting the 1991 Priestley Award, spoke for many chemical educators when he said: "It is truly remarkable that I have been able to make a living doing something that is so much fun" [*14*].

REFERENCES

1. R. W. Ramette, *J. Chem. Educ.* 57:68 (1980).
2. H. N. Alyea and F. B. Dutton, Eds., *Tested Demonstrations in Chemistry,* 6th ed., Journal of Chemical Education: Easton, Pennsylvania (1965).
3. B. Z. Shakhashiri, *Chemical Demonstrations: A Handbook for Teachers of Chemistry,* Vols. 1–3, University of Wisconsin Press: Madison (1983, 1985, 1989).
4. L. R. Summerlin and J. L. Ealy, *Chemical Demonstrations: A Sourcebook for Teachers,* American Chemical Society: Washington, D.C. (1988).
5. E. J. Hartung, *The Screen Projection of Chemical Experiments,* Cambridge University Press: London (1953).
6. D. Kolb, *J. Chem. Educ.* 64:348 (1987).
7. H. N. Alyea, *TOPS in General Chemistry,* 3d ed., Journal of Chemical Education: Easton, Pennsylvania (1967).
8. H. N. Alyea, *J. Chem. Educ.* 66:765 (1989).

9. A. J. Ihde, *J. Chem. Educ.* 57:11 (1980).
10. I. Asimov, *Biographical Encyclopedia of Science and Technology,* Doubleday: Garden City, New York (1964).
11. S. Harris, *What's So Funny about Science?* Cartoons from *American Scientist,* W. Kaufmann: Los Altos, California (1977).
12. G. Kistiakowsky, quoted in E. G. Rochow, *Modern Descriptive Chemistry,* p. iv, Saunders College Publishing Co.: Philadelphia, Pennsylvania (1977).
13. R. E. Oesper, *The Human Side of Scientists,* University Publications, University of Cincinnati: Cincinnati, Ohio (1975).
14. H. Gray, *Chem. Eng. News,* 69 No. 15 (Apr. 15): 40 (1987).

Introduction†

Bassam Z. Shakhashiri

Lecture demonstrations help to focus students' attention on chemical behavior and chemical properties, and to increase students' knowledge and awareness of chemistry. To approach them simply as a chance to show off dramatic chemical changes or to impress students with the "magic" of chemistry is to fail to appreciate the opportunity they provide to teach scientific concepts and descriptive properties of chemical systems. The lecture demonstration should be a process, not a single event.

In lecture demonstrations, the teacher's knowledge of the behavior and properties of the chemical system is the key to successful instruction, and the way in which the teacher manipulates chemical systems serves as a model not only of technique but also of attitude. The instructional purposes of the lecture dictate whether a phenomenon is demonstrated or whether a concept is developed and built by a series of experiments. Lecture experiments, which some teachers prefer to lecture demonstrations, generally involve more student participation and greater reliance on questions and suggestions, such as "What will happen if you add more of . . .?" Even in a lecture demonstration, however, where the teacher is in full control of directing the flow of events, the teacher can ask the same sort of "what if" questions and can proceed with further manipulation of the chemical system. In principle and in practice, every lecture demonstration is a situation in which teachers can convey their attitudes about the experimental basis of chemistry, and can thus motivate their students to conduct further experimentation, and lead them to understand the interplay between theory and experiment.

Lecture demonstrations should not, of course, be considered a substitute for laboratory experiments. In the laboratory, students can work with the chemicals and equipment at their own pace and make their own discoveries. In the lecture hall, students witness chemical changes and chemical systems as manipulated by the teacher. The teacher controls the pace and explains the purposes of each step. Both kinds of instruction are integral parts of the education we offer students.

In teaching and in learning chemistry, teachers and students engage in a complex series of intellectual activities. These activities can be arranged in a hierarchy which indicates their increasing complexity [*1*]:

(1) observing phenomena and learning facts
(2) understanding models and theories
(3) developing reasoning skills
(4) examining chemical epistemology

This hierarchy provides a framework for the purposes of including lecture demonstrations in teaching chemistry.

At the first level, we observe chemical phenomena and learn chemical facts. For example, we can observe that, at room temperature, sodium chloride is a white crys-

†Reprinted with minor modifications from Volume 1.

talline solid and that it dissolves in water to form a solution with characteristic properties of its own. One such property, electrical conductivity, can be readily observed when two wire electrodes connected to a light bulb and a source of current are dipped into the solution. There are additional phenomena and facts that can be introduced: the white solid has a very high melting point; the substance is insoluble in ether; its chemical formula is NaCl; etc.

At the second level, we explain observations and facts in terms of models and theories. For example, we teach that NaCl is an ionic solid compound and that its aqueous solution contains hydrated ions: sodium cations, $Na^{+}(aq)$, and chloride anions, $Cl^{-}(aq)$. The solid, which consists of Na^{+} and Cl^{-} particles, is said to have ionic bonds, that is, there are electrostatic forces between the oppositely charged particles. The ions are arranged throughout the solid in a regular three-dimensional array called a face-centered cube. Here, the teacher can introduce a discussion of the ionic bond model, bond energy, and bond distances. Similarly, a discussion of water as a molecular covalent substance can be presented. The ionic and covalent bonding models can be compared and used to explain the observed properties of a variety of compounds.

At the third level, we develop skills which involve both mathematical tools and logic. For example, we use equilibrium calculations in devising the steps of an inorganic qualitative analysis scheme. We combine solubility product, weak acid dissociation, and complex ion formation constants for competing equilibria which are exploited in analyzing a mixture of ions. The logical sequence of steps is based on understanding the equilibrium aspects of solubility phenomena.

At the fourth level, we are concerned with chemical epistemology. We examine the basis of our chemical knowledge by asking questions such as, "How do we know that the cation of sodium is monovalent rather than divalent?" and "How do we know that the crystal structure of sodium chloride can be determined from x-ray data?" At this level we deal with the limits and validity of our fundamental chemical knowledge.

Across all four levels, the attitudes and motivations of both teacher and student are crucial. The attitude of the teacher is central to the success of interactions with students. Our motivation to teach is reflected in what we do and, as well, in what we do not do, both in and out of the classroom. Our modes of communicating with students affect their motivation to learn. All aspects of our behavior influence students' confidence and their trust in what we say. Our own attitudes toward chemicals and toward chemistry itself are reflected in such matters as how we handle chemicals, adhere to safety regulations, approach chemical problems, and explain and illustrate chemical principles. In my opinion, the single most important purpose that lectures serve is to give teachers the opportunity to convey an attitude toward chemistry—to communicate to students an appreciation of chemistry's diversity and usefulness, its cohesiveness and value as a central science, its intellectual excitement and challenge.

PRESENTING EFFECTIVE DEMONSTRATIONS

In planning a lecture demonstration, I always begin by analyzing the reasons for presenting it. Whether a demonstration is spectacular or quite ordinary I undertake to use the chemical system to achieve specific teaching goals. I determine what I am going to say about the demonstration and at what stage I should say it. Prior to the lecture, I practice doing the demonstration. By doing the demonstration in advance, I often see aspects of the chemical change which help me formulate both statements and questions that I then use in class.

Because one of the purposes of demonstrations is to increase the students' ability to make observations, I try to avoid saying, "Now I will demonstrate the insolubility of barium sulfate by mixing equal volumes of 0.1M barium chloride and 0.1M sodium sulfate solutions." Instead, I say, "Let us mix equal volumes of 0.1M barium chloride and 0.1M sodium sulfate solutions and observe what happens." Rather than announcing what should happen, I emphasize the importance of observing all changes. Often, I ask two or three students to state their observations to the entire class before I proceed with further manipulations. In addition, I help students to sort out observations so that relevant ones can be used in formulating conclusions about the chemical system. Some valid observations may not be relevant to the main purpose of the demonstration. For example, when the above-mentioned solutions are mixed, students may observe that the volumes are additive. However, this observation is not germane to the main purpose of the demonstration, which is to show the insolubility of barium sulfate. However, this observation is relevant if the purposes include teaching about the additive properties of liquids.

Every demonstration that I present in lectures is aimed at enhancing the understanding of chemical behavior. In all cases, the chemistry speaks for itself more eloquently than anything I can describe in words, write on a chalk board, or show on a slide.

Wesley Smith of Ricks College, who was a visiting faculty member at the University of Wisconsin–Madison from 1974 to 1977, has outlined six characteristics of effective demonstrations which best promote student understanding [*2*]:

1. *Demonstrations must be timely and appropriate.* Demonstrations should be done to meet a specific educational objective. For best results, plan demonstrations that are immediately germane to the material in the lesson. Demonstrations for their own sake have limited effectiveness.

2. *Demonstrations must be well prepared and rehearsed.* To ensure success, you need to be thoroughly prepared. *All* necessary material and equipment should be collected well in advance so that they are ready at class time. You should rehearse the entire demonstration from start to finish. Do not just go through the motions or make a dry run. Actually mix the solutions, throw the switches, turn on the heat, and see if the demonstration really works. Only then will you know that all the equipment is present and that all the solutions have been made up correctly. Always practice your presentation.†

3. *Demonstrations must be visible and large-scale.* A demonstration can help only those students who experience it. Hence, you need to set up the effect for the whole class to see. If necessary, rig a platform above desktop level to ensure visibility.

Perhaps the most important factor to consider is the size of what you are presenting. Only in the very tiniest of classes can the students see phenomena on the milligram and milliliter scale. Many situations require the use of oversized glassware and specialized equipment. Solutions and liquids should be shown in full-liter volumes, and solids should be displayed in molar or multi-molar amounts.

Contrasting backgrounds help emphasize chemical changes. A collection of large white and black cards to place behind beakers and other equipment is a valuable addition to your demonstration equipment. These are inexpensive, easy to use, and can provide an extra bit of polish to your demonstration.

4. *Demonstrations must be simple and uncluttered.* A common source of distraction is clutter on the lecture bench. Make sure that the demonstration area is neat and

† As Fred Juergens likes to say, "Prior practice prevents poor presentation."

free of extraneous glassware, scattered papers, and other disorder. All attention must be focused on the demonstration itself.

5. *Demonstrations must be direct and lively.* Action is an important part of a good demonstration. It is the very ingredient that makes demonstrations such efficient attention-grabbers. Students are eager to see something happen, but if nothing perceptible occurs within a few seconds you may lose their attention. The longer they have to wait for results, the less likely it is that the demonstration will have maximum educational value.

6. *Demonstrations must be dramatic and striking.* Usually, a demonstration can be improved by its mode of presentation. A lecture demonstration, according to Alfred T. Collette, is like a stage play. "A demonstration is 'produced' much as a play is produced. Attention must be given to many of the same factors as stage directors consider: visibility, audibility, single centers of attention, audience participation, contrasts, climaxes" [*3*]. The presentation of effective demonstrations is such an important part of good education that "no instructor is doing his best unless he can use this method of teaching to its fullest potential" [*4*].

USING THIS BOOK

The demonstrations in this volume are grouped in topical chapters dealing with the physical behavior of gases, the chemical behavior of gases, and chemical oscillating reactions. Each chapter has an introduction which covers the chemical background for the demonstrations that follow. We confine the discussion of relevant terminology and concepts to the introduction rather than repeating it in the discussion section of each demonstration. Accordingly, when teachers read the discussion section of any particular demonstration, they may find it necessary to refer to the chapter introduction for background information. For additional information teachers may wish to consult the sources listed at the end of each chapter's introduction.

Each demonstration has seven sections: a brief summary, a materials list, a step-by-step account of the procedure to be used, an explanation of the hazards involved, information on how to store or dispose of the chemicals used, a discussion of the phenomena displayed and principles illustrated by the demonstration, and a list of references. The brief summary provides a succinct description of the demonstration. The materials list for each procedure specifies the equipment and chemicals needed. Where solutions are to be used, we give directions for preparing stock amounts larger than those required for the procedure. The teacher should decide how much of each solution to prepare for practicing the demonstration and for doing the actual presentation. The availability and cost of chemicals may also affect decisions about the volumes to be prepared.

The procedure section often contains more than one method for presenting a demonstration. In all cases, the first procedure is the one the authors prefer to use. However, the alternative procedures are also effective and valid pedagogically.

The hazards and disposal sections include information compiled from sources believed to be reliable. We have enumerated many potentially adverse health effects and have called attention to the fact that many of the chemicals should be used only in well-ventilated areas. In all instances teachers should inquire about and follow local disposal practices and should act responsibly in handling potentially hazardous material. We recognize that several chemicals such as silver and mercury can be recovered and reused and have given references to recovery and purification procedures.

The purpose of the discussion section is to provide the teacher with information for explaining each demonstration. We include discussion of chemical equations, relevant data, properties of the materials involved, as well as a theoretical framework through which the chemical processes can be understood. Again, we remind teachers that they should refer to the introduction of each chapter for background information not included in the discussion section of each demonstration. Finally, each demonstration contains a list of references used in developing procedures and providing information for the demonstration.

A WORD ABOUT SAFETY

Jearl Walker, professor of physics at Cleveland State University and editor of the Amateur Scientist section in *Scientific American*, has been quoted in newspaper stories as saying, "The way to capture a student's attention is with a demonstration where there is a possibility the teacher may die." Walker is said to get the attention of his students by dipping his hand in molten lead or liquid nitrogen, or by gulping a mouthful of liquid nitrogen, or by lying between two beds of nails and having an assistant with a sledge hammer break a cinder block on top of him. Walker reportedly has been injured twice, once when he used a small brick instead of a cinder block in the bed-of-nails demonstration, and once when he walked on hot coals and was severely burned.

We disagree strongly with this kind of approach. Demonstrations that result in injury are likely to confirm beliefs that chemicals are dangerous and that their effects are bad. In fact, every chemical is potentially harmful if not handled properly. That is why every person who does lecture demonstrations should be thoroughly knowledgeable about the safe handling of all chemicals used in a demonstration and should be prepared to handle any emergency. A first-aid kit, a fire extinguisher, a safety shower, and a telephone must be accessible in the immediate vicinity of the demonstration area. Demonstrations involving volatile material, fumes, noxious gases, or smoke should be rehearsed and presented only in well-ventilated areas.

We recognize that any of the demonstrations in this book can be hazardous. Our procedures are written for experienced chemists who fully understand the properties of the chemicals and the nature of their behavior. We take no responsibility or liability for the use of any chemical or procedure specified in this book. We urge care and caution in handling chemicals and equipment.

REFERENCES

1. I have adapted many ideas from Paul Saltman's address at the Third Biennial Conference on Chemical Education which was sponsored by the American Chemical Society, Division of Chemical Education, and held at Pennsylvania State University, State College, Pennsylvania (1974); see *J. Chem. Educ.* 52:25 (1975).
2. *Chemical Demonstrations Proceedings*, Western Illinois University and Quincy-Keokuk Section of the American Chemical Society: Macomb, Illinois (1978).
3. A. T. Collette, *Science Teaching in the Secondary School*, Allyn and Bacon: Boston (1973).
4. R. Miller, F. W. Culpepper, Jr., *Ind. Arts Voc. Educ.* 60:24 (1971).

Sources Containing Descriptions of Lecture Demonstrations

We call attention to the following sources of information about lecture demonstrations. These lists, updated from Volume 3, are not intended to be comprehensive. Some of the books are out of print but may be available in libraries.

BOOKS

Alyea, H. N. *TOPS in General Chemistry,* 3d ed., Journal of Chemical Education: Easton, Pennsylvania (1967).

Alyea, H. N., and F. B. Dutton, Eds. *Tested Demonstrations in Chemistry,* 6th ed., Journal of Chemical Education: Easton, Pennsylvania (1965).

Ammon, D., D. Clarke, F. Farrell, R. Schibeci, and J. Webb. *Interesting Chemistry Demonstrations,* 2d ed., Murdoch University: Murdoch, Western Australia (1982).

Arthur, P. *Lecture Demonstrations in General Chemistry,* McGraw-Hill: New York (1939).

Blecha, M. T. Ph.D. Dissertation, "The Development of Instructional Aids for Teaching Organic Chemistry," Kansas State University, Manhattan, Kansas (1981).

Brown, R. J. *333 Science Tricks and Experiments,* Tab Books: Blue Ridge Summit, Pennsylvania (1984).

Chemical Demonstrations Proceedings, Western Illinois University and Quincy-Keokuk Section of the American Chemical Society, Macomb, Illinois, May 5–6, 1978.

Chemical Demonstrations Proceedings, Western Illinois University and Quincy-Keokuk Section of the American Chemical Society, Macomb, Illinois, May 4–5, 1979.

Chemical Demonstrations Proceedings, Western Illinois University and Quincy-Keokuk Section of the American Chemical Society, Macomb, Illinois, May 1–2, 1981.

Chemical Demonstrations Proceedings, Western Illinois University and Quincy-Keokuk Section of the American Chemical Society, Normal, Illinois, June 8, 1982.

Chen, P. S. *Entertaining and Educational Chemical Demonstrations,* Chemical Elements Publishing Co.: Camarillo, California (1974).

Davison, H. F. *A Collection of Chemical Lecture Experiments,* Chemical Catalog Co.: New York (1926).

Ehrlich, R. *Turning the World Inside Out and 174 Other Simple Physics Demonstrations,* Princeton University Press: Princeton, New Jersey (1990).

Faraday, M. *The Chemical History of a Candle: A Course of Lectures Delivered Before a Juvenile Audience at the Royal Institution,* The Viking Press: New York (1960).

Ford, L. A. *Chemical Magic,* T. S. Denison & Co.: Minneapolis, Minnesota (1959).

Fowles, G. *Lecture Experiments in Chemistry,* 5th ed., Basic Books, Inc.: New York (1959).

Frank, J. O., assisted by G. J. Barlow. *Mystery Experiments and Problems for Science Classes and Science Clubs,* 2d ed., J. O. Frank: Oshkosh, Wisconsin (1936).

Freier, G. D., and F. J. Anderson. *A Demonstration Handbook for Physics,* 2d ed., American Association of Physics Teachers: Stony Brook, New York (1981).

Gardner, M. *Entertaining Science Experiments with Everyday Objects,* Dover Publications, Inc.: New York (1981).

Gardner, R. *Magic Through Science,* Doubleday & Co., Inc.: Garden City, New York (1978).
Hartung, E. J. *The Screen Projection of Chemical Experiments,* Melbourne University Press: Carlton, Victoria (1953).
Herbert, D. *Mr. Wizard's Supermarket Science,* Random House: New York (1980).
Herbert, D., and H. Ruchlis. *Mr. Wizard's 400 Experiments in Science,* Revised Edition, Book-Lab: North Bergen, New Jersey (1983).
Humphreys, D. A. *Demonstrating Chemistry: 160 Experiments to Show Your Students,* D. A. Humphreys: Hamilton, Ontario (1983).
Joseph, A., P. F. Brandwein, E. Morholt, H. Pollack, and J. Castka. *A Sourcebook for the Physical Sciences,* Harcourt, Brace, and World, Inc.: New York (1961).
Lanners, E. *Secrets of 123 Classic Science Tricks and Experiments,* Tab Books: Blue Ridge Summit, Pennsylvania (1987).
Lippy, J. D., Jr., and E. L. Palder. *Modern Chemical Magic,* The Stackpole Co.: Harrisburg, Pennsylvania (1959).
Mebane, R. C., and T. R. Rybolt. *Adventures with Atoms and Molecules: Chemistry Experiments for Young People,* Enslow Publishers: Hillside, New Jersey (1985).
Meiners, H. F., Ed. *Physics Demonstration Experiments,* Vols. 1 and 2, The Ronald Press Company: New York (1970).
Mullin, V. L. *Chemistry Experiments for Children,* Dover Publications, Inc.: New York (1968).
My Favorite Lecture Demonstrations, A Symposium at the Science Teachers Short Course, W. Hutton, Chairman; Iowa State University, Ames, Iowa, March 6–7, 1977.
Newth, G. S. *Chemical Lecture Experiments,* Longmans, Green and Co.: New York (1928).
Sharpe, S., Ed. *The Alchemist's Cookbook: 80 Demonstrations,* Shell Canada Centre for Science Teachers, McMaster University: Hamilton, Ontario, undated.
Sarquis, M., and J. Sarquis. *Fun with Chemistry: A Guidebook of K-12 Activities,* Institute for Chemical Education, University of Wisconsin-Madison: Madison, Wisconsin (1991).
Siggins, B.A. M.S. Thesis, "A Survey of Lecture Demonstrations/Experiments in Organic Chemistry," University of Wisconsin-Madison, Wisconsin (1978).
Summerlin, L. R., et al. *Chemical Demonstrations: A Sourcebook for Teachers,* Vols. 1 and 2, American Chemical Society: Washington, D.C. (1985, 1987).
Sutton, R. M. *Demonstration Experiments in Physics,* McGraw-Hill Book Co.: New York (1938).
Talesnick, I. *Idea Bank Collation: A Handbook for Science Teachers,* S17 Science Supplies and Services Co.: Kingston, Ontario (1984).
Walker, J. *The Flying Circus of Physics—With Answers,* Interscience Publishers, John Wiley and Sons: New York (1977).
Weisbruch, F. T. *Lecture Demonstration Experiments for High School Chemistry,* St. Louis Education Publishers: St. Louis, Missouri (1951).
Wilson, J. W., J. W. Wilson, Jr., and T. F. Gardner. *Chemical Magic,* J. W. Wilson: Los Alamitos, California (1977).

ARTICLES

Bailey, P. S., C. A. Bailey, J. Anderson, P. G. Koski, and C. Rechsteiner. Producing a chemistry magic show. *J. Chem. Educ.* 52:524–25 (1975).
Castka, J. F. Demonstrations for high school chemistry. *J. Chem. Educ.* 52:394–95 (1975).
Chem 13 News No. 81. This November issue contained a collection of chemical demonstrations. (1976).
Gilbert, G. L., Ed. Tested demonstrations. Regular column in *J. Chem. Educ.*
Hanson, R. H. Chemistry is fun, not magic. *J. Chem. Educ.* 53:577–78 (1976).
Hughes, K. C. Some more intriguing demonstrations. *Chem. in Australia* 47:458–59 (1980).
Kolb, D. K., Ed. Overhead projector demonstrations. Regular column in *J. Chem. Educ.*

McNaught, I. J., and C. M. McNaught. Stimulating students with colourful chemistry. *School Sci. Review* 62:655–66 (1981).

Rada Kovitz, R. The SSP syndrome. *J. Chem. Educ.* 52:426 (1975).

Schibeci, R. A., J. Webb, and F. Farrel. Some intriguing demonstrations. *Chem. in Australia* 47:246–47 (1980).

Schwartz, A. T., and G. B. Kauffman. Experiments in alchemy, Part I: Ancient arts. *J. Chem. Educ.* 53:136–38 (1976).

Schwartz, A. T., and G. B. Kauffman. Experiments in alchemy, Part II: Medieval discoveries and "transmutations." *J. Chem. Educ.* 53:235–39 (1976).

Shakhashiri, B. Z., G. E. Dirreen, and W. R. Cary. Lecture Demonstrations, in *Sourcebook for Chemistry Teachers,* pp. 3–16, W. T. Lippincott, Ed., American Chemical Society, Division of Chemical Education: Washington, D. C. (1981).

Steiner, R., Ed. Chemistry for kids. Regular column in *J. Chem. Educ.*

Talesnick, I., Ed. Idea bank. Regular column in *The Science Teacher.*

Wilson, J. D., Ed. Favorite demonstrations. Regular column in *J. College Science Teaching.*

Sources of Information on Hazards and Disposal

In preparing the Hazards and Disposal sections of Volume 4, we have used the following sources. The order of listing reflects our degree of utilization.

Laboratory Waste Disposal and Safety Guide, University of Wisconsin–Madison Safety Department: Madison, Wisconsin (1984).

Budavari, S., Ed. *The Merck Index,* 11th ed., Merck & Co., Inc.: Rahway, New Jersey (1989).

Furr, A. K., Ed. *CRC Handbook of Laboratory Safety,* 3d ed., CRC Press: Boca Raton, Florida (1990).

Registry of Toxic Effects of Chemical Substances, Dept. of Health, Education and Welfare (NIOSH): Washington, D.C., revised annually. Available from Superintendent of Documents, U.S. Government Printing Office, Washington, D.C. 20402.

Flinn Chemical Catalog and Reference Manual, Flinn Scientific, Inc.: Batavia, Illinois (1991).

Aldrich Chemical Company Catalog, Aldrich Chemical Co.: Milwaukee, Wisconsin (1990).

Gosselin, R. E., R. P. Smith, H. C. Hodge, and J. E. Braddock. *Clinical Toxicology of Commercial Products,* 5th ed., Williams and Wilkins Co.: Baltimore, Maryland (1984).

Safety in Academic Chemistry Laboratories, 5th ed., American Chemical Society Committee on Chemical Safety: Washington, D.C. (1990). The bibliography lists many journal articles and books.

Bretherick, L., Ed. *Hazards in the Chemical Laboratory,* 4th ed., The Royal Society of Chemistry: London (1986).

Fawcett, H. H. *Hazardous and Toxic Materials: Safe Handling and Disposal,* 2d ed., Wiley Interscience: New York (1988).

Prudent Practices for Handling Hazardous Chemicals in Laboratories, Committee on Hazardous Substances in the Laboratory, National Research Council (1981).

Health and Safety Guidelines for Chemistry Teachers, American Chemical Society Dept. of Educational Activities: Washington, D.C. (1979). The bibliography lists journal articles and books.

Renfrew, M. M., Ed. *Safety in the Chemical Laboratory,* Journal of Chemical Education: Easton, Pennsylvania, Vol. 4 (1981).

Steere, N. V., Ed. *Safety in the Chemical Laboratory,* Journal of Chemical Education: Easton, Pennsylvania, Vol. 1 (1967), Vol. 2 (1971), Vol. 3 (1974).

Guide for Safety in the Chemical Laboratory, 2d ed., Van Nostrand Reinhold Co., Litton Educational Publishing, Inc.: New York (1972).

Sax, N. I. *Dangerous Properties of Industrial Materials,* 3d ed., Van Nostrand Reinhold Co., Litton Educational Publishing, Inc.: New York (1968).

Chemical Demonstrations

Volume 4

10

Clock Reactions

Rodney Schreiner, Bassam Z. Shakhashiri, Earle S. Scott,
Jerry A. Bell, and Mary Ellen Testen

Clock reactions are among the most dramatic and visually pleasing chemical demonstrations. Perhaps the best-known clock reaction is the Landolt iodine clock (Demonstration 10.1) [*1*, *2*]. In this reaction, after two clear, colorless solutions are mixed, the mixture remains colorless for a short time and then suddenly turns dark blue. This behavior is typical of clock reactions. After a clock reaction is initiated, there is a period during which no discernible change takes place in the mixture, and then a change occurs suddenly. This sudden and unexpected change gives clock reactions their charm and visual appeal.

There is a mechanical analogy between the chemical processes in clock reactions and their namesake, the alarm clock. Actually, the analogy works best if we liken clock reactions to an alarm clock with no hands. One cannot tell whether a clock without hands is running, but when its alarm goes off, we infer that it has been running. In a clock reaction, there is no sign of chemical reaction until a sudden change takes place. From the sudden change in the clock reaction we should conclude that some reaction has been occurring. Some of the terminology used to describe clock reactions is based on the alarm clock analogy. The time elapsed between the mixing of the reactants and the sudden change in the mixture is called the *clock period*. The sudden change itself is the *alarm*.

The process occurring in a clock reaction is chemically analogous to a titration. In a titration, one substance is added to another until a critical condition triggers a visible change in the solution (the *end-point*). In the clock reaction, chemical processes generate a substance which reacts with another until a critical condition triggers the alarm. As in a titration, the condition that triggers the alarm is usually a rapid increase or decrease in the concentration of some substance in the mixture. In the titration of a strong acid with a strong base, the condition that triggers the indicator is the sudden increase in the pH of the mixture when the stoichiometric amount of base has just been added. In the Landolt iodine clock reaction, the condition that triggers the alarm is a sudden increase in the concentration of triiodide ions (I_3^-) in the mixture.

The condition that triggers the alarm varies from one clock reaction to another. The Landolt iodine clock reaction can be likened to a titration of bisulfite ions with iodate ions in the presence of starch indicator. The sudden increase in the concentration of triiodide ions that occurs when all the bisulfite has just been consumed triggers the alarm, the formation of a blue complex from starch and triiodide ions. In the case of the hydrolysis of 2-chloro-2-methylpropane (Demonstration 10.7) [*3*], sodium hydroxide is titrated with hydrochloric acid in the presence of a pH indicator. When an excess of hydrochloric acid has been generated, the pH of the mixture suddenly decreases, changing the color of the indicator. Both of these reactions have sharp alarms, as their analogous titrations have sharp end-points. In the "Old Nassau" clock reaction (Demonstration 10.3) [*4*], iodide ions are formed in a solution containing mercury(II). The concentration of iodide ions increases until the solubility of mercury(II) iodide is exceeded, whereupon an orange precipitate forms. This formation of a precipitate can be used to mark the end-point of a titration of mercury(II) with iodide ions. The alarm of a clock reaction can be any of the types of reactions used to detect the end-point of a titration.

The chemical processes involved in different clock reactions and the manner in which this chemistry produces the clock effect vary considerably from one clock reaction to another. The oxidation of iodide by hydrogen peroxide (Demonstration 10.4) produces the same visual effect as the Landolt iodine clock reaction [*5*]. When two clear, colorless solutions are mixed, the mixture remains colorless for a short time, and then suddenly turns dark blue. One of the two solutions that are combined contains sulfuric acid and hydrogen peroxide. The other solution contains potassium iodide, sodium thiosulfate, and starch. The process that occurs when these solutions are mixed can be pictured as occurring in several steps. For each of these steps, a chemical equation can be written. (Although these equations are helpful in imagining how the clock reaction might work, they are not necessarily a complete description of everything that happens in the mixture.) When iodide ions are combined with hydrogen peroxide in an acidic solution, hydrogen peroxide oxidizes the iodide ions to triiodide ions [*6*].

$$3\ I^-(aq) + H_2O_2(aq) + 2\ H^+(aq) \longrightarrow I_3^-(aq) + 2\ H_2O(l) \quad (1)$$

When triiodide ions and thiosulfate ions are combined, thiosulfate ions reduce triiodide ions to iodide ions:

$$I_3^-(aq) + 2\ S_2O_3^{2-}(aq) \longrightarrow 3\ I^-(aq) + S_4O_6^{2-}(aq) \quad (2)$$

The effect of this second reaction is the regeneration of the iodide ions that were consumed in the first reaction. The overall effect of the two reactions is the oxidation of thiosulfate ions by hydrogen peroxide:

$$2\ S_2O_3^{2-}(aq) + H_2O_2(aq) + 2\ H^+(aq) \longrightarrow S_4O_6^{2-}(aq) + 2\ H_2O(aq) \quad (3)$$

Reaction 2 is much faster than reaction 1. Therefore, the triiodide ions are consumed as fast as they are produced, and their concentration remains very low. Throughout the clock period, the concentration of iodide ions remains essentially constant, while the concentrations of hydronium ions (represented here as H^+), hydrogen peroxide, and thiosulfate ions decrease. This overall reaction can continue only as long as there are H^+, H_2O_2, and $S_2O_3^{2-}$ in the mixture. The amount of thiosulfate in the mixture is chosen so that it is the limiting reagent. Eventually, all the thiosulfate ions will be consumed. When this occurs, reaction 2 stops, and another reaction occurs:

$$2\ I_3^- + \text{starch} \rightleftarrows \text{blue starch–}I_5^- \text{ complex} + I^- \quad (4)$$

This reaction causes the mixture to become very deep blue, nearly black [*7*]. This is the alarm reaction. Because reaction 3 keeps the concentration of triiodide ions low until all the thiosulfate ions have been consumed, the equilibrium position of reaction 4 lies to the left until the thiosulfate ions are depleted.

The hydrolysis of 2-chloro-2-methylpropane (Demonstration 10.7) is a clock reaction with a visual effect quite different from that of the oxidation of iodide by hydrogen peroxide [*3, 9*]. In this demonstration, a small amount of colorless liquid is mixed with a blue solution. The solution remains blue for a short time, and then suddenly turns yellow. The chemical reactions involved are also completely different from those in the oxidation of iodide by hydrogen peroxide. The colorless liquid contains 2-chloro-2-methylpropane (tertiary-butyl chloride, *t*-BuCl). The blue solution contains sodium hydroxide and bromothymol blue. Bromothymol blue is a pH indicator with a pK of about 6.8; it is blue in solutions having a pH above its pK and yellow in solutions with a pH below its pK [9]. During the clock period, the pH of the mixture slowly decreases. At the alarm the pH suddenly decreases by several units. When the pH value suddenly crosses the pK value of bromothymol blue, the indicator changes color. The observation that the color of the solution does not change immediately upon addition of *t*-BuCl indicates that the reaction that eventually does cause the pH to change is slow. This reaction is the hydrolysis of *t*-BuCl:

$$(CH_3)_3CCl + H_2O \longrightarrow (CH_3)_3COH + H^+ + Cl^- \qquad (5)$$

This slow reaction is followed by the very rapid reaction of H^+ with the hydroxide ion of the sodium hydroxide:

$$H^+ + OH^- \longrightarrow H_2O \qquad (6)$$

Because reaction 6 is much faster than reaction 5, the H^+ ions are consumed virtually as fast as they are formed, and the pH of the mixture changes only slightly. Once all the sodium hydroxide has been consumed, the excess H^+ produced by reaction 5 reacts rapidly with the indicator, and the color of the solution changes:

$$\underset{}{H^+} + \underset{\text{(blue)}}{Ind^-} \rightleftarrows \underset{\text{(yellow)}}{HInd} \qquad (7)$$

Reaction 7 is the alarm reaction. Because reaction 6 keeps the concentration of H^+ low until all the sodium hydroxide has been consumed, the equilibrium position of reaction 7 lies to the left until the sodium hydroxide is depleted.

GENERAL FEATURES OF A CLOCK REACTION

The two examples described in the previous section reveal some characteristic features of clock reactions. Clock reactions involve three processes. The first process is a relatively slow production of some intermediate chemical species. In terms of the titration analogy, this intermediate is the titrant. In the oxidation of iodide ions by hydrogen peroxide, this titrant species is triiodide ion (I_3^-). In the hydrolysis of *t*-BuCl, the titrant species is hydronium ion (H^+). The second process is the very rapid consumption of this intermediate (titrant) species by the limiting reagent. In the oxidation of iodide by hydrogen peroxide, triiodide ions are rapidly consumed by thiosulfate ions, the limiting reagent. In the hydrolysis of *t*-BuCl, the limiting reagent sodium hydroxide rapidly consumes the hydronium ions. The third process produces the visible change, once all the limiting reagent has been consumed. This process is an equilibrium whose position is shifted when the titrant concentration begins to build. In the oxidation of iodide by hydrogen peroxide, the alarm reaction is the production of the deep blue pentaiodide ion complex with starch. In the hydrolysis of *t*-BuCl, the alarm reaction is the reaction of hydronium ion with the pH indicator.

These three processes can be represented by the simple scheme presented in equations 8, 9, and 10†.

† We thank Professor Roland Stout, Department of Chemistry, Drake University, Des Moines, Iowa, for sharing this succinct treatment of the clock reaction scheme.

$$A + B \longrightarrow T \qquad \text{slow} \qquad (8)$$

$$T + L \longrightarrow X \qquad \text{fast} \qquad (9)$$

$$T + I \rightleftharpoons S \qquad \text{fast} \qquad (10)$$

Initially, the reaction mixture contains A, B, L, and I; these are the reagents combined in the initial mixture. Specifically, L is the limiting reagent, and I is the indicator species. In the oxidation of iodide by thiosulfate, for example, A is potassium iodide, B is hydrogen peroxide, L is sodium thiosulfate, and I is starch. In the hydrolysis of *t*-BuCl, A is *t*-BuCl, B is water, L is sodium hydroxide, and I is bromothymol blue. Reactants A and B produce product T (and perhaps other products) in the relatively slow reaction 8. This product, T, reacts in fast reaction 9 with the limiting reagent, L. (In terms of the titration analogy, T is the titrant.) T is also in rapid equilibrium with indicator I to form S, the signal responsible for the alarm (equation 10). However, as long as T is rapidly removed by L, the concentration of T remains low, the equilibrium position of reaction 10 lies far to the left, and there is very little S in the mixture. Once all of L has been consumed, the concentration of T in the mixture increases, and this affects the equilibrium position of reaction 10. The reaction of hydrogen ions with pH indicator is such an equilibrium. The reaction of starch with pentaiodide ions is another. Although it is thermodynamically favorable, the reaction of L with A (or B) must be much slower than reaction 8. If it were not, then the reaction would consume the limiting reagent prematurely, at best shortening the clock period, at worst eliminating it altogether.

QUALITATIVE RELATIONSHIP BETWEEN CLOCK PERIOD AND INITIAL CONCENTRATIONS

The clock period is related to the rate of the slow process that generates the titrant species in the clock reaction (or, more accurately, to the rate of the rate-determining step in the mechanism that produces the titrant). The more quickly the titrant is produced in this reaction, the sooner the limiting reagent will be used up, and the shorter the clock period will be. In the clock scheme represented by equations 8, 9, and 10, reaction 8 is slow relative to reactions 9 and 10. Reaction 8 is the timer reaction; it is the clockworks of the clock reaction. If reaction 8 were faster than 9, then an excess of titrant could build up *before* all the limiting reagent is consumed. This excess titrant would produce a premature alarm via reaction 10. However, reaction 8 must not be very much slower than reactions 9 and 10. If reaction 8 were very slow, the production of S would be slow as well, because reaction 10 can produce S only as fast as T is produced. This would mean the alarm would appear gradually, diminishing the dramatic impact and appeal of the demonstration.

The clock period tells us how long it takes the slow reaction to produce enough titrant to consume the limiting reagent totally. The appearance of the alarm indicates that the limiting reagent has just been exhausted. Therefore, clock reactions can be used to investigate the effects of several factors on the rates of some chemical reactions. The effects of changing initial reactant concentrations, temperature, and solvent composition can be explored. The clock period, because it depends on the rate at which the titrant is generated, is inversely proportional to the average reaction rate: the faster the titrant is generated, the shorter the clock period. Variations in the clock period with changes in the initial reagent concentrations, temperature, and solvent composition reveal the dependence of the rate of titrant formation on these factors.

Changes in the clock period can be correlated to the initial concentrations of reactants or to the initial amount of the limiting reagent. When the initial amount of limiting reagent in the clock mixture is decreased, less time is required to produce enough titrant to consume it. Therefore, the clock period decreases. If the concentration of the limiting reagent has no effect on the rate of production of the titrant, then the clock period is directly proportional to the initial amount of limiting reagent in the mixture. When the initial amount of limiting reagent is doubled, the clock period will double. If this is not observed, then the concentration of the limiting reagent must have an effect on the rate of titrant production.

In a series of clock reactions, the initial concentration of limiting reagent can be kept constant, and some other factor (e.g., initial concentration of one of the reactants that generates the titrant) can be varied. Changes in this other factor may produce variations in the clock period. If this occurs, the variations in the clock period reflect changes in the rate of production of titrant. A shortened clock period means that less time is required to produce an amount of titrant equivalent to the initial amount of limiting reagent. This indicates that the titrant is produced more rapidly, and the rate of its generation has increased. On the other hand, if the clock period lengthens, then the rate has decreased. (A more extensive discussion of the relationship between reactant concentrations and clock periods is presented in the last section of this introduction, page 12.)

Several demonstrations in this chapter include instructions for investigating the rates of the reactions. Demonstrations 10.4, 10.5, and 10.13 show the effect of initial concentration of reactants on the clock period. The effects of both concentration and temperature are investigated in Demonstration 10.1. The composition of the solvent, as well as initial concentrations of reactants, affects the rate of the reaction in Demonstration 10.7.

ORGANIZATION OF THE CHAPTER

The chapter begins with six "iodine clock" reactions, Demonstrations 10.1 through 10.6. These demonstrations are commonly called iodine clock reactions because they use as their alarm the dark blue color that appears when iodine, iodide ions, and starch are combined. For this reason, most of them produce the same visual effect as the Landolt reaction (Demonstration 10.1). After two colorless solutions are mixed, the mixture remains colorless for a time, and then suddenly turns dark blue. The indicator used in these demonstrations is starch, which contains amylose. Amylose is a polymer of glucose, and it consists of repeating units of the cyclic form of glucose, 1,4-glucopyranose. In an aqueous solution of molecular iodine and iodide ions, the amylose polymer chain forms a spiral strand, wrapped around pentaiodide ions (I_5^-) [7]. These pentaiodide ions are linear chains of five iodine atoms. The starch-pentaiodide complex strongly absorbs yellow light, giving the complex a deep blue color. This starch complex is also used as the end-point indicator in iodometric titrations. It also produces one of the colors in the Briggs-Rauscher oscillating reaction, which is Demonstration 7.1 in Volume 2 of this series.

In Demonstration 10.4, the slow reaction is the oxidation of potassium iodide by hydrogen peroxide. The clock period in this demonstration depends on the initial concentrations of iodide ions and of hydrogen peroxide. Doubling the initial concentration of either reactant will cut the clock period in half. In Demonstration 10.5, the slow reaction is the oxidation of iodide ions by peroxydisulfate ions. In this demonstration, too, the clock period depends on the initial concentrations of the reactants, namely iodide ions and peroxydisulfate ions. Here, too, doubling the initial concentration of either reactant will shorten the clock period by half.

Demonstration 10.6 involves the oxidation of iodide ions by iron(III) ions. Procedure A in this demonstration has a feature unique among these clock reactions. In this procedure, the visual effect is different from the other iodine clocks. Immediately upon mixing the reactants, the mixture turns purple. Then, it gradually fades to colorless and suddenly turns dark blue. This fading reveals that something is indeed happening during the clock period, during which nothing visible happens in the other iodine clock reactions. Although this gradual fading makes the demonstration less visually impactful, it also makes it more instructive.

Demonstration 10.7 involves an organic reaction, the hydrolysis of 2-chloro-2-methylpropane (tertiary-butyl chloride, *t*-BuCl). The sequence of reactions in this clock reaction is the simplest of those in this chapter. The clock period depends on the initial concentration of only one of the reactants, namely *t*-BuCl. This reveals the simplicity of the reaction. The clock period is also sensitive to the composition of the solvent. This fact is related to the nature of the slow reaction in the sequence. The reactions involved in this demonstration are well understood, making the demonstration a good one for relating clock period to reaction scheme. However, the simplicity of the scheme limits the number of factors which can be illustrated with this demonstration.

Demonstration 10.8 also involves an organic reaction, a condensation of aldehydes with acetone. This reaction proceeds in two stages—the first slow, the second fast. The fast stage produces a product which is not very soluble in the reaction medium. The concentration of this product increases so rapidly that the solution becomes supersaturated before a solid forms. Once crystallization begins, it proceeds rapidly, producing a sudden alarm. The series of reactions, as described in the discussion section, is quite involved. Therefore, the relationship between the concentration of reactants and the clock period is complex.

The clockworks of Demonstration 10.9 involve the formation of a complex between formaldehyde and sulfite. The formation of the complex increases the pH of the mixture and, in this regard, can be considered as generating hydroxide ion titrant. The increasing hydroxide ion concentration eventually changes the color of a pH indicator, providing the alarm. Because the alarm is produced by a pH indicator, quite a variety of colors can be obtained by using a variety of indicators.

Demonstrations 10.10 and 10.11 are chemiluminescent clock reactions. In both of these, the alarm is a flash of light. These are described in Chapter 2 on chemiluminescence in Volume 1 of this series.

The clock reaction in Demonstration 10.12 is the disproportionation of thiosulfate ions in an acidic medium. This reaction produces sulfur, which is insoluble in the reaction medium and precipitates, producing the alarm. Demonstration 10.13 is similar, but adds arsenite to the mixture. The arsenite interferes with the production of sulfur, and produces a bright yellow arsenic sulfide precipitate as the alarm.

In Demonstration 10.14, manganese(II) is oxidized to manganese(III) by bromate. Manganese(III) forms a purple complex as the alarm. Demonstration 10.15 is another case where a reaction changes the pH of the mixture and the color of indicators. The reaction involves the oxidation of thiosulfate by periodate.

Because many of the demonstrations use the same solutions, their preparations have been gathered together into the section that immediately follows this introduction.

CHEMISTRY OF AQUEOUS SULFITE SOLUTIONS

Many clock reactions involve compounds of sulfur in the +4 oxidation state, that is, S(IV). These include sulfites, bisulfites, metabisulfites, and sulfur dioxide. The aqueous chemistry of S(IV) is quite complicated, and this complexity affects the behavior of the clock reactions that involve these sulfur compounds. Therefore, a brief description of the chemistry of aqueous S(IV) is included here, before the discussion of quantitative aspects of clock reactions.

The first three demonstrations use sodium bisulfite ($NaHSO_3$), a salt of so-called sulfurous acid. Bubbling gaseous sulfur dioxide through water produces an acidic solution commonly called sulfurous acid. Such solutions have been dealt with as though they contain H_2SO_3, which dissociates to produce hydronium ions (H^+) and bisulfite ions (HSO_3^-). Sulfurous acid has never been isolated. Spectroscopic studies of aqueous solutions of sulfur dioxide have provided no evidence for H_2SO_3 molecules, although they have revealed hydrates of sulfur dioxide, bisulfite ions, some sulfite ions (SO_3^{2-}), and some metabisulfite ions ($S_2O_5^{2-}$) [*10*]. The relative concentrations of these species in the solution depend on the pH of the solution and on the concentration of the total amount of sulfur dioxide dissolved.

Hydrated sulfur dioxide is in equilibrium with hydronium ions and bisulfite ions.

$$SO_2(aq) + H_2O(l) \rightleftarrows HSO_3^-(aq) + H^+(aq)$$

This equation is functionally equivalent to the equation that represents aqueous sulfur dioxide as sulfurous acid.

$$H_2SO_3(aq) \rightleftarrows HSO_3^-(aq) + H^+(aq)$$

The value of the equilibrium constant for these equations is the same. Therefore, aqueous sulfur dioxide is still often represented as sulfurous acid. The first ionization constant for sulfurous acid is 1.6×10^{-2}. This makes sulfurous acid (aqueous sulfur dioxide) a relatively strong member of the class of weak acids. The second ionization constant for sulfurous acid (the ionization constant of bisulfite ion) is 1.0×10^{-7}.

$$HSO_3^-(aq) \rightleftarrows SO_3^{2-}(aq) + H^+(aq)$$

Despite the fact that sulfurous acid does not exist as such, its salts, bisulfites and sulfites, do exist in solution and in solids. The bisulfite ion is unusual in that the hydrogen atom is bonded to the sulfur atom rather than to one of the oxygen atoms, as is more common among oxy-acids. This indicates that the sulfur atom in the sulfite ion possesses an unshared pair of electrons, which means that it can function as a ligand in coordination compounds. Such coordination may occur in Demonstration 10.6, where a yellow complex forms when a solution of bisulfite ions is added to a solution of iron(III) ions.

The bisulfite ion is unusual also in that it can exist in equilibrium with a dimer, disulfite ion (or metabisulfite ion in older nomenclature), which contains a sulfur-sulfur bond.

$$2\ HSO_3^-(aq) \rightleftarrows S_2O_5^{2-}(aq) + H_2O(l)$$

This is an equilibrium which lies far to the left in dilute solutions of bisulfites. However, if a solution of a bisulfite salt is evaporated to isolate a solid salt, the solid obtained is substantially the disulfite. Consequently, when buying material from which to prepare a bisulfite solution, it is often economically advantageous to purchase a disulfite salt in place of a bisulfite. The disulfite splits into the bisulfite when it is dissolved, the solid contains a higher concentration of sulfur(IV), and the disulfites are frequently less expensive because they can be prepared by evaporation.

Both sulfite and bisulfite ions in solution are good reducing agents. The Landolt reaction and its derivatives (Demonstrations 10.1 through 10.3) depend upon the reducing ability of the bisulfite ion and the speed with which the reaction occurs. Unfortunately, this makes these solutions subject to air oxidation to such a degree that they cannot be preserved for long. The reactions of molecular oxygen are normally quite slow. However, a number of metal ions, especially those which can exist in solution in oxidation states separated by one electron (e.g., Cu^{+}/Cu^{2+} and Fe^{2+}/Fe^{3+}), catalyze these oxidations. In some cases, it is possible to inhibit this oxidation reaction by adding to the solution a complexing agent, such as the disodium salt of EDTA (ethylenediaminetetraacetic acid), which binds the metal ions in a manner that prevents their acting as catalysts.

When a bisulfite salt is dissolved to make a dilute solution, that solution will contain primarily bisulfite ions. However, when a bisulfite is dissolved in a highly acidic solution, as it is in some of the demonstrations, a number of different species containing sulfur in the +4 oxidation state are formed. Each of these species may have its own reactions with the oxidizing agents. This greatly complicates the problem of determining reaction mechanisms in such solutions. It certainly prohibits drawing any definitive conclusions about reaction mechanisms from the observation of how clock periods depend on initial concentrations.

CHEMISTRY OF AQUEOUS THIOSULFATE SOLUTIONS

Thiosulfates are salts of thiosulfuric acid. Thiosulfuric acid ($H_2S_2O_3$) has a molecular structure in which the two sulfur atoms are not equivalent.

$$\begin{array}{ccccccc} & & & \mathrm{O} & & & \\ & & & | & & & \\ \mathrm{H} & — & \mathrm{S} — & \mathrm{S} & — \mathrm{O} & — & \mathrm{H} \\ & & & | & & & \\ & & & \mathrm{O} & & & \end{array}$$

The central sulfur atom is similar to that in the sulfate ion (SO_4^{2-}). The peripheral sulfur is similar to that in hydrogen sulfide (H_2S). If these analogous structures are used to assign oxidation numbers to the individual sulfur atoms, the central sulfur atom in thiosulfuric acid has an oxidation number of +6, and the peripheral sulfur atom has an oxidation number of −2. The average of these two numbers is +2, which is the number that would be obtained from the standard rules for oxidation number assignments.

A characteristic reaction of an H—S— group (mercapto group) is oxidation to a disulfide.

$$\begin{array}{c} \mathrm{O} \\ | \\ {}^{-}\mathrm{O—S—S—H} \\ | \\ \mathrm{O} \end{array} + \begin{array}{c} \mathrm{O} \\ | \\ \mathrm{H—S—S—O^{-}} \\ | \\ \mathrm{O} \end{array} + I_2 \longrightarrow \begin{array}{c} \mathrm{O}\qquad\quad\;\mathrm{O} \\ |\qquad\qquad | \\ {}^{-}\mathrm{O—S—S—S—S—O^{-}} \\ |\qquad\qquad | \\ \mathrm{O}\qquad\quad\;\mathrm{O} \end{array} + 2\,I^{-} + 2H^{+}$$

The peripheral sulfur also acts as a mercapto group in its ability to coordinate with transition metal ions and heavy metal ions. The sulfur atom tends to coordinate to larger, lower valence ions. Perhaps the best-known example of such coordination is the use of sodium thiosulfate (photographers' "hypo") in dissolving silver halides during the fixing step of photographic processing. In Demonstration 10.6, thiosulfate ions form a colored complex with iron(III). This complex formation permits a direct observation of the change in the concentration of thiosulfate ions during the course of the clock period of this reaction.

All attempts to synthesize thiosulfuric acid from aqueous solution have failed. This failure can be explained by noting that, when pure thiosulfuric acid is exposed to water, it undergoes a variety of internal oxidation-reduction reactions (disproportionations). Among the products of these reactions are elemental sulfur (some as cyclic S_8), sulfur dioxide, hydrogen sulfide, hydrogen polysulfides, sulfuric acid, and polythionates. This illustrates dramatically the difficulties which may be encountered in dealing with the addition of acid to a solution of sodium thiosulfate. Demonstration 10.12 depends upon this instability of thiosulfate ions in acidic aqueous solution and upon the formation of elemental sulfur which precipitates to form a cloudy suspension. It is very likely that the process that produces this effect in Demonstration 10.12 is not a single reaction but an assortment of reactions. Nevertheless, the process is frequently described by a single equation, which represents a disproportionation of thiosulfate ions.

$$S_2O_3^{2-}(aq) + H^+(aq) \longrightarrow S(s) + HSO_3^-(aq)$$

The complexity of thiosulfate ion chemistry can be illustrated with the methods for synthesizing sodium thiosulfate. Bubbling hydrogen sulfide into an aqueous solution of sodium sulfite produces sodium thiosulfate in the solution.

$$2\ H_2S(g) + 4\ SO_3^{2-}(aq) \longrightarrow 3\ S_2O_3^{2-}(aq) + H_2O(l) + 2\ OH^-(aq)$$

The reaction is a complicated one which involves the formation of sulfur. The sulfur reacts with more bisulfite ions to produce thiosulfate ions.

$$2\ H_2S(g) + SO_3^{2-}(aq) \longrightarrow 3\ S(s) + H_2O(l) + 2\ OH^-(aq)$$

$$3\ S(s) + 3\ SO_3^{2-}(aq) \longrightarrow 3\ S_2O_3^{2-}(aq)$$

When the hydrogen sulfide used in the synthesis of thiosulfate contains isotopically labelled sulfur, the sulfur produced by the acidification of the product thiosulfate solution is two-thirds labelled sulfur and one-third unlabelled. This ratio is the one predicted on the basis of the above equations. These equations suggest that sulfur will react with sulfite or bisulfite solutions. In fact, boiling a solution of sodium sulfite or sodium bisulfite with sulfur does yield thiosulfate ions in the solution. The thiosulfate system is even more complicated, because hydrogen sulfide and sulfur dioxide react in aqueous solution to form a wide range of products. These products include tetrathionate ions ($S_4O_6^{2-}$) and pentathionate ions ($S_5O_6^{2-}$), which contain short chains of sulfur atoms with SO_3^- groups on each end. Similarly, hydrogen polysulfides (H_2S_x) react with sulfur dioxide to produce a series of species containing chains of sulfur atoms with a hydrogen atom at one end and an SO_3^- group at the other.

The variety of these reactions indicates that a great range of chemical transformations can occur in an aqueous solution containing thiosulfate ions. Because of this, it is very difficult to make any definitive statement about what happens in a clock reaction involving thiosulfate ions.

Because the titration of thiosulfate ions by iodine has been a useful reaction in quantitative chemical analysis, a great deal is known about the preservation of the thiosulfate solutions. Such solutions must not be acidic, even to the extent that develops when the solution becomes saturated with atmospheric carbon dioxide. It is common practice to guard against the slow decomposition of thiosulfate solutions by adding to the solution a very small amount (50–100 mg/liter) of pure sodium bicarbonate, which makes the solution slightly basic and buffers against contamination by atmospheric carbon dioxide.

Sodium thiosulfate solutions may also decompose by the action of bacteria. The visible sign of such decomposition is the deposition of solid sulfur. This problem may be minimized by boiling the distilled water from which the solution is to be prepared. This treatment will also remove any carbon dioxide dissolved in the water.

Air oxidation of pure sodium thiosulfate solutions is not usually a problem. However, metal ions in the solution can catalyze the oxidation. Copper ions are an effective catalyst and a common contaminant in distilled water. Water which has been deionized using ion exchange resins will contain no such contaminants.

QUANTITATIVE RELATIONSHIP BETWEEN THE CLOCK PERIOD AND INITIAL CONCENTRATIONS

An analysis of reaction kinetics reveals the sort of information about the rate of a reaction that can be obtained from a clock reaction. Suppose the rate-determining process in the clock reaction involves the combination of a molecule of A with a molecule of B to form T.

$$A + B \longrightarrow T$$

Because this process depends on the collision of A and B, its rate at any time is proportional to the concentrations of A and B at that time.

$$\text{rate} = \frac{d[T]}{dt} = k[A]_t[B]_t$$

In this equation, k is a proportionality constant called the *rate constant*. If at some time the total amount of T produced up to that time is x moles/liter, then, at that time, the concentration of A is $[A]_0 - x$, and the concentration of B is $[B]_0 - x$, where $[A]_0$ and $[B]_0$ are the initial concentrations of A and B. At that time, the rate of the reaction will be given by

$$\frac{dx}{dt} = k([A]_0 - x)([B]_0 - x)$$

The clock period is the length of time required for the amount of T generated by the slow process to be stoichiometrically equivalent to the amount of limiting reagent in the mixture. The amount of time required for this much T to be generated can be related to $[A]_0$ and $[B]_0$ by integrating the previous equation. If the amount of T needed to react with all the limiting reagent is c moles/liter, and $[A]_0 \neq [B]_0$ (which is the case in the demonstrations), the clock period, t_c, is related to $[A]_0$ and $[B]_0$ by

$$t_c = \frac{1}{k([B]_0 - [A]_0)} \ln \frac{([B]_0 - c)[A]_0}{([A]_0 - c)[B]_0}$$

This equation reveals that, in general, the relationship between the clock period and initial concentrations of A and B is anything but simple. However, under certain circumstances, the relationship can be quite simple. If the amount of limiting reagent is much less than the initial amounts of A and B, then c is much smaller than $[A]_0$ and $[B]_0$. The logarithmic terms in the equation above can be rearranged.

$$\ln([B]_0 - c) = \ln\left[[B]_0\left(1 - \frac{c}{[B]_0}\right)\right]$$
$$= \ln[B]_0 + \ln\left(1 - \frac{c}{[B]_0}\right)$$

When $c << [B]_0$, the ratio $c/[B]_0$ is much less than 1. In these circumstances, $\ln(1 - c/[B]_0)$ is approximately equal to $-c/[B]_0$. This can be used to reveal a simple relationship between the clock period and $[A]_0$, and the amount of limiting reagent.

$$t_c = \frac{1}{k([B]_0 - [A]_0)}\{\ln([B]_0 - c) - \ln([A]_0 - c) + \ln[A]_0 - \ln[B]_0\}$$
$$\approx \frac{1}{k([B]_0 - [A]_0)}\left\{\ln[B]_0 - \frac{c}{[B]_0} - \ln[A]_0 + \frac{c}{[A]_0} + \ln[A]_0 - \ln[B]_0\right\}$$
$$= \frac{c}{k[A]_0[B]_0}$$

When the amount of limiting reagent is small compared with the initial amounts of A and B, the clock period is inversely proportional to the initial concentrations of the reagents involved in the rate-determining process of the reaction. Furthermore, under these conditions, the clock period is directly proportional to the amount of limiting reagent. In most of the demonstrations in this chapter, the amount of limiting reagent is small compared with the initial amounts of reactants. Therefore, for most of these demonstrations, when the clock period is inversely proportional to the initial concentration of a reagent, that reagent is involved in the rate-determining process. In addition, if the clock period is directly proportional to the initial concentration of one of the reagents, that reagent is the limiting reagent, and it is not involved in the rate-determining process.

REFERENCES

1. H. Landolt, *Chem. Ber.* 19:1317 (1886).
2. H. Landolt, *Chem. Ber.* 20:745 (1887).
3. J. A. Landgrebe, *J. Chem. Educ.* 41:567 (1964).
4. H. N. Alyea and F. B. Dutton, *Tested Demonstrations in Chemistry,* 6th ed., p. 19, Journal of Chemical Education: Easton, Pennsylvania (1965).
5. W. Seger, *J. Chem. Educ.* 8:166 (1931).
6. R. K. McAlpine, *J. Chem. Educ.* 22:387 (1945).
7. R. C. Teitelbaum, S. L. Ruby, and T. J. Marks, *J. Am. Chem. Soc.* 102:3322 (1980).
8. J. T. Riley, *J. Chem. Educ.* 54:29 (1977).
9. M. Windholz, Ed., *The Merck Index,* 10th ed., Merck and Co.: Rahway, New Jersey (1983).
10. F. A. Cotton and G. Wilkinson, *Advanced Inorganic Chemistry,* 5th ed., John Wiley and Sons: New York (1988).

Preparation of Stock Solutions

The following instructions are for the preparation of 1.0 liter of each of the stock solutions.

potassium iodate, KIO_3

0.10M KIO_3 Dissolve 21.4 g of KIO_3 in 800 mL of distilled water and dilute the resulting solution to 1.0 liter.

Caution! Potassium iodate is a strong oxidizing agent, and mixtures of KIO_3 with combustible materials can be flammable or explosive.

potassium iodide, KI

1.0M KI Dissolve 166 g of KI in 800 mL of distilled water and dilute the resulting solution to 1.0 liter. (The iodide ions in this solution will be oxidized slowly by oxygen in the air. The product is molecular iodine, which darkens the solution. This air oxidation can be retarded by preparing the solution with water that has been deoxygenated by boiling or by passing nitrogen gas through it, and then storing the solution in an air-tight container.)

sodium bisulfite, $NaHSO_3$

0.25M $NaHSO_3$ Dissolve 26 g of $NaHSO_3$ (or 24 g of NaS_2O_5, sodium metabisulfite) in 600 mL of distilled water and dilute the resulting solution to 1.0 liter. (This solution should be prepared within 3 days of use. Sodium bisulfite solutions are oxidized by the oxygen of the air, and slowly decrease in concentration. If this stock solution is to be used only for presentation of qualitative or relative concentration effects, its concentration can be increased by adding a suitable amount of solid $NaHSO_3$. The suitable amount can be determined by performing trial runs of the demonstration, adding very small amounts of solid $NaHSO_3$ to the stock solution until the appropriate clock period is obtained. For more information on $NaHSO_3$ solutions, see page 0 of the introduction to this chapter.)

Caution! Sodium bisulfite is a strong reducing agent and the solid or concentrated solutions should not be mixed with oxidizing agents. Concentrated solutions are irritating to the skin and mucous membranes. When mixed with acid, solutions of $NaHSO_3$ release gaseous sulfur dioxide, which can irritate the respiratory system. Some persons are hypersensitive to sulfites and should avoid direct contact with them.

sodium thiosulfate, $Na_2S_2O_3$

2.0M $Na_2S_2O_3$ Dissolve 496 g of $Na_2S_2O_3 \cdot 5H_20$ in 500 mL of distilled water and dilute the resulting solution to 1.0 liter.

0.20M $Na_2S_2O_3$ Dissolve 49.6 g of $Na_2S_2O_3 \cdot 5H_20$ in 600 mL of distilled water and dilute the resulting solution to 1.0 liter, *or* pour 100 mL of 2.0M $Na_2S_2O_3$ into 600 mL of distilled water and dilute the resulting solution to 1.0 liter.

(Sodium thiosulfate solutions are subject to decomposition as a result of acidification, bacterial action, or air oxidation catalyzed by metal ions. This decomposition can be minimized by boiling the water used to prepare the solution, to remove dissolved carbon dioxide and sterilize the water. Deionize the water to remove any metal ions. For more information on $Na_2S_2O_3$ solutions, see page 10 of the introduction to this chapter.)

starch solution

1% starch Bring 500 mL of distilled water to a boil in a 1-liter beaker. In a 50-mL beaker, make a slurry of 10 g of soluble starch in 20 mL of distilled water. Pour the slurry into the boiling water and boil the mixture for 5 minutes. Place 450 g of ice or 450 mL of very cold water in a 2-liter beaker and pour the hot starch mixture into the 2-liter beaker. After the ice has melted, dilute the mixture to 1 liter. (If kept for several months, starch solutions may be attacked by mold. This can be prevented by adding 10 drops of a 2% solution of thymol in ethanol. Prepare the thymol solution by dissolving 1 g thymol in 60 mL of 95% ethanol.)

sulfuric acid, H_2SO_4

2.0M H_2SO_4 Set a 2-liter beaker containing 600 mL of distilled water in a pan of ice water. While stirring the water, slowly pour 110 mL of concentrated (18M) H_2SO_4 into the beaker. Cool the resulting solution and dilute it to 1.0 liter.

1.0M H_2SO_4 Set a 2-liter beaker containing 600 mL of distilled water in a pan of ice water. While stirring the water, slowly pour 55 mL of concentrated (18M) H_2SO_4 into the beaker. Cool the resulting solution and dilute it to 1.0 liter. Alternatively, pour 500 mL of 2.0M H_2SO_4 into 400 mL of distilled water and dilute the resulting solution to 1.0 liter.

Caution! Because sulfuric acid is both a strong acid and a powerful dehydrating agent, it must be handled with great care. The dilution of concentrated H_2SO_4 is a highly exothermic process and releases sufficient heat to cause burns. Therefore, when preparing dilute solutions from the concentrated acid, always add the acid to water, slowly and with stirring.

10.1

The Landolt Iodine Clock: Oxidation of Bisulfite by Iodate

In each of several sets of beakers arranged in a row, two clear, colorless solutions are mixed to form a clear colorless solution which, after several seconds, suddenly changes to deep blue (Procedure A). Changing the concentrations of the solutions (Procedure B) or changing the temperature (Procedure C) affects the clock period, the time before the mixture changes color. Ten sets of solutions are mixed at the start of a musical selection, and the mixtures suddenly turn blue at various points during the music (Procedure D). When a different set of clear, colorless solutions is mixed, the mixture remains colorless for about 15 seconds, suddenly turns deep blue, and over another 15 seconds fades back to colorless (Procedure E).

MATERIALS FOR PROCEDURE A†

For preparation of stock solutions, see pages 14–15.

500 mL 0.10M potassium iodate, KIO_3 (stock solution)

250 mL 1% starch solution (stock solution)

1150 mL distilled water

100 mL of 0.25M sodium bisulfite, $NaHSO_3$ (stock solution)

100-mL graduated cylinder

5 400-mL beakers

5 600-mL beakers

MATERIALS FOR PROCEDURE B

For preparation of stock solutions, see pages 14–15.

300 mL 0.10M potassium iodate, KIO_3 (stock solution)

200 mL 1% starch solution (stock solution)

1020 mL distilled water

80 mL 0.25M sodium bisulfite, $NaHSO_3$ (stock solution)

4 400-mL beakers, with labels

100-mL graduated cylinder

† To enhance the impact of the demonstration, instructions in Procedure A are for five sequential presentations of the Landolt clock reaction. If only one such presentation is desired, only a fifth of the amount of each listed material is needed.

4 600-mL beakers
timing device capable of measuring seconds

MATERIALS FOR PROCEDURE C

For preparation of stock solutions, see pages 14–15.

600 mL 0.10M potassium iodate, KIO_3 (stock solution)
300 mL 1% starch solution (stock solution)
1380 mL distilled water
120 mL 0.25M sodium bisulfite, $NaHSO_3$ (stock solution)
100-mL graduated cylinder
6 400-mL beakers
6 600-mL beakers
ice bath large enough to accommodate 2 600-mL and 2 400-mL beakers
hot plate large enough to accommodate 2 400-mL beakers
thermometer, −10°C to +110°C
timing device capable of measuring seconds

MATERIALS FOR PROCEDURE D

For preparation of stock solutions, see pages 14–15.

150 mL 0.10M potassium iodate, KIO_3 (stock solution)
ca. 785 mL distilled water
60 mL 0.25M sodium bisulfite, $NaHSO_3$ (stock solution)
30 mL 1% starch solution (stock solution)
60–65-second audio recording (See Procedure D for description.)
2 400-mL beakers, with labels
100-mL graduated cylinder
20 250-mL beakers, with labels
50-mL or 25-mL measuring pipette or graduated cylinder
50-mL graduated cylinder

MATERIALS FOR PROCEDURE E

For preparation of stock solutions, see pages 14–15.

2.5 g malonic acid, $CH_2(CO_2H)_2$
ca. 450 mL distilled water
25 mL 0.10M potassium iodate, KIO_3 (stock solution)

4 mL 0.25M sodium bisulfite, $NaHSO_3$ (stock solution)

25 mL 1% starch solution (stock solution)

250-mL volumetric flask or graduated cylinder

2 600-mL beakers

PROCEDURE A

Preparation

Pour 100 mL of 0.10M KIO_3, 50 mL of 1% starch, and 100 mL of distilled water into each of five 400-mL beakers. Pour 20 mL of 0.25M $NaHSO_3$ and 130 mL of distilled water into each of the five 600-mL beakers.

Arrange the 600-mL beakers in a row. Place a 400-mL beaker behind each of the 600-mL beakers, forming a back row of beakers.

Presentation

Select a 400-mL beaker from one end of the back row of beakers. Pour the contents of this beaker into the 600-mL beaker immediately in front of it, and begin counting up from 1 at a rate of about one count per second. The mixture will turn suddenly from colorless to deep blue at a count of about 10. When this color change occurs, pour the solution from the next 400-mL beaker into the 600-mL beaker in front of it and start counting again. Repeat the pouring and counting, progressing down the row of beakers. If the rate of counting is constant, the color change in each beaker will occur on the same count. (Immediately upon mixing, the concentration of KIO_3 is 0.025M, and that of $NaHSO_3$ is 0.013M.)

PROCEDURE B

Preparation

Label two of the 400-mL beakers with "0.04M KIO_3." Into each of these beakers, pour 100 mL of 0.10M KIO_3 solution, 50 mL of starch solution, and 100 mL distilled water. Label the remaining two 400-mL beakers with "0.02M KIO_3." Into each of these beakers, pour 50 mL of 0.10M KIO_3 solution, 50 mL of starch solution, and 150 mL of distilled water.

Pour 20 mL of 0.25M $NaHSO_3$ and 130 mL of distilled water into each of the four 600-mL beakers.

Arrange the four 600-mL beakers in a row. Place a labelled 400-mL beaker in front of each of the 600-mL beakers.

Presentation

Pour the contents of one of the 400-mL beakers labelled "0.04M KIO_3" into the 600-mL beaker behind it and start the timer. (Upon mixing, the concentration of the solution is 0.013M $NaHSO_3$ and 0.025M KIO_3.) The mixture will remain colorless for about 10 seconds and then suddenly turn deep blue. Stop the timer as soon as the mixture turns deep blue. Record the time elapsed between mixing and the color change (the clock period). Repeat this with the other beaker of 0.04M KIO_3 and note the extent to which the clock periods are reproducible.

Repeat the procedure described in the previous paragraph using the beakers of 0.02M KIO_3. (Upon mixing, the concentration of the solutions is 0.013M $NaHSO_3$ and 0.013M KIO_3.) Note how the clock period depends on the concentration of KIO_3.

PROCEDURE C

Preparation

Pour 100 mL of 0.10M KIO_3, 50 mL of 1% starch, and 100 mL of distilled water into each of six 400-mL beakers. Pour 20 mL of 0.25M $NaHSO_3$ and 130 mL of distilled water into each of the six 600-mL beakers.

Set two of the 600-mL beakers and two of the 400-mL beakers in the ice bath. Allow at least 15 minutes for the temperature of the solutions to equilibrate with that of the ice bath.

Place two of the 400-mL beakers on the hot plate, insert the thermometer in one of the beakers, and adjust the plate to warm the solutions to about 60°C. Do not keep these solutions warm for more than 5 minutes before presenting the demonstration; water will evaporate and change their concentrations. Do not warm the $NaHSO_3$ solutions in the 600-mL beakers; warming will drive off sulfur dioxide (SO_2), changing the concentration.

Presentation

Pour the contents of one of the room-temperature 400-mL beakers into one of the room temperature 600-mL beakers and start the timer. The mixture will remain colorless for about 10 seconds and then suddenly turn deep blue. Stop the timer as soon as the mixture turns deep blue. Record the time elapsed between mixing and the color change (the clock period). Repeat this with the second pair of room-temperature beakers and note the extent to which the clock periods are reproducible.

Record the temperature of the solutions in the ice bath. Then, after removing the beakers from the ice bath, repeat the procedure of the previous paragraph with the two sets of cold solutions. Compare the clock periods of these cold mixtures with those of the room-temperature mixtures. Note the effect of the lower temperature on the clock period.

Place the thermometer in one of the remaining 600-mL beakers. Quickly pour one of the 60°C solutions into the beaker and start the timer. The mixture will remain colorless for a short time and then suddenly turn deep blue. Stop the timer as soon as the mixture turns deep blue. Record the time elapsed between mixing and the color change. Also record the temperature of the mixture. Repeat this with the remaining solutions. Note the effect of temperature differences on the clock period.

PROCEDURE D†

Preparation

Make an audio recording containing 60–65 seconds of lively music. (Excerpts from Rossini's *William Tell Overture,* Tchaikovsky's *1812 Overture,* or Beethoven's *Wellington's Victory* are all suitable. For example, effective selections from the *William Tell Overture* include the opening trumpet call [about 15 seconds], the "Lone Ranger" theme [beginning about 1 minute from the end and lasting about 35 seconds], and the final 10 seconds of the recording.)

Label one of the 400-mL beakers "KIO_3" and the other "$NaHSO_3$." Pour 150 mL of 0.10M KIO_3 and 150 mL of distilled water into the "KIO_3" beaker. Stir the mixture. (This solution is 0.05M in KIO_3.) Pour 60 mL of 0.25M $NaHSO_3$ into the "$NaHSO_3$" beaker. Add 30 mL of 1% starch solution and 210 mL of distilled water and stir the mixture. (This solution is 0.05M in $NaHSO_3$ and about 0.1% starch.)

Label ten of the 250-mL beakers "A1" through "A10" and ten more "B1" through "B10." Use the measuring pipette or graduated cylinder to dispense into each of the "A" beakers the volumes of 0.05M KIO_3 and distilled water indicated in the chart below. In each of the "B" beakers, combine the indicated volumes of 0.05M $NaHSO_3$–starch solution and distilled water.

Beaker	mL of 0.05M KIO_3	mL of distilled water
A1	50	0
A2	40	10
A3	35	15
A4	30	20
A5	27	23
A6	25	25
A7	23	27
A8	21	29
A9	19	31
A10	18	32

Beaker	mL of 0.05M $NaHSO_3$	mL of distilled water
B1	50	0
B2	40	10
B3	35	15
B4	30	20
B5	27	23
B6	25	25
B7	23	27
B8	21	29
B9	19	31
B10	18	32

Presentation

Recruit ten assistants from the audience and give each assistant a pair of beakers, one from the "A" set and the correspondingly numbered one from the "B" set. Instruct the assistants to pour simultaneously the contents of one of their beakers into the other at a signal from you. Start the prepared audio recording and give the signal to the assistants. The mixtures in the beakers will remain colorless for a time and one by one will suddenly turn deep blue, seemingly in time with the music. With motions similar to those of an orchestra conductor, point to each beaker as it changes color.

† This procedure was developed by L. K. Brice of the Virginia Polytechnic Institute and State University, Blacksburg, Virginia.

PROCEDURE E [1]

Preparation

Dissolve 2.5 g of malonic acid in 150 mL of distilled water. Add 25 mL of 0.10M KIO_3 to this solution. In the 250-mL volumetric flask or graduated cylinder, dilute this mixture to 250 mL with distilled water. Rinse the flask thoroughly. Pour the solution into one of the 600-mL beakers.

Add 4 mL of 0.25M $NaHSO_3$ to 25 mL of 1% starch solution. In the 250-mL volumetric flask or graduated cylinder, dilute this mixture to 250 mL with distilled water. Pour the solution into the other 600-mL beaker.

Presentation

Pour the solution from one beaker into the other. The mixture will remain colorless for about 15 seconds, then suddenly turn deep blue. The deep blue color will immediately begin to fade, and within another 20 seconds the mixture will have returned to colorless.

HAZARDS

Potassium iodate is a strong oxidizing agent, and mixtures of potassium iodate with combustible materials can be flammable or explosive.

Sodium bisulfite is a strong reducing agent, and the solid or concentrated solutions should not be mixed with oxidizing agents. Concentrated solutions are irritating to the skin and mucous membranes. When mixed with acid, solutions of sodium bisulfite release gaseous sulfur dioxide, which can irritate the respiratory system. Some persons are hypersensitive to sulfites and should avoid direct contact with them.

DISPOSAL

Combine the deep blue waste solutions in a large beaker. While stirring the mixture, slowly add solid sodium thiosulfate ($Na_2S_2O_3 \cdot 5H_2O$) until the mixture is no longer blue. Flush this mixture down the drain with large quantities of water.

DISCUSSION

The reaction in this demonstration is the one that many chemists think of first when they hear the term *clock reaction*. The sudden, uniform change from a colorless, clear solution to the deep blue, nearly opaque, mixture makes the reaction one of the most visually effective of all clock reactions. Furthermore, the number of factors affecting reaction rates which can be demonstrated with this reaction (reactant concentrations and temperature) make it versatile as well.

In Procedure A, the visual effect alone is presented. Here, the impact of the effect is multiplied by presenting the reaction five times in sequence. (To conserve materials and preparation time, however, the reaction may be presented only once.) In Procedure B, the initial concentration of potassium iodate is varied, and this has an effect on the observed clock period (the time between mixing of the solutions and the appearance of the deep blue color). The effect of temperature on the clock period is presented in Procedure C. In Procedure D, ten solutions of varying concentrations are mixed at the start of a musical selection, and the solutions turn blue at various points during the selection. In Procedure E, a mixture remains colorless for a time, suddenly turns deep blue, and gradually fades back to colorless.

When a solution of bisulfite ions is mixed with a solution of iodate ions and a starch solution, the mixture remains colorless for a time and then suddenly turns deep blue. This observation was first reported by Landolt in the 1880s; this is why many chemists know this clock reaction as the Landolt reaction [*2, 3*]. The time during which the mixture remains colorless is called the clock period. The clock period depends upon the initial concentrations of bisulfite ions and iodate ions, on the temperature of the mixture, and on its pH. The clock period is inversely proportional to the initial concentrations of iodate ions and bisulfite ions. The clock period becomes shorter as the temperature of the mixture increases. A decrease in the pH of the mixture, such as would be caused by the addition of sulfuric acid to the mixture, also shortens the clock period.

At a temperature of 25°C and with no added acid, the empirical relationship between the clock period, P, and the initial concentrations of bisulfite ions and iodate ions is

$$P = \frac{0.003 \text{ s} \cdot \text{mol}^2\text{liter}^{-2}}{[IO_3^-]_0[HSO_3^-]_0}$$

This indicates that a mixture containing an initial iodate ion concentration of 0.025M and an initial bisulfite ion concentration of 0.013M, as used in Procedure A, has a clock period of about 9 seconds at 25°C. Decreasing either the initial iodate ion concentration or the initial bisulfite ion concentration will lengthen the clock period, and increasing either will shorten the clock period. However, there is a limit to the application of this relationship. If the initial concentration of bisulfite ions is more than three times the initial iodate ion concentration, the mixture will not undergo a color change.

The ratio of initial bisulfite ion concentration to initial iodate ion concentration must be less than 3 for the solution to undergo a sudden color change. This means that sodium bisulfite is the limiting reagent at an initial mole ratio less than 3, and that potassium iodate is the limiting reagent at initial mole ratios greater than 3. Furthermore, it seems that the color change occurs at the moment when all the bisulfite has just been consumed. The overall process can be represented by the following sequence of equations [*4, 5*]:

$$IO_3^-(aq) + 3\ HSO_3^-(aq) \longrightarrow I^-(aq) + 3\ SO_4^{2-}(aq) + 3\ H^+(aq) \quad (1)$$

$$IO_3^-(aq) + 8\ I^-(aq) + 6\ H^+(aq) \longrightarrow 3\ I_3^-(aq) + 3\ H_2O(aq) \quad (2)$$

$$I_3^-(aq) + HSO_3^-(aq) + H_2O(l) \longrightarrow 3\ I^-(aq) + SO_4^{2-}(aq) + 3\ H^+(aq) \quad (3)$$

$$2\ I_3^-(aq) + \text{starch} \rightleftharpoons \text{blue starch–}I_5^-\text{ complex} + I^-(aq) \quad (4)$$

In the first reaction in this sequence, bisulfite ions reduce iodate ions to iodide ions. In reaction 2 these iodide ions are oxidized by iodate ions to triiodide ions. At this point, the solution contains triiodide ions and starch, which are the components of the blue complex whose formation occurs in reaction 4. However, reaction 3 occurs so rapidly that it prevents the formation of the blue complex. In reaction 3, triiodide ions are reduced by bisulfite ions to iodide ions until all the bisulfite ions have been consumed. At that point, triiodide ions and starch combine to form the blue starch–I_5^- complex. Thus, the solution remains colorless while the bisulfite is being consumed, but, once it is gone, the blue color of the complex appears. The overall result of reactions 1 through 3 is the oxidation of three bisulfite ions for each iodate ion reduced.† If the initial concentration of bisulfite ions is more than three times that of iodate ions, no color change will occur, because the excess bisulfite ions will consume all the I_3^- produced in reaction 2.

In Procedure E, malonic acid is added to the mixture of potassium iodate, sodium bisulfite, and starch. This causes the deep blue of the starch-pentaiodide complex to fade gradually after it suddenly appears. The blue color fades because malonic acid reacts slowly with aqueous triiodide ions.

$$\mathrm{HOC(=O)\!-\!CH_2\!-\!C(=O)OH} + I_3^- \longrightarrow \mathrm{HOC(=O)\!-\!CHI\!-\!C(=O)OH} + 2\,I^- + H^+ \qquad (5)$$

The products of this reaction are colorless. Both iodate ions and malonic acid are in large stoichiometric excess over bisulfite ions. Once all the bisulfite ions in the mixture have been consumed, reaction 2 produces triiodide ions, which react to form the blue complex (equation 4). However, triiodide ions also react with malonic acid (equation 5). This reaction is quite slow compared with reaction 4, and does not prevent the sudden appearance of the blue complex. As triiodide ions react with malonic acid, the equilibrium of reaction 4 gradually shifts to the left, releasing triiodide ions from the complex. As the complex dissociates, the blue color fades. Eventually, all the triiodide ions react with malonic acid, and the solution becomes colorless.

The reaction sequence presented in equations 1 through 4 is only a convenient way of describing how the iodine clock reaction can occur. It is not a complete description of all the reactions occurring in the mixture. A sequence of reactions that explains the observations made with this clock reaction has been suggested [*6*]. However, more information than can be obtained from clock reactions is needed to determine precisely what reactions occur in the mixture. This information would come from studies of the various component reactions in the absence of other materials, for example, how iodide ions react with bisulfite ions. Such studies would include how the rates of these reactions depend on the concentrations of the components. There have been few reports of recent studies on the rates of the individual reactions involved in the Landolt reactions. Some early studies of the rate of the reaction between iodate ions and bisulfite ions produced the following rate law for reaction 1 [*7–9*].

$$-\frac{d[HSO_3^-]}{dt} = k_1[IO_3^-][HSO_3^-][H^+]$$

More recent studies of the oxidation of iodide ions by iodate ions yielded a rate law for reaction 2 [*10*].

$$-\frac{d[IO_3^-]}{dt} = k_2[IO_3^-][I^-]^2[H^+]^2$$

Reaction 3 is much faster than reaction 2, consuming the triiodide ions as fast as they are formed. A rate law for this reaction has not yet been determined. As these rate laws

† To see this, multiply equation 1 by 5, equation 2 by 1, and equation 3 by 3, add the three resulting equations, and divide the sum by 6.

indicate, the dependence of the rate of the reactions on the concentrations of iodate ions and bisulfite ions is rather complex, and relating the clock period either to the rate laws or to the mechanism of the reaction is very difficult. One should not expect to devise a mechanism only from data obtained by studying how the clock period varies with the initial concentrations of iodate ions and bisulfite ions. Additional studies of the component reactions are also required.

One of the factors complicating the study of the rates of the reactions that occur in the Landolt clock is the complexity of the aqueous chemistry of bisulfite. In aqueous solution, bisulfite ions can dissociate into hydrogen ions and sulfite ions.

$$HSO_3^-(aq) \rightleftarrows H^+(aq) + SO_3^{2-}(aq) \qquad (6)$$

They can also dissociate in water to form aqueous sulfur dioxide.

$$HSO_3^-(aq) \rightleftarrows SO_2(aq) + OH^-(aq) \qquad (7)$$

Thus, an aqueous solution of sodium bisulfite can contain sulfite ions and aqueous sulfur dioxide, as well as bisulfite ions. All these contain sulfur in the +4 oxidation state and are referred to collectively as S(IV) species. Each of these can react with iodate ions at different rates. Furthermore, the concentration of each of these species in the solution depends on, among other things, the pH of the solution. As the clock reaction proceeds, the various reactions produce and consume hydrogen ions, with a net decrease in pH. Therefore, the various equilibria will shift, changing the relative concentrations of the S(IV) species, and the rates of the reactions change with changing hydrogen ion concentration.

Procedure C illustrates that the clock period of the reaction decreases as the temperature of the reactants increases. This means that the rate of the reaction increases with increasing temperature. Data from this procedure can be used to calculate an activation energy for the reaction using the Arrhenius equation [*11*].

$$k = Ae^{-E_a/RT}$$

In this equation, k represents the rate constant of the reaction at the absolute temperature T, E_a is the activation energy of the reaction, R is the gas constant ($8.33\ J \cdot mol^{-1}K^{-1}$), and A is a proportionality constant called the frequency factor. Taking the logarithm of this equation yields

$$\ln(k) = -\frac{E_a}{RT} + \ln(A)$$

When ln(k) is plotted versus 1/T over a small temperature range, the plot is a straight line. The value of the slope of this line is $-E_a/R$. The rate constant for the overall reaction in this clock reaction is inversely proportional to the clock period. Therefore, plotting ln(1/P) versus 1/T will yield a straight line with a slope of $-E_a/R$, where E_a is the apparent activation energy of the overall reaction.

Because the sudden color change is so dramatic, many teachers have adopted methods of presenting this demonstration that emphasize the change. One method is to point at the mixture immediately before the color change takes place. This can add an aura of magic to the demonstration, but it requires careful planning and testing of the solutions to assure proper timing. (The clock period can be lengthened by adding several milligrams of sodium bisulfite to the bisulfite/starch mixture. The clock period can be shortened by diluting the potassium iodate solution.) A method that emphasizes the time factor is to count during the clock period so that the color change occurs on the count of, perhaps, 10. A particularly virtuosic presentation, which was developed by

L. K. Brice of the Virginia Polytechnic Institute and State University, uses a number of concentration combinations which students mix at predetermined points during the last 40 seconds of a recording of Rossini's *William Tell Overture,* so that the solutions have all changed color by the end of the music. This method is described in Procedure D.

REFERENCES

1. M. A. Autuori, A. G. Brolo, and A. L. M. L. Mateus, *J. Chem. Educ.* 66:852 (1989).
2. H. Landolt, *Chem. Ber.* 19:1317 (1886).
3. H. Landolt, *Chem. Ber.* 20:745 (1887).
4. J. A. Church and S. A. Dreskin, *J. Phys. Chem.* 72:1387 (1968).
5. C. H. Sorum, F. S. Charlton, J. A. Neptune, and J. O. Edwards, *J. Am. Chem. Soc.* 74:219 (1952).
6. J. L. Lambert and G. T. Fina, *J. Chem. Educ.* 61:1037 (1984).
7. A. Skrabal, *Z. Elektrochem.* 28:224 (1922).
8. R. Rieder and A. Skrabal, *Z. Elektrochem.* 30:109 (1924).
9. A. Skrabal and A. Zahorka, *Z. Elektrochem.* 33:42 (1927).
10. K. J. Morgan, M. G. Peard, and C. F. Cullis, *J. Chem. Soc.* 1951:1865 (1951).
11. B. Z. Shakhashiri and G. E. Dirreen, *Manual for Laboratory Investigations in General Chemistry,* Stipes Publishing Co.: Champaign, Illinois (1982).

10.2

Color Variations of the Landolt Reaction

One pair of colorless solutions is mixed; the mixture remains colorless for a period of about 10 seconds, and then suddenly turns red. A second pair of solutions is mixed and, after 10 seconds, the mixture suddenly turns yellow. A third pair is mixed, and it suddenly turns blue after 10 seconds [*1*].

MATERIALS

For preparation of stock solutions, see pages 14–15.

200 mL 0.25M sodium bisulfite, $NaHSO_3$ (stock solution)

1 liter distilled water

ca. 200 mL 0.10M potassium iodate, KIO_3 (stock solution)

1 mL 1% starch solution (stock solution)

50.0-mL volumetric pipette

500-mL Erlenmeyer flask

100-mL graduated cylinder

50-mL buret, with stand

3 400-mL beakers, with labels

stirring rod

3 600-ml beakers

PROCEDURE

Preparation

Use the volumetric pipette to measure 50.0 mL of 0.25M $NaHSO_3$ into the 500-mL Erlenmeyer flask. Add 100 mL of distilled water to the flask. Fill the buret with 0.10M KIO_3 and clamp it over the flask. Record the initial buret reading. Titrate the $NaHSO_3$ solution in the flask with the KIO_3 solution in the buret until the mixture in the flask just turns and stays yellow. (This will require between 37 and 42 mL of KIO_3 solution.) Subtract the initial buret reading from the final reading, and record this titer amount.

Label the three 400-mL beakers with the letters A, B, and C. Use the buret to measure the titer amount of 0.10M KIO_3 solution into all three beakers. Add an additional 5.0 mL of 0.10M KIO_3 solution to beakers A and C, and add an additional

1.0 mL to beaker B. Add 1 mL of 1% starch solution to the contents of beaker C. Pour 200 mL of distilled water into all three beakers and stir their contents. Arrange the beakers in a row.

Use the volumetric pipette to measure 50.0 mL of 0.25M $NaHSO_3$ into each of the 600-mL beakers. Add 100 mL of distilled water to each beaker and swirl them to mix their contents. Arrange the beakers in a row in front of the 400-mL beakers.

Presentation

Pour the contents of beaker A into the 600-mL beaker immediately in front of it, and begin counting up from 1 at a rate of about one count per second. The mixture will turn suddenly from colorless to red at a count of about 10. When this color occurs, pour the solution from beaker B into the 600-mL beaker in front of it and start counting again. This mixture will turn suddenly from colorless to yellow at a count of about 10. When this color change occurs, pour the solution from beaker C into the 600-mL beaker in front of it and start counting again. This mixture will turn suddenly from colorless to blue at a count of about 10.

HAZARDS

Potassium iodate is a strong oxidizing agent, and mixtures of potassium iodate with combustible materials can be flammable or explosive.

Sodium bisulfite is a strong reducing agent, and the solid or concentrated solutions should not be mixed with oxidizing agents. Concentrated solutions are irritating to the skin and mucous membranes. Some persons are hypersensitive to sulfites and should avoid direct contact with them.

DISPOSAL

Combine the waste solutions in a large beaker. While stirring the mixture, slowly add solid sodium thiosulfate ($Na_2S_2O_3 \cdot 5H_2O$) until the mixture becomes colorless. Flush this mixture down the drain with large quantities of water.

DISCUSSION

In this demonstration, three different colors—red, yellow, and blue—are produced by clock reactions in three separate beakers. The reactions in the three beakers are all the reduction of iodate by bisulfite. (The same reaction is used in Demonstration 10.1, and it is discussed there.) The process can be described by the sequence of reactions represented by equations 1 through 3 [2].

$$IO_3^-(aq) + 3\ HSO_3^-(aq) \longrightarrow I^-(aq) + 3\ SO_4^{2-}(aq) + 3\ H^+(aq) \quad (1)$$

$$IO_3^-(aq) + 8\ I^-(aq) + 6\ H^+(aq) \longrightarrow 3\ I_3^-(aq) + 3\ H_2O(l) \quad (2)$$

$$I_3^-(aq) + HSO_3^-(aq) + H_2O(l) \longrightarrow 3\ I^-(aq) + SO_4^{2-}(aq) + 3\ H^+(aq) \quad (3)$$

Iodate ions react with bisulfite ions to form iodide ions (equation 1), and these iodide ions react with more iodate ions to produce triiodide (equation 2). However, as quickly as the triiodide ions are formed, they are consumed by a reaction with bisulfite, returning them to iodide ions, as indicated in equation 3. This prevents the concentration of triiodide ions from becoming significant as long as there is bisulfite in the solution. For each iodate ion reduced to triiodide ions, three bisulfite ions are consumed. If the iodate ions are in excess, eventually all the bisulfite ions will be consumed. When this happens, the triiodide ions produced in reaction 2 remain in the solution, and their concentration increases. These triiodide ions color the solution yellow-orange. If only a small amount of triiodide ions are formed, the solution appears yellow. As the concentration of triiodide ions increases, the color deepens and appears orange-red.

The titration is performed to determine the precise volume of potassium iodate solution that is equivalent to 50.0 mL of sodium bisulfite solution. When the precise equivalence is known, it is possible to assure that beaker B contains only a slight excess of iodate ions over bisulfite ions. Then, when all the bisulfite ions have been consumed, there are only enough iodate ions remaining to produce a small amount of triiodide ions, which color the solution yellow. In beaker A, a larger excess of iodate ions is used. When all the bisulfite ions have been consumed, a greater number of triiodide ions are formed, and the solution turns orange-red. Beaker C contains the same excess of iodate ions as beaker A. Beaker C also contains soluble starch. This starch reacts with triiodide ions to form a dark blue complex.

$$2\ I_3^-(aq) + \text{starch} \longrightarrow \text{starch–}I_5^-\ \text{complex} + I^-(aq) \qquad (5)$$

There must be only a slight excess of potassium iodate over sodium bisulfite for the mixture to turn yellow. If there is a larger excess, the mixture turns red. If sodium bisulfite is in excess, no color change will occur. The titration assures that a small excess of potassium iodate is used with the particular solutions used in the demonstration. The titration is needed because commercial sodium bisulfite is not pure sodium bisulfite, but a mixture containing sodium bisulfite and sodium metabisulfite ($Na_2S_2O_5$). The composition of the mixture can vary from batch to batch of sodium bisulfite. Metabisulfite ions react with water to form bisulfite ions.

$$S_2O_5^{2-}(aq) + H_2O(l) \rightleftarrows 2\ HSO_3^-(aq)$$

Therefore, their effect in the solution is the same as that of the bisulfite ions, and each metabisulfite ion is equivalent to two bisulfite ions. However, the molar mass of sodium metabisulfite (190 g/mol) is not twice that of sodium bisulfite (104 g/mol). Therefore, when a solution is prepared by weighing an amount of "sodium bisulfite," the exact amount of potassium iodate required to react with it will not be known. Furthermore, if the sodium bisulfite has been exposed to atmospheric oxygen, it has undergone some oxidation, and the concentration of a solution prepared by weighing the solid cannot be known accurately. For these reasons, a titration is used to determine the amount of potassium iodate which will provide a small excess over sodium bisulfite.

REFERENCES

1. J. W. Wilson, J. W. Wilson, Jr., and T. F. Gardner, *Chemical Magic*, pp. 51–57, Chemical Magic: Los Alamitos, California (1977).
2. C. H. Sorum, F. S. Charlton, J. A. Neptune, and J. O. Edwards, *J. Am. Chem. Soc.* 74:219 (1952).

10.3

Old Nassau Orange and Black: The Landolt Reaction with Mercury Indicator

Three colorless solutions are mixed in sequence, and the mixture remains colorless for several seconds. Then, an orange precipitate forms suddenly throughout the mixture. The mixture remains orange for several seconds and then turns suddenly black (Procedure A) [*1*]. Another combination of solutions produces a change from colorless to orange and back to colorless (Procedure B). When differing volumes of three colorless solutions are combined, the mixtures undergo various combinations of the orange and black color changes (Procedure C).

MATERIALS FOR PROCEDURE A

For preparation of stock solutions, see pages 14–15.

65 mL 0.25M sodium bisulfite, $NaHSO_3$ (stock solution)

85 mL 1% starch solution (stock solution)

150 mL 0.01M mercury(II) chloride, $HgCl_2$ (To prepare 1 liter of solution, dissolve 2.7 g of $HgCl_2$ in 600 mL of distilled water and dilute the resulting solution to 1.0 liter.)

110 mL 0.10M potassium iodate, KIO_3 (stock solution)

40 mL distilled water

250-mL beaker

400-mL beaker

600-mL beaker

timing device capable of measuring seconds (optional)

MATERIALS FOR PROCEDURE B

For preparation of stock solutions, see pages 14–15.

90 mL 0.25M sodium bisulfite, $NaHSO_3$ (stock solution)

90 mL 0.01M mercury(II) chloride, $HgCl_2$ (For preparation, see Materials for Procedure A.)

65 mL 0.10M potassium iodate, KIO_3 (stock solution)
255 mL distilled water
100-mL graduated cylinder
250-mL beaker
stirring rod
400-mL beaker
600-mL beaker
timing device capable of measuring seconds (optional)

MATERIALS FOR PROCEDURE C

For preparation of stock solutions, see pages 14–15.

400 mL 0.25M sodium bisulfite, $NaHSO_3$ (stock solution)
100 mL 1% starch solution (stock solution)
ca. 2 liters distilled water
2.7 g mercury(II) chloride, $HgCl_2$
500 mL 0.10M potassium iodate, KIO_3 (stock solution)
3 1-liter volumetric flasks, with labels
400-mL beaker
3 250-mL graduated cylinders, with labels
7 600-mL beakers

PROCEDURE A

Preparation

In the 250-mL beaker, combine 65 mL of 0.25M $NaHSO_3$ and 85 mL of 1% starch solution. (This solution is 0.11M in $NaHSO_3$ and about 0.6% starch.)

Into the 400-mL beaker, pour 150 mL of 0.01M $HgCl_2$.

In the 600-mL beaker, combine 110 mL of 0.10M KIO_3 and 40 mL of distilled water. (This solution is 0.073M in KIO_3.)

Presentation

Pour the solution from the 250-mL beaker into the 400-mL beaker. Then, quickly pour the contents of the 400-mL beaker into the 600-mL beaker and start the timer. (The initial concentrations of the solutes in this mixture are 0.036M $NaHSO_3$, 0.0033M $HgCl_2$, 0.024M KIO_3, and about 0.2% starch.) The mixture will remain colorless for several seconds, and then a bright orange precipitate will suddenly form throughout the solution. Note the time at which the precipitate forms. The mixture will remain orange for several seconds more and then turn suddenly black. Note the time at which the mixture becomes black.

PROCEDURE B

Preparation

Combine 90 mL of 0.25M $NaHSO_3$ and 60 mL of distilled water in the 250-mL beaker. Stir the mixture. (This solution is 0.15M in $NaHSO_3$.)

Pour 90 mL of 0.01M $HgCl_2$ and 60 mL of distilled water into the 400-mL beaker. Swirl the mixture. (This solution is 0.0060M in $HgCl_2$.)

In the 600-mL beaker, combine 65 mL of 0.10M KIO_3 and 135 mL of distilled water. Swirl the mixture. (This solution is 0.043M in KIO_3.)

Presentation

Pour the solution from the 250-mL beaker into the 400-mL beaker. Then, quickly pour the contents of the 400-mL beaker into the 600-mL beaker and start the timer. (The initial concentrations of the solutes in this mixture are 0.050M $NaHSO_3$, 0.0020M $HgCl_2$, and 0.014M KIO_3.) The mixture will remain colorless for about 5 seconds, and then a bright orange precipitate will suddenly form throughout the solution. The mixture will remain orange for about another 5 seconds, and then the orange precipitate will dissolve.

PROCEDURE C†

Preparation

Label one of the 1-liter volumetric flasks "A." Pour 400 mL of 0.25M $NaHSO_3$ and 100 mL of 1% starch solution into this flask. Dilute the mixture to 1 liter with distilled water. (This is solution A, and it is 0.10M $NaHSO_3$ and about 0.1% starch.)

Label a second 1-liter volumetric flask "B." Dissolve 2.7 g of $HgCl_2$ in 200 mL of distilled water in the 400-mL beaker and pour the solution into the volumetric flask. Rinse the beaker with distilled water and pour the rinse into the flask. Dilute the solution in the flask to 1 liter. (This is solution B, and it is 0.010M $HgCl_2$.)

Label the third volumetric flask "C." Pour 500 mL of 0.10M KIO_3 into the flask and dilute it to 1 liter with distilled water. (This is solution C, and it is 0.050M in KIO_3.)

Label one of the 250-mL graduated cylinders "A," another "B," and the third "C."

Presentation

Trial 1. Pour 100 mL of solution A into cylinder A, 100 mL of solution B into cylinder B, and 100 mL of solution C into cylinder C. Pour the solution from cylinder

† This procedure was developed by Professor John J. Fortman of Wright State University, Dayton, Ohio.

A into one of the 600-mL beakers, followed by the solution in cylinder B, and finally by the solution in cylinder C. Swirl the beaker to mix the solutions. The mixture will remain colorless for about 5 seconds, and then an orange precipitate will form suddenly throughout the mixture. After another 10 seconds, the mixture will turn dark blue.

Trial 2. Pour 100 mL of solution A into cylinder A and 100 mL of solution C into cylinder C. Pour the solution from cylinder A into one of the 600-mL beakers, followed by the solution in cylinder C. Swirl the beaker to mix the solutions. The mixture will remain colorless for several seconds, and then suddenly turn dark blue.

Trial 3. Pour 50 mL of solution A into cylinder A, 100 mL of solution B into cylinder B, and 100 mL of solution C into cylinder C. Pour the solution from cylinder A into another 600-mL beaker, followed by the solution in cylinder B, and finally by the solution in cylinder C. Swirl the beaker to mix the solutions. The mixture will remain colorless for about 5 seconds, and then an orange precipitate will form suddenly throughout the mixture. However, the mixture will not later turn dark blue.

Trial 4. Pour 200 mL of solution A into cylinder A, 100 mL of solution B into cylinder B, and 100 mL of solution C into cylinder C. Pour the solution from cylinder A into another of the 600-mL beakers, followed by the solution in cylinder B, and finally by the solution in cylinder C. Swirl the beaker to mix the solutions. The mixture will remain colorless for about 5 seconds, and then an orange precipitate will form suddenly throughout the mixture. After another 10 seconds, the precipitate will dissolve, and the mixture will be clear and colorless. The solution will not turn dark blue.

Trial 5. Pour 100 mL of solution A into cylinder A, 100 mL of solution B into cylinder B, and 50 mL of solution C into cylinder C. Pour the solution from cylinder A into another 600-mL beaker, followed by the solution in cylinder B, and finally by the solution in cylinder C. Swirl the beaker to mix the solutions. The mixture will remain colorless for about 5 seconds, and then an orange precipitate will form suddenly throughout the mixture. The dark blue color will not appear.

Trial 6. Pour 100 mL of solution A into cylinder A, 100 mL of solution B into cylinder B, and 200 mL of solution C into cylinder C. Pour the solution from cylinder A into one of the 600-mL beakers, followed by the solution in cylinder B, and finally by the solution in cylinder C. Swirl the beaker to mix the solutions. The mixture will remain colorless for about 5 seconds, and then an orange precipitate will form suddenly throughout the mixture. After another 10 seconds, the mixture will turn dark blue.

Trial 7. Pour 100 mL of solution A into cylinder A, 200 mL of solution B into cylinder B, and 100 mL of solution C into cylinder C. Pour the solution from cylinder A into one of the 600-mL beakers, followed by the solution in cylinder B, and finally by the solution in cylinder C. Swirl the beaker to mix the solutions. The mixture will remain colorless for about 5 seconds, and then an orange precipitate will form suddenly throughout the mixture. The dark blue color will not appear.

HAZARDS

Mercury and all of its compounds are highly toxic by ingestion, inhalation, and skin absorption, and may be fatal. Chronic effects can result from exposure to small concentrations over an extended period of time. The dust from salts of mercury is quite poisonous and can irritate the skin and eyes.

Potassium iodate is a strong oxidizing agent, and mixtures of potassium iodate with combustible materials can be flammable or explosive.

Sodium bisulfite is a strong reducing agent, and the solid or concentrated solutions should not be mixed with oxidizing agents. Concentrated solutions are irritating to the skin and mucous membranes. Some persons are hypersensitive to sulfites and should avoid direct contact with them.

DISPOSAL

Mercury wastes should be converted to the virtually insoluble mercury(II) sulfide, and this should be buried in a landfill approved for the disposal of toxic heavy metals. To form the sulfide of mercury, combine the contents of the 600-mL beakers in a pan. (The excess $HgCl_2$ solution from Procedure C should also be poured into this pan.) For each beakerful of solution in the pan, add 1 g of sodium sulfide ($Na_2S \cdot 9H_2O$) or 1 mL of ammonium sulfide ($(NH_4)_2S$) to the contents of the pan. Stir the mixture occasionally for 1 hour. Add aqueous ammonia until the mixture is just basic to litmus. Filter off the insoluble materials and allow them to dry. Consult local authorities to locate an approved toxic-waste disposal site for the dry materials which contain mercury(II) sulfide. To destroy excess sulfide in the filtrate solution, add 50 mL of 5% sodium hypochlorite solution (liquid laundry bleach) and stir occasionally for 1 hour. Flush this solution down the drain with plenty of water.

DISCUSSION

This demonstration uses a modification of the Landolt reaction (Demonstration 10.1) devised by H. N. Alyea [*1–3*]. In Procedure A, the solution remains clear and colorless for a short time, then suddenly turns opaque and bright orange. The mixture remains bright orange for a similar period, then suddenly turns dark blue, virtually black. Orange and black are the school colors of Princeton University and of the House of Nassau, which gave its name to Nassau Hall, built in 1798 and site of the first undergraduate chemistry laboratory in the New World. Thus, this reaction is often called the Old Nassau reaction. Procedure B involves a modification of the reaction in which the orange precipitate dissolves, and the mixture returns to clear and colorless instead of turning dark blue. In Procedure C, the amounts of the reagents are varied in several trials, thereby changing the identity of the limiting reagent in the trials, altering the course of the reactions, and changing their visible results.

The clock reaction in all the procedures of this demonstration is the same as the one in Demonstration 10.1, the Landolt reaction. The net reaction is the reduction of iodate ions to iodide ions by an excess of bisulfite ions.

$$IO_3^-(aq) + 3\ HSO_3^-(aq) \longrightarrow I^-(aq) + 3\ SO_4^{2-}(aq) + 3\ H^+(aq) \qquad (1)$$

Whenever the mole ratio of bisulfite ions to iodate ions is greater than 3, this will be the complete process involving these reactants. If the mole ratio of bisulfite ions to iodate ions is less than 3, there will still be some iodate ions in the solution after the bisulfite ions are exhausted. These excess iodate ions react with the iodide ions to form triiodide ions.

$$IO_3^-(aq) + 8\ I^-(aq) + 6\ H^+(aq) \longrightarrow 3\ I_3^-(aq) + 3\ H_2O(aq) \qquad (2)$$

The triiodide ions combine with starch to produce the dark blue starch-pentaiodide complex.

$$2\ I_3^-(aq) + \text{starch} \rightleftarrows \text{blue starch–}I_5^-\text{ complex} + I^-(aq) \qquad (3)$$

In all but one of the mixtures used in this demonstration, a small amount of mercury(II) chloride is added. Mercury(II) ions form soluble complexes with chloride ions, with sulfite ions, and with iodide ions [*4*].

$$Hg^{2+}(aq) + 2\ Cl^-(aq) \rightleftarrows HgCl_2(aq) \qquad K_f = 1.7 \times 10^{13}$$

$$Hg^{2+}(aq) + 2\ SO_3^{2-}(aq) \rightleftarrows Hg(SO_3)_2^{2-}(aq) \qquad K_f = 1.2 \times 10^{24}$$

$$Hg^{2+}(aq) + 2\ I^-(aq) \rightleftarrows HgI_2(aq) \qquad K_f = 6.6 \times 10^{23}$$

In the initial mixture, the mercury(II) ions are complexed by chloride and sulfite ions. The concentration of iodide ion increases as a result of reaction 1. As this occurs, the mercury(II) ions become complexed by iodide ions. Eventually, the concentration becomes large enough to precipitate mercury(II) iodide, which is orange.

$$Hg^{2+}(aq) + 2\ I^-(aq) \rightleftarrows HgI_2(s) \qquad 1/K_{sp} = 9.1 \times 10^{27}$$

As iodide ions continue to be produced by reaction 1, mercury(II) iodide continues to precipitate, until all the mercury(II) ions have been removed from the solution, or until iodide ion production stops.

If the ratio of moles of bisulfite ions to moles of mercury(II) ions is less than 6, then the number of iodide ions produced is insufficient to precipitate all the mercury(II). Even if there are excess iodate ions in the solution, there will be no iodide ions available to be oxidized to triiodide ions and produce the blue complex.

If the mole ratio of bisulfite ions to mercury(II) ions is greater than 6, bisulfite ions will be available to reduce iodate ions to iodide ions, after all the mercury(II) has precipitated. These iodide ions can react with more iodate ions to produce triiodide ions and the blue complex. However, even if the mole ratio of bisulfite ions to mercury(II) ions is greater than 6, no triiodide ions or blue color will form if the mole ratio of iodate ions to mercury(II) ions is less than 2. In this situation all the iodate ions will be consumed before all the mercury(II) has precipitated, and there can be no oxidation of iodide ions.

If the mole ratio of bisulfite ions to mercury(II) ions is greater than 12 and the mole ratio of iodate ions to mercury(II) ions is greater than 4, the reactions produce enough iodide ions to complex all the mercury(II) iodide precipitate and to dissolve it [*4*].

$$HgI_2(s) + I^-(aq) \rightleftarrows HgI_3^-(aq) \qquad K_{eq} = 0.45$$

$$HgI_3^-(aq) + I^-(aq) \rightleftarrows HgI_4^{2-}(aq) \qquad K_{eq} = 63$$

The processes occurring in the mixtures in this demonstration can be summarized by the scheme given in equations 4 through 7.

$$IO_3^-(aq) + 3\ HSO_3^-(aq) \longrightarrow I^-(aq) + 3\ SO_4^{2-}(aq) + 3\ H^+(aq) \qquad (4)$$

$$Hg^{2+}(aq) + 2\ I^-(aq) \rightleftarrows HgI_2(s) \qquad (5)$$

$$IO_3^-(aq) + 8\ I^-(aq) + 6\ H^+(aq) \longrightarrow 3\ I_3^-(aq) + 3\ H_2O(aq) \qquad (6)$$

$$2\ I_3^-(aq) + \text{starch} \rightleftarrows \text{blue starch–}I_5^-\text{ complex} + I^-(aq) \qquad (7)$$

In the first reaction of this scheme (equation 4), bisulfite ions reduce iodate ions to iodide ions. In the reaction of equation 5 these iodide ions combine with mercury(II) ions to form the bright orange mercury(II) iodide (HgI_2) precipitate. The precipitate does not form immediately, because the first iodide ions produced form a soluble 1:1 complex with $Hg^{2+}(aq)$. Only after the mole ratio of I^- to Hg^{2+} exceeds 1:1 does the mercury(II) iodide precipitate form. After all the mercury has precipitated, the excess iodide ions are oxidized by iodate to triiodide ions (equation 6). The triiodide ions react with starch, forming the blue starch–pentaiodide complex (equation 7).

The initial amounts of sodium bisulfite, mercury(II) chloride, and potassium iodate used in Procedure A are listed in the following chart:

Substance	Initial amount
$NaHSO_3$	0.0163 mole
$HgCl_2$	0.0015 mole
KIO_3	0.011 mole

As equation 5 reveals, two iodide ions are required to precipitate each mercury(II) ion. Equation 4 shows that to produce these two iodide ions, two iodate ions and six bisulfite ions are required. The initial amounts in the chart above produce an initial ratio of bisulfite ions to mercury(II) ions over 10, and of iodate ions to mercury(II) ions over 7. Even after all the mercury(II) has been precipitated, there are iodate ions and bisulfite ions remaining. Therefore, after all the mercury(II) has been precipitated, more iodide ions are produced. Equation 4 shows that for all the bisulfite ions to be consumed, the ratio of bisulfite ions to iodate ions must be less than 3. That is the case here, so eventually all the bisulfite ions are consumed. When this happens, the reaction of equation 6 occurs, in which the remaining iodate ions react with the excess iodide ions. This leads to the production of the deep blue starch-pentaiodide complex (equation 7).

In Procedure B, the ratios of the initial amounts of sodium bisulfite, mercury(II) chloride, and potassium iodate are changed from those in Procedure A. The initial amounts of these substances are given in the following chart:

Substance	Initial amount
$NaHSO_3$	0.0225 mole
$HgCl_2$	0.00090 mole
KIO_3	0.0065 mole

Because the initial ratio of bisulfite ions to iodate ions is greater than 3, the bisulfite ions will consume all the iodate ions in the solution, so the reaction of equation 6 cannot occur, and the solution will not turn dark blue (equation 7). All the iodate ions are converted to iodide ions. The initial amount of iodate ions is more than 7 times that of mercury(II) ions, so the ultimate ratio of iodide ions to mercury(II) ions is over 7. This creates a sufficient excess of iodide ions to cause the mercury(II) iodide precipitate to dissolve as the soluble $[HgI_4]^{2-}$ complex, as shown by equation 8.

$$HgI_2(s) + 2\,I^-(aq) \longrightarrow [HgI_4]^{2-}(aq) \tag{8}$$

Therefore, as the reaction in Procedure B progresses, the mixture remains colorless for a short time, then the orange mercury(II) iodide precipitate forms, then the precipitate dissolves, and the mixture returns to a colorless solution.

Procedure C shows how the relative amounts of the three reactants affect the results of the reaction. The results of this procedure are summarized in the table, where A is 0.10M sodium bisulfite, B is 0.010M mercury(II) chloride, and C is 0.050M potassium iodate.

Effect of Relative Amounts of Reactants on Color Changes

		Initial amounts			Ratio of initial amounts		Outcome		
Trial	Volume ratio A:B:C	Moles $NaHSO_3$	Moles $HgCl_2$	Moles KIO_3	HSO_3^- $/IO_3^-$	HSO_3^- /Hg(II)	Orange precipitate appears	Precipitate dissolves	Dark blue appears
1	2:2:2	0.010	0.0010	0.0050	2	10	yes	no	yes
2	2:0:2	0.010	0	0.0050	2	—	no	—	yes
3	1:2:2	0.0050	0.0010	0.0050	1	5	yes	no	no
4	4:2:2	0.020	0.0010	0.0050	4	20	yes	yes	no
5	2:2:1	0.010	0.0010	0.0025	4	10	yes	no	no
6	2:2:4	0.010	0.0010	0.0010	1	10	yes	no	yes
7	2:4:2	0.010	0.0020	0.0050	2	5	yes	no	no

As revealed by equation 4, if the ratio of bisulfite ions to iodate ions is greater than 3, then bisulfite will remain after all iodate ions have been consumed. This means that the reaction of equation 6 cannot occur, and no dark blue will appear. This is the case in trials 4 and 5. In trial 4, the mercury(II) iodide precipitate dissolves, because the excess of iodate ions over mercury(II) ions is more than 4. In trial 5, the precipitate does not dissolve, because the ratio of iodate ions to mercury(II) ions is less than 3, which is not large enough to generate a sufficient number of iodide ions to dissolve the precipitate.

Trials 3 and 7 do not turn deep blue either. In these, the ratio of bisulfite ions to mercury(II) is less than 6. Therefore, all the iodide ions produced by the bisulfite ions are tied up in the mercury(II) iodide precipitate. There are no iodide ions remaining to react with iodate ions for the ultimate generation of the blue starch complex.

In trial 1, the ratio of bisulfite ions to mercury(II) ions is greater than 6, which means that iodide ions will form after all the mercury(II) iodide has precipitated. However, the ratio of bisulfite ions to iodate ions is less than 3, so iodate ions will remain after all the bisulfite has been consumed. Therefore, iodate ions can react with iodide ions to produce the blue complex. (This trial is similar to the situation in Procedure A.) In trial 2 the absence of mercury(II) chloride in the mixture reveals that it is essential for the production of the orange precipitate.

REFERENCES

1. H. N. Alyea and F. B. Dutton, *Tested Demonstrations in Chemistry,* 6th ed., p. 19, Journal of Chemical Education: Easton, Pennsylvania (1965).
2. H. N. Alyea, *J. Chem. Educ.* 32:9 (1955).
3. H. N. Alyea, *J. Chem. Educ.* 54:166 (1977).
4. R. M. Smith and A. E. Martell, *Critical Stability Constants,* Vol. 4, Plenum Press: New York (1976).

10.4

Hydrogen Peroxide Iodine Clock: Oxidation of Potassium Iodide by Hydrogen Peroxide

Two colorless solutions are mixed, and after several seconds the mixture suddenly turns deep blue (Procedure A). The time required for the color to change can be varied by changing the concentrations of iodide ions (Procedure B), of hydrogen peroxide (Procedure C), and of thiosulfate ions (Procedure D) [*1*].

MATERIALS FOR PROCEDURE A

For preparation of stock solutions, see pages 14–15.

25 mL 2.0M sulfuric acid, H_2SO_4 (stock solution)
25 mL 3% hydrogen peroxide, H_2O_2
ca. 450 mL distilled water
0.10 g sodium thiosulfate pentahydrate, $Na_2S_2O_3 \cdot 5H_2O$
13 mL 1.0M potassium iodide, KI (stock solution)
10 mL 1% starch solution (stock solution)
250-mL graduated cylinder
250-mL beaker
2 600-mL beakers
stirring rod
timing device capable of measuring seconds (optional)

MATERIALS FOR PROCEDURE B

For preparation of stock solutions, see pages 14–15.

35 mL 0.20M sodium thiosulfate, $Na_2S_2O_3$ (stock solution)
ca. 2 liters distilled water
125 mL 1.0M potassium iodide, KI (stock solution)
100 mL 2.0M sulfuric acid, H_2SO_4 (stock solution)
100 mL 3% hydrogen peroxide, H_2O_2
50 mL 1% starch solution (stock solution)

2 250-mL volumetric flasks
100-mL graduated cylinder
1-liter volumetric flask
5 600-mL beakers, with labels
5 stirring rods
5 400-mL beakers
timing device capable of measuring seconds

MATERIALS FOR PROCEDURE C

For preparation of stock solutions, see pages 14–15.

35 mL 0.20 sodium thiosulfate, $Na_2S_2O_3$ (stock solution)
ca. 2 liters distilled water
125 mL 1.0M potassium iodide, KI (stock solution)
25 mL 1% starch solution (stock solution)
45 mL 2.0M sulfuric acid, H_2SO_4 (stock solution)
105 mL 3% hydrogen peroxide, H_2O_2
2 250-mL volumetric flasks
500-mL volumetric flask
100-mL graduated cylinder
3 400-mL beakers
3 stirring rods
3 250-mL beakers, with labels
timing device capable of measuring seconds

MATERIALS FOR PROCEDURE D

Same as Materials for Procedure B

PROCEDURE A

Preparation

In a 250-mL graduated cylinder, combine 25 mL of 2.0M H_2SO_4 and 25 mL of 3% H_2O_2. Dilute the solution to 250 mL with distilled water. Pour the solution back and forth between the graduated cylinder and one of the 600-mL beakers to mix the solution. Leave the solution in the beaker. (This solution is 0.20M in H_2SO_4 and 0.09M in H_2O_2.)

In a 250-mL beaker, dissolve 0.10 g $Na_2S_2O_3 \cdot 5H_2O$ in 50 mL of distilled water.

Rinse the 250-mL graduated cylinder with distilled water. Add the $Na_2S_2O_3$ solution, 13 mL 1.0M KI, and 10 mL of 1% starch solution to the cylinder. Dilute the mixture to 250-mL with distilled water. Pour the mixture back and forth several times between the cylinder and the remaining 600-mL beaker. Leave the solution in the beaker. (This solution is 0.052M in KI and 0.0016M in $Na_2S_2O_3$.)

Place the two 600-mL beakers side by side on the display table and put a stirring rod in one of the beakers.

Presentation

Quickly pour the contents of one beaker into the other, stir the mixture, and begin timing. The mixture will remain colorless for about 10 seconds and then suddenly turn deep blue. (In this mixture, the initial concentrations are 0.10M H_2SO_4, 0.045M H_2O_2, 0.026M KI, and 0.0008M $Na_2S_2O_3$.)

PROCEDURE B

Preparation

Solution B-1. In a 250-mL volumetric flask, dilute 35 mL of 0.20M $Na_2S_2O_3$ to 250 mL with distilled water. (This solution is 0.028M in $Na_2S_2O_3$.)

Solution B-2. In another 250-mL volumetric flask, dilute 125 mL of 1.0M KI to 250-mL with distilled water. (This solution is 0.50M in KI.)

Solution B-3. In a 1-liter volumetric flask combine 100 mL of 2.0M H_2SO_4, 100 mL of 3% H_2O_2, and about 500 mL of distilled water. Swirl the flask to mix the solutions and dilute the mixture to 1.0 liter with distilled water. (This solution is 0.20M in H_2SO_4 and 0.09M in H_2O_2.)

Label five 600-mL beakers with the numbers 1 through 5 and place them in a row on the display table. In each of these beakers combine the volumes of solutions as follows:

Beaker	mL of solution B-1	mL of solution B-2	mL of 1% starch solution	mL of distilled water
1	40	50	10	100
2	40	40	10	110
3	40	30	10	120
4	40	20	10	130
5	40	10	10	140

Place a stirring rod in each beaker and stir the mixtures.

Set a 400-mL beaker behind each of the numbered beakers. Pour 200 mL of solution B-3 into each of the 400-mL beakers.

Presentation

Pour the contents of the first 400-mL beaker into beaker 1, stir the mixture, and start timing. The mixture will remain colorless for several seconds and then suddenly turn deep blue. Note the time required for the color to appear.

Repeat the procedure described in the preceding paragraph with the remaining four sets of beakers. Note how the initial concentration of KI affects the time required for the color to change. The initial concentrations of the mixtures in beakers 1 through 5 are

Beaker	$[H_2SO_4]_0$	$[H_2O_2]_0$	$[Na_2S_2O_3]_0$	$\mathbf{[KI]_0}$
1	0.10	0.045	0.0028	**0.063**
2	0.10	0.045	0.0028	**0.050**
3	0.10	0.045	0.0028	**0.038**
4	0.10	0.045	0.0028	**0.025**
5	0.10	0.045	0.0028	**0.013**

PROCEDURE C

Preparation

Prepare solutions B-1 and B-2 as described in Procedure B.

Solution C. In a 500-mL volumetric flask, combine 100 mL of solution B-1, 50 mL of solution B-2, and 25 mL of 1% starch solution. Swirl the flask to mix the solutions and dilute the mixture to 500 mL with distilled water.

Set three 400-mL beakers in a row on the display table, pour 150 mL of solution C into each, and place a stirring rod in each one. Label three 250-mL beakers with the numbers 1 through 3 and set one behind each of the 400-mL beakers. In the 250-mL beakers, mix the solutions as follows:

Beaker	mL of 2.0M H_2SO_4	mL of 3% H_2O_2	mL of distilled water
1	15	60	75
2	15	30	105
3	15	15	120

Presentation

Pour the contents of beaker 1 into its corresponding 400-mL beaker, stir the mixture, and start timing. The mixture will remain colorless for about 10 seconds and then suddenly turn deep blue. Note the time required for the color to appear.

Repeat the procedure described in the preceding paragraph with the remaining two sets of beakers. Note how the initial concentration of hydrogen peroxide affects the time required for the color to appear. The initial concentrations of the mixtures in beakers 1 through 3 are

Beaker	$[H_2SO_4]_0$	$\mathbf{[H_2O_2]_0}$	$[Na_2S_2O_3]_0$	$[KI]_0$
1	0.10	**0.18**	0.0028	0.025
2	0.10	**0.090**	0.0028	0.025
3	0.10	**0.045**	0.0028	0.025

PROCEDURE D

Preparation

Prepare solutions B-1, B-2, and B-3 as described in Procedure B.

Label five 600-mL beakers with the numbers 1 through 5 and place them in a row on the display table. In each of these beakers combine the volumes of solutions as follows:

Beaker	mL of solution B-1	mL of solution B-2	mL of 1% starch solution	mL of distilled water
1	10	20	10	160
2	20	20	10	150
3	30	20	10	140
4	40	20	10	130
5	50	20	10	120

Place a stirring rod in each beaker and stir the mixtures.

Set a 400-mL beaker behind each of the numbered beakers. Pour 200 mL of solution B-3 into each of the 400-mL beakers.

Presentation

Pour the contents of the first 400-mL beaker into beaker 1, stir the mixture, and start timing. The mixture will remain colorless for several seconds and then suddenly turn deep blue. Note the time required for the color to appear.

Repeat the procedure described in the preceding paragraph with the remaining four sets of beakers. Note how the initial concentration of $Na_2S_2O_3$ affects the time required for the color to appear. The initial concentrations of the mixtures in beakers 1 through 5 are

Beaker	$[H_2SO_4]_0$	$[H_2O_2]_0$	$\mathbf{[Na_2S_2O_3]_0}$	$[KI]_0$
1	0.10	0.045	**0.0007**	0.025
2	0.10	0.045	**0.0014**	0.025
3	0.10	0.045	**0.0021**	0.025
4	0.10	0.045	**0.0028**	0.025
5	0.10	0.045	**0.0035**	0.025

HAZARDS

The 3% hydrogen peroxide solution can irritate the skin and eyes.

DISPOSAL

Waste solutions should be flushed down the drain with water.

DISCUSSION

Each of the four procedures of this demonstration illustrates a particular feature of the reaction. Procedure A presents an example of a typical clock reaction in which a mixture of clear, colorless solutions remains colorless for a while and then suddenly turns deep blue. Procedure B shows the effect of initial iodide ion concentration on the time required for the color change, that is, on the clock period. Procedure C shows the effect of initial hydrogen peroxide concentration on the clock period. The last procedure, Procedure D, shows how the initial concentration of thiosulfate ions affects the clock period.

The sudden change from colorless to deep blue solutions in this demonstration can be explained with the following sequence of equations [*2*]:

$$3\ I^-(aq) + H_2O_2(aq) + 2\ H^+(aq) \longrightarrow I_3^-(aq) + 2\ H_2O(l) \quad (1)$$

$$I_3^-(aq) + 2\ S_2O_3^{2-}(aq) \longrightarrow 3\ I^-(aq) + S_4O_6^{2-}(aq) \quad (2)$$

$$2\ I_3^-(aq) + \text{starch} \rightleftharpoons \text{starch–}I_5^-\ \text{complex} + I^-(aq) \quad (3)$$

The first equation indicates that, in an acidic solution, iodide ions are oxidized by hydrogen peroxide to triiodide ions. These triiodide ions are reduced back to iodide ions by thiosulfate ions, as indicated in equation 2. This reaction is much faster than the reaction of equation 1; it consumes triiodide ions as fast as they are formed. This prevents any readily apparent reaction of equation 3. However, after all the thiosulfate ions have been consumed by the reaction of equation 2, triiodide ions react with starch to form the blue starch–pentaiodide complex [*3*].

Procedure B illustrates the effect of initial iodide ion concentration on the clock period. The results of this procedure show that doubling the initial concentration halves the clock period. Therefore, doubling the initial concentration of iodide ions must double the initial rate of the reaction. This means that the rate-determining reaction is first order in iodide ions.

In Procedure C, the effect of varying the initial hydrogen peroxide concentration is observed. Doubling the initial concentration of hydrogen peroxide cuts the clock period in half. This indicates that doubling the initial concentration of hydrogen peroxide doubles the initial rate of the reaction, and therefore, the rate-determining reaction is first order in hydrogen peroxide.

In Procedure D, the initial concentration of thiosulfate ions is varied. The results indicate that the clock period is inversely proportional to their initial concentrations. When their initial concentration is doubled, the clock period is doubled. This reveals that the thiosulfate ions do not have an effect on the rate of the rate-determining reaction. If thiosulfate ions were involved in the rate-determining reaction, then an increase in their concentration would increase the rate of this reaction and shorten the clock period.

Procedures B and C of this demonstration indicate that the rate-determining process in this reaction is first order in both iodide ions and hydrogen peroxide. This suggests that the slow step in the mechanism of the reaction involves both an iodide ion and a hydrogen peroxide molecule. Such a step is represented in equation 4.

$$H_2O_2 + I^- + H^+ \longrightarrow HOI + H_2O \quad (4)$$

The hypoiodous acid produced in this step can react rapidly with iodide ions to form iodine.

$$HOI + 2\ I^- + H^+ \rightleftharpoons I_3^- + H_2O \quad (5)$$

Equation 1 is the sum of equations 4 and 5. The slowness of reaction 4 controls the overall rate of reaction 1. Reaction 4 is first order in both iodide ions and hydrogen peroxide, which makes the entire process in equation 1 first order in these species, too.

Other reactions could occur in a mixture of iodide ions and hydrogen peroxide, but these reactions require conditions that are not met in this reaction mixture. For example, hydrogen peroxide can oxidize iodine to iodate ions, and iodide ions catalyze the decomposition of hydrogen peroxide to oxygen and water. However, these processes require neutral or basic conditions, and the solutions used in this demonstration are acidic.

The reactions of hydrogen peroxide and iodide ions have been investigated by Liebhafsky over a period of half a century [*4*]. The Briggs-Rauscher oscillating reaction, which is Demonstration 7.1 in Volume 2 of this series, also involves reactions of hydrogen peroxide and iodide ions. Investigations of the Briggs-Rauscher reaction have revealed quite a few details of the reactions involving these substances [*5*].

Reactions 1–3 do not represent all that is occurring during this demonstration. In this demonstration, hydrogen peroxide oxidizes iodide ions to triiodide ions, which in turn oxidize thiosulfate ions to tetrathionate ions. Thus, hydrogen peroxide is energetically capable of oxidizing thiosulfate ions. Equation 6 indicates how this could happen.

$$H_2O_2 + 2\ S_2O_3^{2-} + 2\ H^+ \longrightarrow S_4O_6^{2-} + 2\ H_2O \qquad (6)$$

That this reaction does not interfere with the demonstration shows that it is much slower than the oxidation of iodide ions by hydrogen peroxide.

REFERENCES

1. W. Seger, *J. Chem. Educ.* 8:166 (1931).
2. R. K. McAlpine, *J. Chem. Educ.* 22:387 (1945).
3. R. C. Teitelbaum, S. L. Ruby, and T. J. Marks, *J. Am. Chem. Soc.* 102:3322 (1980).
4. H. A. Liebhafsky, W. C. McGavock, R. J. Reyes, G. M. Roe, and L. S. Wu, *J. Am. Chem. Soc.* 100:87 (1978), and references cited therein.
5. K. R. Sharma and R. M. Noyes, *J. Amer. Chem. Soc.* 98:4345 (1976), and references cited therein.

10.5

Thiosulfate-Countered Oxidation of Iodide by Peroxydisulfate

In each of several sets of beakers arranged in a row, two clear, colorless solutions are mixed to form a clear colorless solution which, after several seconds, suddenly changes to deep blue (Procedure A). The time between mixing and the appearance of the deep blue depends on the initial concentrations of potassium iodide (Procedure B) and potassium peroxydisulfate (Procedure C). Varying the amount of sodium thiosulfate in the mixture also changes the time required for the color to appear (Procedure D).

MATERIALS FOR PROCEDURE A†

For preparation of stock solutions, see pages 14–15.

200 mL 1.0M potassium iodide, KI (stock solution)

5.0 mL 0.20M sodium thiosulfate, $Na_2S_2O_3$ (stock solution)

100 mL 1% starch solution (stock solution)

ca. 700 mL distilled water

500 mL 0.10M potassium peroxydisulfate (potassium persulfate), $K_2S_2O_8$ (This solution should be prepared within 24 hours of use. To prepare 1.0 liter of solution, dissolve 27 g of $K_2S_2O_8$ in 700 mL of distilled water and dilute the solution to 1.0 liter.)

1-liter volumetric flask

5 400-mL beakers

100-mL graduated cylinder

5 600-mL beakers

MATERIALS FOR PROCEDURE B

For preparation of stock solutions, see pages 14–15.

300 mL 1.0M potassium iodide, KI (stock solution)

650 mL distilled water

† To enhance the impact of the demonstration, instructions in Procedure A are for five sequential presentations of the clock reaction. If only one such presentation is desired, only a fifth of the amounts of the listed materials are needed.

50 mL 1% starch solution (stock solution)

5 mL 0.20M sodium thiosulfate, $Na_2S_2O_3$ (stock solution)

500 mL 0.10M potassium peroxydisulfate, $K_2S_2O_8$ (For preparation, see Materials for Procedure A.)

5 400-mL beakers, with labels

100-mL graduated cylinder

5 stirring rods

5 600-mL beakers

timing device capable of measuring seconds

MATERIALS FOR PROCEDURE C

For preparation of stock solutions, see pages 14–15.

200 mL 1.0M potassium iodide, KI (stock solution)

5.0 mL 0.20M sodium thiosulfate, $Na_2S_2O_3$ (stock solution)

100 mL 1% starch solution (stock solution)

ca. 1100 mL distilled water

600 mL 0.10M potassium peroxydisulfate (potassium persulfate), $K_2S_2O_8$ (For preparation, see Materials for Procedure A.)

1-liter volumetric flask

5 400-mL beakers

5 stirring rods

5 600-mL beakers, with labels

100-mL graduated cylinder

timing device capable of measuring seconds

MATERIALS FOR PROCEDURE D

For preparation of stock solutions, see pages 14–15.

200 mL 1.0M potassium iodide, KI (stock solution)

100 mL 1% starch solution (stock solution)

ca. 700 mL distilled water

7 mL 0.20M sodium thiosulfate, $Na_2S_2O_3$ (stock solution)

500 mL 0.10M potassium peroxydisulfate (potassium persulfate), $K_2S_2O_8$ (For preparation, see Materials for Procedure A.)

1-liter volumetric flask

5 400-mL beakers, with labels

5 stirring rods
5 600-mL beakers
100-mL graduated cylinder
10-mL graduated cylinder
timing device capable of measuring seconds

PROCEDURE A

Preparation

Combine 200 mL of 1.0M KI solution, 5.0 mL of 0.20 M $Na_2S_2O_3$, and 100 mL of 1% starch solution in a 1-liter volumetric flask, and dilute the solution to 1.0 liter with distilled water. (This solution is 0.20M in KI and 0.0010M in $Na_2S_2O_3$.) Pour 200 mL of this solution into each of five 400-mL beakers. Arrange the beakers in a row.

Pour 100 mL of 0.10M $K_2S_2O_8$ into each of five 600-mL beakers. Place a 600-mL beaker in front of each of the 400-mL beakers.

Presentation

Select a 400-mL beaker from one end of the back row of beakers. Pour the contents of this beaker into the 600-mL beaker immediately in front of it, and begin counting up from 1 at a rate of about one count per second. The mixture will turn suddenly from colorless to deep blue at a count of about 10. When this color change occurs, pour the solution from the next 400-mL beaker into the 600-mL beaker in front of it and start counting again. Repeat the pouring and counting, progressing down the row of beakers. If the rate of counting is constant, the color change in each beaker will occur on the same count. (The initial concentration of each solution is 0.13M KI, 0.0007M $Na_2S_2O_3$, and 0.033M $K_2S_2O_8$.)

PROCEDURE B

Preparation

Label each of five 400-mL beakers with one of the following labels: "0.25M," "0.20M," "0.15M," "0.10M," and "0.05M." In each beaker, combine the liquids indicated in the following chart:

Beaker	mL of 1.0M KI	mL of distilled water
"0.25M"	100	90
"0.20M"	80	110
"0.15M"	60	130
"0.10M"	40	150
"0.05M"	20	170

Add 10 mL of 1% starch and 1.0 mL of 0.20M $Na_2S_2O_3$ to each beaker. Stir the mixtures. Arrange the beakers in a row, leaving the stirring rod in each beaker.

Pour 100 mL of 0.10M $K_2S_2O_8$ into each of five 600-mL beakers. Place a 600-mL beaker beside each of the 400-mL beakers.

Presentation

Quickly add the contents of the 400-mL beaker labelled "0.15M" to the 600-mL beaker adjacent to it, stir the mixture, and start the timer. The mixture will remain colorless for 10–15 seconds and suddenly turn dark blue. Record the time elapsed between mixing and the appearance of the blue color (the clock period).

Repeat the procedure described in the previous paragraph with the remaining four sets of beakers. Note how the clock period depends on the initial concentration of KI.

The initial concentrations of the solutions in each beaker are

Beaker	$[K_2S_2O_8]_0$	$[Na_2S_2O_3]_0$	$\mathbf{[KI]_0}$
"0.25M"	0.033	0.0007	**0.25**
"0.20M"	0.033	0.0007	**0.20**
"0.15M"	0.033	0.0007	**0.15**
"0.10M"	0.033	0.0007	**0.10**
"0.05M"	0.033	0.0007	**0.05**

PROCEDURE C

Preparation

Combine 200 mL of 1.0M KI solution, 5.0 mL of 0.20M $Na_2S_2O_3$, and 100 mL of 1% starch solution in a 1-liter volumetric flask, and dilute the solution to 1.0 liter with distilled water. (This solution is 0.20M in KI.)

Pour 200 mL of the solution in the volumetric flask into each of the five 400-mL beakers. Arrange the beakers in a row and place a stirring rod in each beaker.

Label each of five 600-mL beakers with one of the following labels: "0.050M," "0.040M," "0.030M," "0.020M," and "0.010M." In each beaker, combine the liquids indicated in the following chart.

Beaker	mL of 0.10M $K_2S_2O_8$	mL of distilled water
"0.050M"	200	0
"0.040M"	160	40
"0.030M"	120	80
"0.020M"	80	120
"0.010M"	40	160

Place a 600-mL beaker beside each of the 400-mL beakers.

Presentation

Quickly add the contents of one of the 400-mL beakers to the 600-mL beaker labelled "0.030M," stir the mixture, and start the timer. The mixture will remain colorless for 10–15 seconds and suddenly turn dark blue. Record the time elapsed between mixing and the appearance of the blue color (the clock period).

Repeat the procedure described in the previous paragraph with the remaining four sets of beakers. Note how the clock period depends on the initial concentration of $K_2S_2O_8$.

The initial concentrations of the solutions in each beaker are

Beaker	$[K_2S_2O_8]_0$	$[Na_2S_2O_3]_0$	$[KI]_0$
"0.050M"	**0.050**	0.0007	0.13
"0.040M"	**0.040**	0.0007	0.13
"0.030M"	**0.030**	0.0007	0.13
"0.020M"	**0.020**	0.0007	0.13
"0.010M"	**0.010**	0.0007	0.13

PROCEDURE D

Preparation

Combine 200 mL of 1.0M KI solution and 100 mL of 1% starch solution in a 1-liter volumetric flask, and dilute the solution to 1.0 liter with distilled water. (This solution is 0.20M in KI.)

Label each of five 400-mL beakers with one of the following labels: "none," "1.0 mL," "1.5 mL," "2.0 mL," and "2.5 mL." Pour 200 mL of the solution in the volumetric flask into each of the 400-mL beakers. Arrange the beakers in a row and place a stirring rod in each beaker.

Pour 100 mL of 0.10M $K_2S_2O_8$ into each of five 600-mL beakers. Place a 600-mL beaker beside each of the 400-mL beakers.

Presentation

To each of the 400-mL beakers, add the amount of 0.20M $Na_2S_2O_3$ indicated on its label, and stir the mixtures.

Quickly add the contents of the 400-mL beaker labelled "none" to the 600-mL beaker adjacent to it. The mixture will turn deep blue immediately.

Pour the contents of the beaker labelled "1.0 mL" into the adjacent 600-mL beaker, stir the mixture, and start timing. Record how many seconds elapse before this mixture turns deep blue.

Repeat the procedure described in the previous paragraph with the remaining three sets of beakers. Note how the clock period depends on the initial concentration of $Na_2S_2O_3$.

The initial concentrations of the solutions in each beaker are

Beaker	$[K_2S_2O_8]_0$	$[Na_2S_2O_3]_0$	$[KI]_0$
"none"	0.033	**0**	0.13
"1.0 mL"	0.033	**0.0007**	0.13
"1.5 mL"	0.033	**0.0010**	0.13
"2.0 mL"	0.033	**0.0013**	0.13
"2.5 mL"	0.033	**0.0016**	0.13

HAZARDS

Potassium peroxydisulfate is a powerful oxidizing agent, and mixtures of potassium peroxydisulfate with combustible materials can be flammable or explosive. Solid potassium peroxydisulfate is irritating to mucous membranes.

DISPOSAL

The solutions should be flushed down the drain with water.

DISCUSSION

The procedures of this demonstration use the oxidation of iodide ions by peroxydisulfate in the presence of starch and thiosulfate to produce a clock reaction. Procedure A shows the typical iodine-clock phenomenon: a mixture of two colorless solutions remains colorless for a time, and then suddenly turns blue. Here, the impact of the effect is multiplied by presenting the reaction five times in sequence. (To conserve materials and preparation time, however, the reaction may be presented only once.) All five mixtures have the same initial composition. In Procedure B the initial concentration of potassium iodide varies, and in Procedure C the initial concentration of potassium peroxydisulfate varies. Both of these initial concentrations affect the time between mixing of the solutions and the appearance of the blue color (the clock period). Procedure D shows that the clock period is proportional to the amount of thiosulfate in the mixture. The oxidation of iodide ions by peroxydisulfate ions has been used as a laboratory experiment for many years, and is also suitable as a lecture demonstration or experiment [*1–4*].

The sudden production of the blue color in this demonstration can be explained with the following series of reactions [*4*]. Iodide ions are oxidized to triiodide ions by peroxydisulfate ions, as represented in equation 1.

$$3\ I^-(aq) + S_2O_8^{2-}(aq) \longrightarrow 2\ SO_4^{2-}(aq) + I_3^-(aq) \quad (1)$$

The triiodide ions produced in the first reaction are quickly returned to iodide ions by a reaction with thiosulfate ions, as equation 2 indicates.

$$I_3^-(aq) + 2\ S_2O_3^{2-}(aq) \longrightarrow 3\ I^-(aq) + S_4O_6^{2-}(aq) \quad (2)$$

Once all the thiosulfate ions have been consumed, the triiodide ions react with starch to form the blue starch–pentaiodide complex, as represented in equation 3 [*5*].

$$2\ I_3^-(aq) + \text{starch} \longrightarrow \text{blue starch–}I_5^-\text{ complex} + I^-(aq) \quad (3)$$

Reaction 2 is considerably faster than reaction 1, and as long as there is thiosulfate in the mixture, I_3^- is consumed as fast as it is produced. If there is an excess of peroxydisulfate ions over thiosulfate ions, the thiosulfate ions will eventually be completely consumed. When all the thiosulfate ions are consumed, the excess peroxydisulfate oxidizes iodide ions to triiodide ions, which combine with starch to form the blue complex throughout the solution.

Procedure B illustrates the effect of initial iodide ion concentration on the clock period. The results of this procedure show that doubling the concentration halves the clock period. Therefore, doubling the initial concentration of iodide ions must double the initial rate of the reaction.

In Procedure C, the effect of varying the peroxydisulfate ion concentration is observed. Doubling the initial concentration of $S_2O_8^{2-}$ cuts the clock period in half. This indicates that doubling the initial concentration of $S_2O_8^{2-}$ doubles the initial rate of the reaction.

In Procedure D, the initial concentration of thiosulfate ions is varied. The results indicate that the clock period is proportional to their initial concentration. When the initial concentration of thiosulfate ions is doubled, the clock period is doubled. This reveals that the thiosulfate ions do not have an effect on the rate of the rate-determining reaction. If thiosulfate ions were involved in the rate-determining reaction, then an increase in their concentration would increase the rate of this reaction and shorten the clock period.

Reactions 1–3 do not represent all that is occurring during this demonstration. Peroxydisulfate ions can oxidize thiosulfate ions, as indicated in equation 4 [*6*].

$$S_2O_8^{2-} + 2\ S_2O_3^{2-} \longrightarrow S_4O_6^{2-} + 2\ SO_4^{2-} \quad (4)$$

This reaction, however, is slower than the oxidation of iodide ions by peroxydisulfate ions. Therefore, it does not compete significantly with other reactions in the mixture, and it has little effect on the observations in this demonstration [*3*].

Because the sudden color change is so dramatic, many teachers have adopted methods of presenting this demonstration that emphasize the change. One method is to point at the mixture immediately before the color change takes place. This can add an aura of magic to the demonstration, but it requires careful planning and testing of the solutions to assure proper timing. A method that emphasizes the time factor is to count during the time before the color change, so that the color change occurs on the count of, perhaps, 10.

REFERENCES

1. E. Mack, Jr., and W. G. France, *A Laboratory Manual of Elementary Physical Chemistry,* 2d ed., Van Nostrand: New York (1934).
2. G. G. Evans, *J. Chem. Educ.* 29:139 (1952).
3. P. C. Moews, Jr., and R. H. Petrucci, *J. Chem. Educ.* 41:549 (1964).
4. B. Z. Shakhashiri and G. E. Dirreen, *Manual for Laboratory Investigations in General Chemistry,* Stipes Publishing Co.: Champaign, Illinois (1982).
5. R. C. Teitelbaum, S. L. Ruby, and T. J. Marks, *J. Am. Chem. Soc.* 102:3322 (1980).
6. C. H. Sorum and J. O. Edwards, *J. Am. Chem. Soc.* 74:1204 (1952).

10.6

Oxidation of Iodide by Iron(III)

When two colorless solutions are mixed, the mixture immediately turns purple. Over a period of about 10 seconds, the mixture gradually fades to colorless, and then suddenly turns dark blue (Procedure A). When another two colorless solutions are mixed, the mixture is nearly colorless for about 10 seconds, and then suddenly turns dark blue (Procedure B).

MATERIALS FOR PROCEDURE A

For preparation of stock solutions, see pages 14–15.

ca. 750 mL distilled water

10 mL concentrated (16M) nitric acid, HNO_3

2.5 g iron(III) nitrate nonahydrate, $Fe(NO_3)_3 \cdot 9H_2O$

1.8 g potassium iodide, KI

10 mL 1% starch solution (stock solution)

4.0 mL 0.20M sodium thiosulfate, $Na_2S_2O_3$ (stock solution)

1-liter beaker

stirring rod

250-mL beaker

250-mL volumetric flask or 250-mL graduated cylinder

400-mL beaker

100-mL graduated cylinder

600-mL beaker

MATERIALS FOR PROCEDURE B

For preparation of stock solutions, see pages 14–15.

ca. 750 mL distilled water

10 mL concentrated (16M) nitric acid, HNO_3

2.5 g iron(III) nitrate nonahydrate, $Fe(NO_3)_3 \cdot 9H_2O$

1.8 g potassium iodide, KI

10 mL 1% starch solution (stock solution)

1.7 mL 0.25M sodium bisulfite, $NaHSO_3$ (stock solution)

1-liter beaker
stirring rod
250-mL beaker
250-mL volumetric flask or 250-mL graduated cylinder
400-mL beaker
100-mL graduated cylinder
600-mL beaker

PROCEDURE A

Preparation

Pour 500 mL of distilled water into the 1-liter beaker, add 10 mL of concentrated HNO_3, and stir the mixture. (This solution is 0.3M in HNO_3.) Pour about 150 mL of this 0.3M HNO_3 into the 250-mL beaker. Add 2.5 g of $Fe(NO_3)_3 \cdot 9H_2O$ to the solution in the 250-mL beaker and stir the mixture to dissolve the solid. Transfer the solution to the 250-mL volumetric flask or graduated cylinder. Pour some of the 0.3M HNO_3 remaining in the 1-liter beaker into the solution to dilute it to 250 mL. (This solution is 0.025M in $Fe(NO_3)_3$ and 0.3M in HNO_3.) Pour the solution into the 400-mL beaker. Rinse the volumetric flask or graduated cylinder with distilled water.

Rinse the 250-mL beaker with distilled water and pour about 150 mL of distilled water into it. Dissolve 1.8 g of KI in the water. Add 10 mL of 1% starch solution and 4.0 mL of 0.20M $Na_2S_2O_3$ to this mixture. Pour the mixture into the 250-mL volumetric flask or graduated cylinder. Dilute the mixture to 250-mL with distilled water. (This solution is 0.043M in KI, 0.0032M in $Na_2S_2O_3$, and about 0.04% starch.) Pour this solution into the 600-mL beaker.

Presentation

Quickly pour the contents of the 400-mL beaker into the 600-mL beaker. (Upon mixing, the concentrations in the solution are 0.15M HNO_3, 0.012M $Fe(NO_3)_3$, 0.021M KI, and 0.0016M $Na_2S_2O_3$.) Immediately upon mixing, the solution will turn purple. The purple will gradually fade, and after the solution becomes colorless, it will suddenly turn dark blue.

PROCEDURE B

Preparation

Prepare a 0.025M $Fe(NO_3)_3$ solution in 0.3M HNO_3 as described in the first paragraph of Procedure A.

Pour about 150 mL of distilled water into the 250-mL beaker. Dissolve 1.8 g of KI in the water. Add 10 mL of 1% starch solution and 1.7 mL of 0.25M $NaHSO_3$ to this mixture. Pour the mixture into the 250-mL volumetric flask or graduated cylinder.

Dilute the mixture to 250-mL with distilled water. (This solution is 0.043M in KI, 0.0017M in $NaHSO_3$, and about 0.04% starch.) Pour this solution into the 600-mL beaker.

Presentation

Quickly pour the contents of the 400-mL beaker into the 600-mL beaker. (Upon mixing, the concentrations in the solution are 0.15M HNO_3, 0.012M $Fe(NO_3)_3$, 0.021M KI, and 0.0008M $Na_2S_2O_3$.) Immediately upon mixing, the solution will turn pale yellow. The yellow will gradually fade, and about 10 seconds after mixing, it will suddenly turn dark blue.

HAZARDS

Concentrated nitric acid is both a strong acid and a powerful oxidizing agent. Contact with the skin can result in severe burns. The vapor irritates the respiratory system, eyes, and other mucous membranes; therefore, concentrated nitric acid should be handled only in a well-ventilated area.

Solid iron(III) nitrate is a strong oxidizing agent and a skin irritant. Contact with combustible materials can cause fires.

Sodium bisulfite is a strong reducing agent, and the solid or concentrated solutions should not be mixed with oxidizing agents. Concentrated solutions are irritating to the skin and mucous membranes. Some persons are hypersensitive to sulfites and should avoid direct contact with them.

DISPOSAL

The excess nitric acid solution and the waste solution should be flushed down the drain with water.

DISCUSSION

In Procedure A, a colorless solution of iron(III) nitrate is mixed with a colorless solution containing thiosulfate ions, iodide ions, and starch. Immediately upon mixing, the solution turns purple. The purple color gradually fades, and the mixture becomes colorless. Once the solution has become colorless, it suddenly turns deep blue. In Procedure B, the iron(III) nitrate solution is mixed with an iodide solution containing bisulfite ions and starch. Here, the solution turns pale yellow upon mixing, gradually fades to colorless, and then suddenly turns deep blue.

The mixture of the two solutions in Procedure A turns purple because a colored complex forms between iron(III) ions and thiosulfate ions [*1, 2*].

$$[Fe(H_2O)_6]^{3+}(aq) + S_2O_3^{2-}(aq) \rightleftarrows [Fe(H_2O)_5(S_2O_3)]^{+}(aq) + H_2O(l) \quad (1)$$

The color of the mixture fades because the thiosulfate ions in the solution are consumed by the reactions that occur in the mixture, and the concentration of the complex gradually decreases. In Procedure B, iron(III) forms a complex with sulfite ions, but this complex is pale yellow and its color is not as visible as that of the thiosulfate complex in Procedure A.

The mixtures eventually turn deep blue because reactions in them produce molecular iodine. The molecular iodine, iodide ions, and starch form a complex which is blue. The complex contains pentaiodide ions (I_5^-) surrounded by coils of the starch polymer [*3*].

$$2\ I_3^-(aq) + \text{starch} \rightleftarrows \text{blue starch–}I_5^-\text{ complex} + I^-(aq) \tag{2}$$

The specific reactions that produce the clock effect (i.e., the sudden appearance of the blue color) in this demonstration are not known. However, it is easy to determine that iron(III) ions oxidize iodide ions to molecular iodine. This can be observed by mixing a solution containing $Fe^{3+}(aq)$ with another containing $I^-(aq)$. The mixture will immediately turn red-orange from the production of triiodide ions.

$$2\ Fe^{3+}(aq) + 3\ I^-(aq) \longrightarrow 2\ Fe^{2+}(aq) + I_3^-(aq) \tag{3}$$

If starch is added to the mixture, it will turn deep blue as a result of the formation of the starch-pentaiodide complex (equation 2).

That the mixtures in this demonstration do not turn deep blue immediately indicates that the thiosulfate ions (or bisulfite ions in Procedure B) interfere with the oxidation of iodide ions by iron(III). A simple explanation for this phenomenon is that any triiodide ions produced by the reaction of iodide ions with iron(III) are immediately reduced to iodide ions by thiosulfate ions (or bisulfite ions). This is the explanation provided for the thiosulfate-countered oxidation of iodide by peroxydisulfate, Demonstration 10.5 (and for the Landolt reaction, Demonstration 10.1). Applying this explanation to this demonstration by analogy yields the following sequence of reactions for Procedure A.

$$2\ Fe^{3+}(aq) + 3\ I^-(aq) \longrightarrow 2\ Fe^{2+}(aq) + I_3^-(aq) \tag{4}$$

$$I_3^-(aq) + 2\ S_2O_3^{2-}(aq) \longrightarrow 3\ I^-(aq) + S_4O_6^{2-}(aq) \tag{5}$$

$$2\ I_3^-(aq) + \text{starch} \rightleftharpoons \text{blue starch–}I_5^-\text{ complex} + I^-(aq) \tag{6}$$

Iron(III) slowly oxidizes iodide ions to triiodide ions (equation 4). As fast as triiodide ions are produced, they are consumed by thiosulfate ions (equation 5). Once all the thiosulfate has been consumed, the starch-pentaiodide complex forms, and the mixture turns blue (equation 6). For Procedure B, equation 7 would replace equation 5 in the set above.

$$I_3^-(aq) + HSO_3^-(aq) + H_2O(l) \longrightarrow 3\ I^-(aq) + SO_4^{2-}(aq) + 3\ H^+(aq) \tag{7}$$

In Procedures A and B, the same amounts of iron(III) and iodide ions are used. The amount of bisulfite ions used in Procedure B is half the amount of the thiosulfate ions used in Procedure A. As equations 5 and 7 show, it takes twice as many thiosulfate ions as bisulfite ions to consume a given number of triiodide ions. Therefore, the effective amounts of thiosulfate and bisulfite ions used in the two procedures are the same. The clock period (the length of time between mixing of the solutions and the sudden appearance of the deep blue color) in the two procedures is also the same. This indicates that the reaction that determines the length of the clock period does not involve thiosulfate or bisulfite ions.

The clock period depends on the initial concentrations of thiosulfate ions (or bisulfite ions), iodide ions, and iron(III). The clock period is directly proportional to the initial concentration of thiosulfate ions in Procedure A (and of bisulfite ions in Procedure B). If the initial concentration of thiosulfate ions is doubled, the clock period will double. This indicates that these are the limiting reagents in the reactions (see page 7 in the introduction to the chapter). Decreasing either the iodide ion concentration or the iron(III) concentration will increase the clock period. The clock period is inversely proportional to the square of the initial iodide ion concentration. Doubling the iodide ion concentration will decrease the clock period by a factor of 4. However, the relationship between the initial concentration of iron(III) and the clock period is not a simple one.

REFERENCES

1. F. M. Page, *Trans. Faraday Soc.* 49:635 (1953).
2. F. M. Page, *Trans. Faraday Soc.* 50:120 (1954).
3. R. C. Teitelbaum, S. L. Ruby, and T. J. Marks, *J. Am. Chem. Soc.* 102:3322 (1980).

10.7

Hydrolysis of 2-Chloro-2-Methylpropane

A small amount of a colorless liquid is stirred into a blue solution. After a short time, the mixture suddenly turns yellow (Procedure A). Increasing the initial concentration of sodium hydroxide in the mixture lengthens the time until the color changes (Procedure B). Increasing the initial concentration of 2-chloro-2-methylpropane in the mixture shortens the time until the color changes (Procedure C). Varying the fraction of acetone in the solvent affects the time required for the color change (Procedure D). The solvent effect is also shown using an overhead projector (Procedure E) [*1*, *2*].

MATERIALS FOR PROCEDURE A

1.0 mL 2-chloro-2-methylpropane (tertiary-butyl chloride, *t*-BuCl), $(CH_3)_3CCl$

75 mL acetone, CH_3COCH_3

20.0 mL 0.10M sodium hydroxide, NaOH (To prepare 1.0 liter of stock solution, dissolve 4.0 g of NaOH in about 600 mL of distilled water, and dilute the resulting solution to 1.0 liter with distilled water.)

5 mL bromothymol blue indicator solution (To prepare 100 mL of stock solution, dissolve 0.04 g of the sodium salt of bromothymol blue [3′,3″-dibromothymolsulfonephthalein] in 100 mL of distilled water.)

400 mL distilled water

1.0-mL volumetric pipette, with bulb

250-mL Erlenmeyer flask, with stopper

250-mL graduated cylinder

600-mL beaker

25-mL graduated cylinder

stirring rod

MATERIALS FOR PROCEDURE B

3.0 mL 2-chloro-2-methylpropane (tertiary-butyl chloride, *t*-BuCl), $(CH_3)_3CCl$

240 mL acetone, CH_3COCH_3

60.0 mL 0.10M sodium hydroxide, NaOH (See Materials for Procedure A for preparation.)

15 mL bromothymol blue indicator solution (See Materials for Procedure A for preparation.)

1200 mL distilled water

1.0-mL volumetric pipette, with bulb

3 250-mL Erlenmeyer flasks, with stoppers

3 600-mL beakers, with labels

50-mL graduated cylinder

timing device capable of measuring seconds

3 stirring rods

MATERIALS FOR PROCEDURE C

3.5 mL 2-chloro-2-methylpropane (tertiary-butyl chloride, *t*-BuCl), $(CH_3)_3CCl$

240 mL acetone, CH_3COCH_3

60.0 mL 0.10M sodium hydroxide, NaOH (See Materials for Procedure A for preparation.)

15 mL bromothymol blue indicator solution (See Materials for Procedure A for preparation.)

1200 mL distilled water

3 250-mL Erlenmeyer flasks, with stoppers and labels

5.0-mL graduated pipette, with bulb

25-mL graduated cylinder

3 600-mL beakers

3 stirring rods

timing device capable of measuring seconds

MATERIALS FOR PROCEDURE D

3.0 mL 2-chloro-2-methylpropane (tertiary-butyl chloride, *t*-BuCl), $(CH_3)_3CCl$

300 mL acetone, CH_3COCH_3

1.2 liters distilled water

60.0 mL 0.10M sodium hydroxide, NaOH (See Materials for Procedure A for preparation.)

12 mL bromothymol blue indicator solution (See Materials for Procedure A for preparation.)

3 250-mL Erlenmeyer flasks, with labels and stoppers

1.0-mL volumetric pipette, with bulb

250-mL graduated cylinder

3 600-mL beakers, with labels

25-mL graduated cylinder

timing device capable of measuring seconds

3 stirring rods

MATERIALS FOR PROCEDURE E

overhead projector, with transparency and marking pen

6.0 mL 0.1M 2-chloro-2-methylpropane (tertiary-butyl chloride, *t*-BuCl), $(CH_3)_3CCl$, in acetone (To prepare 100 mL of stock solution, dissolve 1.1 mL of $(CH_3)_3CCl$ in 75 mL of acetone, and dilute the resulting solution to 100 mL.)

15.0 mL distilled water

6.0 mL 0.010M sodium hydroxide, NaOH (To prepare 100 mL of stock solution, dissolve 0.40 g of NaOH in 75 mL of distilled water and dilute the resulting solution to 100 mL.)

3.0 mL acetone, CH_3COCH_3

9 drops bromothymol blue indicator solution (For preparation, see Materials for Procedure A.)

10-mL graduated pipette

3 test tubes, 10-mm × 75-mm, with stoppers

3 50-mL beakers, with labels

timing device capable of measuring seconds

PROCEDURE A

Preparation

Pipette 1.0 mL of *t*-BuCl into the Erlenmeyer flask. Add 75 mL of acetone to the flask. Stopper the flask and swirl it to mix its contents.

Use the graduated cylinder to measure 20 mL of 0.10M NaOH and 5 mL of bromothymol blue indicator solution into the 600-mL beaker. Pour 400 mL of distilled water into the 600-mL beaker.

Presentation

Pour the *t*-BuCl–acetone solution from the Erlenmeyer flask into the beaker and stir the mixture. (The initial concentrations of this mixture are 0.004M NaOH and 0.02M *t*-BuCl.) The mixture will remain blue for about 15 seconds and then suddenly turn yellow.

PROCEDURE B

Preparation

Pipette 1.0 mL of *t*-BuCl into each of the three Erlenmeyer flasks. Add 80 mL of acetone to each flask. Stopper the flasks and swirl them to mix their contents.

Label one of the 600-mL beakers "10 mL NaOH," another "20 mL NaOH," and the third "30 mL NaOH." Use the 50-mL graduated cylinder to measure the appropriate volume of 0.10M NaOH into each of the beakers. Add 5 mL of bromothymol blue indicator solution to each beaker. Add enough distilled water to each beaker to bring the total volume of the solution in each to 425 mL (410 mL in the "10 mL NaOH" beaker, 400 mL in the "20 mL NaOH" beaker, and 390 mL in the "30 mL NaOH" beaker).

Presentation

Pour the *t*-BuCl–acetone solution from one of the Erlenmeyer flasks into the beaker labelled "20 mL NaOH," start the timer, and stir the mixture. (The initial concentrations of this mixture are 0.004M NaOH and 0.02M *t*-BuCl.) The mixture will remain blue for about 15 seconds and then suddenly turn yellow. Stop the timer as soon as the color change occurs. Record the time required for the color to change (the clock period).

Repeat the procedure of the previous paragraph, pouring the solution from one of the remaining flasks into the beaker labelled "10 mL NaOH." (The initial concentrations of this mixture are 0.002M NaOH and 0.02M *t*-BuCl.) This mixture will remain blue for about 8 seconds, and then suddenly turn yellow. Record the time required for the color to change.

Repeat the procedure once more, pouring the solution from the remaining flask into the beaker labelled "30 mL NaOH." (The initial concentrations of this mixture are 0.006M NaOH and 0.02M *t*-BuCl.) This mixture will remain blue for about 25 seconds, and then suddenly turn yellow. Record the time required for the color to change. Note the effect of the initial concentration of NaOH on the clock period.

PROCEDURE C

Preparation

Label one of the 250-mL Erlenmeyer flasks "0.5 mL *t*-BuCl," another "1.0 mL *t*-BuCl," and the third "2.0 mL *t*-BuCl." Pipette 3.5 mL of *t*-BuCl into the flask labelled "2.0 mL *t*-BuCl." Pour 140 mL of acetone into this flask. Stopper the flask and swirl it to mix its contents. Pour 20 mL of this solution into the flask labelled "0.5 mL *t*-BuCl," and 40 mL into the flask labelled "1.0 mL *t*-BuCl." Pour 60 mL of acetone into the flask labelled "0.5 mL *t*-BuCl" and containing 20 mL of solution. Pour 40 mL of acetone into the flask labelled "1.0 mL *t*-BuCl" and containing 40 mL of solution. Stopper all the flasks and swirl them to mix their contents. The flasks each hold 80 mL of solution containing the labelled volume of *t*-BuCl.

Use the 25-mL graduated cylinder to measure 20.0 mL of 0.10M NaOH to add to each of the three 600-mL beakers. Add 5 mL of bromothymol blue indicator solution to each beaker. Pour 400 mL of water into each beaker. Stir the mixture in each beaker.

Presentation

Pour the *t*-BuCl–acetone solution from the Erlenmeyer flask labelled "1.0 mL *t*-BuCl" into one of the beakers, start the timer, and stir the mixture. (The initial concentrations of this mixture are 0.004M NaOH and 0.02M *t*-BuCl.) The mixture will remain blue for about 15 seconds and then suddenly turn yellow. Stop the timer as soon as the color change occurs. Record the time required for the color to change (the clock period).

Repeat the procedure of the previous paragraph, pouring the solution from the flask labelled "2.0 mL *t*-BuCl" into one of the remaining beakers. (The initial concentrations of this mixture are 0.004M NaOH and 0.04M *t*-BuCl.) This mixture will remain blue for about 8 seconds, and then suddenly turn yellow. Record the time required for the color to change.

Repeat the procedure once more, pouring the solution from the remaining flask, labelled "0.5 mL *t*-BuCl" into the remaining beaker. (The initial concentrations of this mixture are 0.004M NaOH and 0.01 M *t*-BuCl.) This mixture will remain blue for about 30 seconds, and then suddenly turn yellow. Record the time required for the color to change. Note the effect of the initial concentration of *t*-BuCl on the clock period.

PROCEDURE D

Preparation

Label one of the 250-mL Erlenmeyer flasks "75 mL acetone," another "100 mL acetone," and the third "125 mL acetone." Pipette 1.0 mL of *t*-BuCl into each flask. Use the 250-mL graduated cylinder to measure into each flask the amount of acetone indicated on its label. Stopper the flasks and swirl them to mix their contents.

Label one of the 600-mL beakers "425 mL water," another "400 mL water," and the third "375 mL water." Use the 25-mL graduated cylinder to measure 20.0 mL of 0.10M NaOH and 4 mL of bromothymol blue indicator solution to add to each beaker. Pour into each beaker the amount of distilled water indicated on its label.

Presentation

Pour the *t*-BuCl–acetone solution from the Erlenmeyer flask labelled "100 mL acetone" into the beaker labelled "400 mL water," start the timer, and stir the mixture. The mixture will remain blue for about 20 seconds and then suddenly turn yellow. Stop the timer as soon as the color change occurs. Record the time required for the color to change (the clock period).

Repeat the procedure of the previous paragraph, pouring the solution from the flask labelled "125 mL acetone" into the beaker labelled "375 mL water." This mixture will remain blue for about 30 seconds, and then suddenly turn yellow. Record the time required for the color to change.

Repeat the procedure once more, pouring the solution from the flask labelled "75 mL acetone" into the beaker labelled "425 mL water." This mixture will remain blue for about 15 seconds, and then suddenly turn yellow. Record the time required for the color to change, and note the effect of the solvent composition on the clock period.

PROCEDURE E

Preparation

Use the pipette to measure 2.0 mL of 0.1M *t*-BuCl into each of the three test tubes. Stopper the test tubes.

Label the three 50-mL beakers from 1 to 3. No more than 10 minutes before presenting the demonstration, combine the materials as listed in the chart below. Mix the contents by swirling the beakers.

Material	Beaker 1	Beaker 2	Beaker 3
water	4.0 mL	5.0 mL	6.0 mL
0.01M NaOH	2.0 mL	2.0 mL	2.0 mL
acetone	2.0 mL	1.0 mL	none
bromothymol blue	3 drops	3 drops	3 drops

Presentation

Place a transparency on the overhead projector and set the three beakers on the transparency. Write "60%" next to beaker 1, "70%" next to beaker 2, and "80%" next to beaker 3. (These indicate the percentage of water in the solvent once the reactions have been initiated.)

Pour 2.0 mL of 0.1M *t*-BuCl from one of the test tubes into beaker 1, swirl the beaker to mix the contents, and begin timing. Swirl the beaker occasionally to prevent local heating of the solution. Record on the transparency the time required for the color of the mixture to change (the clock period).

Repeat the procedure in the previous paragraph with the other two beakers. Note the effect of differing solvent compositions on the clock period.

HAZARDS

2-Chloro-2-methylpropane is flammable and should be kept away from open flames.

Acetone is flammable and should be kept away from open flames. Acetone is toxic by inhalation and ingestion, and it should be handled only in well-ventilated areas. The liquid can cause severe eye damage.

Solid sodium hydroxide and its concentrated solutions can cause severe burns to the eyes, skin, and mucous membranes. Dust from solid sodium hydroxide is very irritating to the eyes and respiratory system.

DISPOSAL

The waste solutions should be flushed down the drain with water.

DISCUSSION

In this demonstration, a colorless solution is mixed with a blue solution. The mixture remains blue for a short while, then suddenly turns yellow. This clock effect is shown in a single instance in Procedure A. In Procedure B, the initial concentration of sodium hydroxide in the mixture is varied. As the concentration is increased, the time between mixing and the color change (the clock period) increases proportionally. In Procedure C, the initial concentration of 2-chloro-2-methylpropane is varied. As the concentration of this component increases, the clock period decreases proportionally. In all of these mixtures, the solvent is a combination of acetone and water. In Procedure D, the composition of the solvent is changed. As the proportion of water in the mixture increases, the clock period decreases. A similar effect is revealed in Procedure E using an overhead projector.

The color change in these mixtures is caused by the pH indicator bromothymol blue. This indicator is blue in basic solutions and yellow in acidic solution. The pK_a of bromothymol blue is 6.8. A color change of bromothymol blue indicates that a reaction which affects the pH of the mixture must be occurring in the mixture. The reaction is the hydrolysis of 2-chloro-2-methylpropane (tertiary-butyl chloride, *t*-BuCl), which is represented by equation 1.

$$CH_3\text{—}\underset{\underset{\displaystyle CH_3}{|}}{\overset{\overset{\displaystyle Cl}{|}}{C}}\text{—}CH_3 + H_2O \longrightarrow CH_3\text{—}\underset{\underset{\displaystyle CH_3}{|}}{\overset{\overset{\displaystyle OH}{|}}{C}}\text{—}CH_3 + H^+ + Cl^- \qquad (1)$$

Because the hydrolysis releases H^+, the pH of the mixture decreases as the reaction proceeds. The mixture is blue initially because it contains sodium hydroxide, which makes the mixture basic. As the hydrolysis of *t*-BuCl proceeds, it generates hydrogen ions which neutralize the hydroxide ions of the sodium hydroxide. All the mixtures contain an excess of *t*-BuCl over sodium hydroxide. Eventually, all the hydroxide ions from sodium hydroxide are consumed. At this point the mixture becomes acidic and the bromothymol blue indicator changes from blue to yellow.

Procedure B reveals the effect of varying the initial amount of sodium hydroxide in the mixture. The three solutions in Procedure B have the same solvent (16% by volume of acetone in water) and the same initial concentration of *t*-BuCl (0.02M). The solutions differ in their initial sodium hydroxide concentrations (0.002M, 0.004M, and 0.006M). The results of Procedure B show that the clock period is directly proportional to the initial concentration of sodium hydroxide. This is not surprising, because the color change occurs when the sodium hydroxide has been neutralized by the hydrolysis of *t*-BuCl. The direct proportionality indicates that sodium hydroxide is not involved in the rate-determining process in the reaction (see page 7 in the introduction to this chapter). If it were involved, increasing its concentration would increase the rate of the rate-determining process. Increasing the rate would tend to shorten the clock period. If sodium hydroxide were involved in the rate-determining process, then the effect on the clock period of increasing the initial concentration of sodium hydroxide would be more complicated than a proportional lengthening. The complication is due to the operation of two opposing factors—the additional sodium hydroxide that must be consumed during the clock period, and the increased rate at which it would be consumed.

Procedure C shows that an increase in the initial concentration of *t*-BuCl produces a proportional shortening of the clock period. This indicates that the initial concentration of 2-chloro-2-methylpropane does affect the rate of the rate-determining process. Therefore, 2-chloro-2-methylpropane is involved in this process.

There are at least two possibilities for the rate-determining process. One is the reaction of a molecule of 2-chloro-2-methylpropane with a water molecule to produce 2-methyl-2-propanol and HCl.

$$H_2O + (CH_3)_3C\text{—}Cl \longrightarrow \underset{\underset{\displaystyle H}{|}}{HO}\cdots\underset{\underset{\displaystyle CH_3}{|}}{\overset{H_3C\ \ \diagdown\ \diagup\ \ CH_3}{C}}\cdots Cl \longrightarrow HO\text{—}C(CH_3)_3 + Cl^- + H^+ \qquad (2)$$

The rate of this process would be proportional to the concentration of 2-chloro-2-methylpropane and independent of the concentration of sodium hydroxide. Another possibility is the slow dissociation of a molecule of 2-chloro-2-methylpropane into a chloride ion and a $(CH_3)_3C^+$ ion, followed by the rapid reaction of the $(CH_3)_3C^+$ ion with water to form the alcohol.

$$(CH_3)_3CCl \longrightarrow (CH_3)_3C^+ + Cl^- \quad \text{(slow)} \quad (3)$$

$$(CH_3)_3C^+ + H_2O \longrightarrow (CH_3)_3COH + H^+ \quad \text{(fast)} \quad (4)$$

The rate of this process would also depend on the concentration of 2-chloro-2-methylpropane and be independent of the concentration of NaOH.

The solvent used in this demonstration is a mixture of acetone and water. Acetone is used because 2-chloro-2-methylpropane is not soluble in pure water. Water is used because sodium hydroxide is not soluble in pure acetone. To dissolve both of these substances in a single solution, a mixed solvent is used. The concentration of each solvent component must be quite high to keep both solutes in solution. In Procedures D and E the effect of solvent composition on the clock period is investigated. These procedures reveal that, as the proportion of water in the solvent increases, the clock period decreases. This might suggest that the process described by equation 2 is the rate-determining process. In equation 2, water is involved as a reactant, and increasing its concentration would increase the rate of the process. However, the concentration of water used in all trials in Procedures D and E is very high. The concentration is much too high to reveal a concentration effect. What is observed is the effect of the dielectric constant of the reaction medium. Increasing the concentration of water in the solvent makes it a better solvent for ions. The process described by equation 2 involves the merging of two neutral molecules. The rate at which this occurs should be unaffected by the ion-dissolving properties of the solvent. Instead, the rate of process of equation 3, which involves the dissociation of a single neutral molecule into ions, would be much more sensitive to the composition of the solvent.

The process represented by equation 2 is called a bimolecular nucleophilic substitution, abbreviated as S_N2. The mechanism is a substitution, because hydroxide is substituted for chloride in the product. It is bimolecular, because it involves the interaction of two particles: a water molecule and a molecule of *t*-BuCl. It is nucleophilic ("nucleus-loving") because the partly negative oxygen atom of water is attracted to the central carbon atom of *t*-BuCl, which has a slight positive (i.e., nuclear) charge. The process described by equations 3 and 4 is called a unimolecular nucleophilic substitution (S_N1).

Another reaction in this mixture that becomes significant only under acidic conditions is the elimination of hydrogen chloride (HCl) from *t*-BuCl (equation 5).

$$CH_3-\underset{\underset{CH_3}{|}}{\overset{\overset{Cl}{|}}{C}}-CH_3 \longrightarrow CH_3-\underset{\underset{CH_3}{|}}{C}=CH_2 + H^+ + Cl^- \quad (5)$$

The elimination reaction produces hydrogen chloride, as does the substitution reaction, and therefore also results in a decrease in the pH of the mixture as the reaction proceeds. The elimination reaction can proceed by either a bimolecular elimination (E2) or unimolecular elimination (E1) mechanism. The E1 mechanism involves the same slow step as the S_N1 mechanism, and is affected in the same way by solvent polarity. Therefore, the proportion of substitution to elimination is of no consequence to this demonstration. The hydroxide ion concentration decreases at the same rate whether substitution, elimination, or some combination of the two is occurring.

Other indicators can be used in place of bromothymol blue in this demonstration. As long as the indicator changes color in the pH range from about 5 to about 9, a sharp color change will occur. If universal indicator is used, the mixture will undergo several gradual color changes.

REFERENCES

1. J. A. Landgrebe, *J. Chem. Educ.* 41:567 (1964).
2. J. T. Riley, *J. Chem. Educ.* 54:29 (1977).

10.8

Aldehyde-Acetone Condensation

Two colorless solutions are mixed, and the mixture becomes yellow but remains clear for 45–90 seconds, after which time a cloud of crystals suddenly forms throughout the solution [*1, 2*].

MATERIALS FOR PROCEDURE A

20.0 mL cinnamaldehyde, $C_6H_5CH{=}CH{-}CHO$

125 mL 95% ethanol, C_2H_5OH

75 mL 2M potassium hydroxide, KOH (To prepare 250 mL of stock solution, dissolve 28 g of KOH in about 200 mL of distilled water and dilute the solution to 250 mL.)

10.0 mL acetone, CH_3COCH_3

400-mL beaker

stirring rod

timing device capable of measuring seconds (optional)

MATERIALS FOR PROCEDURE B

20.0 mL benzaldehyde, C_6H_5CHO

125 mL 95% ethanol, C_2H_5OH

75 mL 4M potassium hydroxide, KOH (To prepare 250 mL of stock solution, dissolve 56 g of KOH in about 200 mL of distilled water and dilute the solution to 250 mL.)

10.0 mL acetone, CH_3COCH_3

600-mL beaker

stirring rod

timing device capable of measuring seconds (optional)

PROCEDURE A

Preparation and Presentation

In a 400-mL beaker, dissolve 20.0 mL of cinnamaldehyde in 125 mL of 95% ethanol. Add 75 mL of 2M KOH and stir the mixture to form a homogeneous solution.

Quickly stir 10.0 mL (0.14 mole) of acetone into the solution in the beaker and begin timing. (The mixture is 0.8M in cinnamaldehyde, 0.6M in acetone, and 0.7M in KOH.) The solution will turn yellow and remain clear for about 30 seconds. Then, a cloud of bright yellow crystals will suddenly appear throughout the solution.

PROCEDURE B

Preparation and Presentation

In a 600-mL beaker, dissolve 20.0 mL of benzaldehyde in 125 mL of 95% ethanol. Add 75 mL of 4M KOH and stir the mixture to form a homogeneous solution.

Quickly stir 10.0 mL (0.14 mole) of acetone into the solution in the beaker and begin timing. (The mixture is 0.9M in benzaldehyde, 0.6M in acetone, and 1.4M in KOH.) The color of the solution will gradually change from pale yellow to deep orange over a span of about 30 seconds, and after this time a cloud of pale yellow crystals will suddenly appear throughout the solution.

HAZARDS

Potassium hydroxide can cause severe burns of the eyes and skin. Dust from solid potassium hydroxide is very caustic.

Benzaldehyde and cinnamaldehyde can cause skin irritations. High concentrations of their vapors may irritate the eyes and respiratory system. Both can be toxic if taken internally at high doses (LD_{50}—lethal dose in 50% of the cases—in rats is 1–2 g/kg).

DISPOSAL

The solutions should be flushed down the drain with water.

DISCUSSION

In the reactions that occur in both procedures of this demonstration, an aldehyde combines with acetone under the influence of hydroxide ions. In Procedure A the aldehyde is cinnamaldehyde, and in Procedure B it is benzaldehyde.

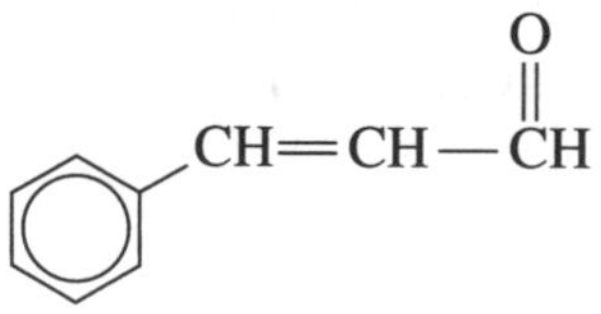

$$C_6H_5{-}CH{=}CH{-}\overset{\overset{\displaystyle O}{\|}}{C}H \qquad C_6H_5{-}\overset{\overset{\displaystyle O}{\|}}{C}H$$

cinnamaldehyde 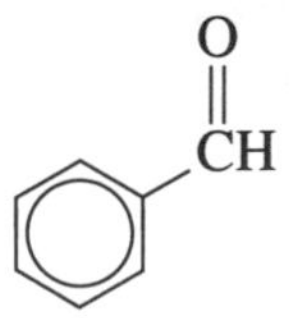 benzaldehyde

The overall reaction is represented in equation 1, where R represents a phenyl group (C_6H_5—) in the case of benzaldehyde or C_6H_5—CH=CH— in the case of cinnamaldehyde.

$$2\,\mathrm{RCH(=O)} + \mathrm{CH_3C(=O)CH_3} \xrightarrow{\mathrm{OH^-}} \mathrm{R{-}CH{=}CH{-}C(=O){-}CH{=}CH{-}R} + 2\,\mathrm{H_2O} \quad (1)$$

When the aldehyde is benzaldehyde, the product is 1,5-diphenyl-1,4-pentadien-3-one (dibenzalacetone); when the aldehyde is cinnamaldehyde, the product is 1,9-diphenyl-1,3,6,8-nonatetraen-5-one (dicinnamalacetone).

The reaction involves two sequential *aldol condensations*. The reaction is called an aldol condensation because it involves an aldehyde as the reactant and it produces an alcohol. With the aldehydes used here, the product alcohol undergoes a further reaction, a dehydration, to produce the final product. The mechanism of aldol condensations is well understood [*3*]. The mechanism for the reaction that occurs in the demonstration begins with the removal by hydroxide ion of a hydrogen ion from acetone. Hydrogen atoms bonded to a carbon atom attached to a carbonyl group (C=O) are somewhat acidic.

$$\mathrm{CH_3C(=O)CH_3} + \mathrm{OH^-} \rightleftarrows \mathrm{CH_3C(=O)CH_2^-} + \mathrm{H_2O} \quad (2)$$

The resulting anion of acetone adds to the aldehyde at its carbonyl carbon atom (the carbon atom doubly bonded to an oxygen atom).

$$\mathrm{CH_3C(=O)CH_2^-} + \mathrm{R{-}CH(=O)} \rightleftarrows \mathrm{CH_3C(=O)CH_2{-}C(O^-)(R){-}H} \quad (3)$$

Then, as indicated in equation 4, the product anion removes a hydrogen ion from water, regenerating the hydroxide ion consumed in the first step. The product is an alcohol, the alcohol of the "aldol" reaction.

$$\mathrm{CH_3C(=O)CH_2{-}C(O^-)(R){-}H} + \mathrm{HOH} \rightleftarrows \mathrm{CH_3C(=O)CH_2{-}C(OH)(R){-}H} + \mathrm{OH^-} \quad (4)$$

Each of the previous steps is an equilibrium. For each of these equilibria, the equilibrium constant is less than 1, and the position of each equilibrium lies to the left. What drives each of these ultimately to the right is the dehydration of the alcohol. The alcohol dehydrates, producing a ketone (equation 5).

$$\mathrm{CH_3C(=O)CH_2{-}C(OH)(R){-}H} \longrightarrow \mathrm{CH_3C(=O)CH{=}CH(R)} + \mathrm{H_2O} \quad (5)$$

The product ketone has more double bonds alternating with single bonds (i.e., it is more conjugated) than does the alcohol. (Recall that the R group in both procedures contains a phenyl group, which is conjugated.) The increased conjugation causes the product to absorb light at longer wavelengths, so the product is a different color from the reactants. As the amount of the product increases, the color of the mixture gradually deepens and changes from yellow to yellow-orange. Because this ketone is soluble in the ethanol solvent, the mixture remains clear while the reaction proceeds.

The product ketone can undergo the same four reaction steps as acetone. Therefore, the reaction continues with the hydroxide ion removing a hydrogen ion from the ketone (equation 6).

$$RCH{=}CH\overset{\overset{\large O}{\|}}{C}CH_3 + OH^- \rightleftarrows RCH{=}CH\overset{\overset{\large O}{\|}}{C}CH_2^- + H_2O \qquad (6)$$

The organic anion formed in this reaction adds to another molecule of aldehyde, as did the acetone anion in equation 3.

$$RCH{=}CH\overset{\overset{\large O}{\|}}{C}CH_2^- + R{-}\overset{\overset{\large O}{\|}}{C}H \rightleftarrows RCH{=}CH\overset{\overset{\large O}{\|}}{C}CH_2{-}\underset{\underset{\large R}{|}}{\overset{\overset{\large O^-}{|}}{C}}{-}H \qquad (7)$$

In a reaction analogous to equation 4, this addition product reacts with water to form an alcohol.

$$RCH{=}CH\overset{\overset{\large O}{\|}}{C}CH_2{-}\underset{\underset{\large R}{|}}{\overset{\overset{\large O^-}{|}}{C}}{-}H + HOH \rightleftarrows RCH{=}CH\overset{\overset{\large O}{\|}}{C}CH_2{-}\underset{\underset{\large R}{|}}{\overset{\overset{\large OH}{|}}{C}}{-}H + OH^- \qquad (8)$$

This alcohol dehydrates to form the ultimate product of the reaction, dibenzalacetone or dicinnamalacetone.

$$RCH{=}CH\overset{\overset{\large O}{\|}}{C}CH_2{-}\underset{\underset{\large R}{|}}{\overset{\overset{\large OH}{|}}{C}}{-}H \longrightarrow RCH{=}CH\overset{\overset{\large O}{\|}}{C}CH{=}\underset{\underset{\large R}{|}}{C}H + H_2O \qquad (9)$$

Because this double addition product is even more conjugated than the single addition product, the color of the mixture changes from yellow-orange to orange. Furthermore, the double addition product is not very soluble in the ethanol-water mixture used as the solvent, and it precipitates out of the solution.

Exactly how this reaction functions as a clock is not known. It has been suggested that the solution gradually becomes supersaturated in the double addition product, the concentration eventually becomes too great to remain in solution, and the excess suddenly precipitates [*1*]. The solution supersaturates because it lacks suitable sites (nucleation sites) for crystals to begin forming. Once the crystals do start forming, the formation is very rapid, because the solution is quite supersaturated.

Because the reaction involves so many steps, the relationship between the clock period and the initial concentrations of the components is complex. However, within limits, the clock period is inversely proportional to the initial concentration of hydroxide ions. Doubling the initial concentration of hydroxide ions decreases the clock period by half. At similar hydroxide ion concentrations, the clock period for the cinnamaldehyde reaction is about half that for the benzaldehyde reaction. To obtain similar clock periods from these two aldehydes, the hydroxide ion concentration used with benzaldehyde is twice that used with cinnamaldehyde. (Potassium hydroxide is used in this demonstration rather than sodium hydroxide, because potassium hydroxide is much more soluble in the alcohol solvent used to dissolve the aldehydes.) Increasing the initial concentration of acetone decreases the clock period only slightly. A similar effect is produced by changing the initial concentration of aldehyde. The composition of the solvent has an effect on the appearance of the clock reaction. If the ratio of ethanol to water is much greater than that used in the demonstration, the appearance of precipitate is gradual. If the ratio is much less, then the reactants will not remain in solution.

As is the case for most volatile aldehydes, cinnamaldehyde and benzaldehyde have strong odors. Cinnamaldehyde is found in oil of cinnamon and has a strong odor of cinnamon. Benzaldehyde is artificial essence of almond oil. Both are used as artificial flavoring in foods and in perfumes.

REFERENCES

1. L. C. King and G. K. Ostrum, *J. Chem. Educ.* 41:A139 (1964).
2. L. R. Summerlin, C. L. Borgford, and J. B. Ealy, *Chemical Demonstrations: A Sourcebook for Teachers,* Vol. 2, p. 157, American Chemical Society: Washington, D.C. (1987).
3. C. D. Gutsche and D. J. Pasto, *Fundamentals of Organic Chemistry,* pp. 923–25, Prentice-Hall: Englewood Cliffs, New Jersey (1975).

10.9

Formaldehyde-Sulfite Complex Formation

Three sets of clear, colorless solutions are mixed. After about 20 seconds, each mixture undergoes a sudden color change, one becoming clear red, another cloudy white, and the third clear blue (Procedure A) [*1*]. Three sets of liquids are mixed, producing a colorless mixture, a yellow mixture, and a red mixture. After about 20 seconds the first mixture changes from colorless to yellow, the second from yellow to red, and the third from red to blue (Procedure B).

MATERIALS FOR PROCEDURE A

22 mL 37% aqueous formaldehyde (formalin), CH_2O

2 liters distilled water

6.3 g sodium sulfite, Na_2SO_3

21 g sodium bisulfite, $NaHSO_3$

0.37 g disodium ethylenediaminetetraacetate dihydrate, $Na_2H_2[CH_2N(CH_2CO_2)_2]_2 \cdot 2H_2O$ (optional)

2–3 mL phenolphthalein indicator solution (To prepare 50 mL of stock solution, dissolve 0.05 g of phenolphthalein in 50 mL of 95% ethanol.)

5 mL 2M aqueous magnesium chloride solution, $MgCl_2$ (To prepare 50 mL of stock solution, dissolve 10 g of $MgCl_2$ in 40 mL of distilled water, and dilute the solution to 50 mL.)

2–3 mL thymolphthalein indicator solution (To prepare 100 mL of stock solution, dissolve 0.04 g of thymolphthalein in 50 mL of 95% ethanol, and dilute the resulting solution with 50 mL of distilled water.)

2 1-liter Erlenmeyer flasks

3 600-mL beakers

3 stirring rods

3 250-mL beakers

250-mL graduated cylinder

10-mL graduated cylinder

MATERIALS FOR PROCEDURE B

22 mL 37% aqueous formaldehyde (formalin), CH_2O

2 liters distilled water

6.3 g sodium sulfite, Na_2SO_3

21 g sodium bisulfite, $NaHSO_3$

0.37 g disodium ethylenediaminetetraacetate dihydrate, $Na_2H_2[CH_2N(CH_2CO_2)_2]_2 \cdot 2H_2O$ (optional)

2–3 mL *m*-nitrophenol indicator solution (To prepare 50 mL of stock solution, dissolve 0.15 g of *m*-nitrophenol in 50 mL of distilled water.)

2–3 mL phenol red indicator solution (To prepare 250 mL of stock solution, mix 0.1 g of phenol red [phenolsulfonephthalein] with 28 mL of 0.01M NaOH, dissolve the mixture in 200 mL of distilled water, and dilute the resulting solution to 250 mL with distilled water.)

1–2 mL 4,4′-bis(4-amino-1-naphthylazo)-2,2′-stilbene disulfonic acid indicator solution (To prepare 100 mL of stock solution, dissolve 0.1 g of 4,4′-bis(4-amino-1-naphthylazo)-2,2′-stilbene disulfonic acid† in 6 mL of 0.05M NaOH, and dilute the mixture to 100 mL with distilled water.)

2 1-liter Erlenmeyer flasks

3 600-mL beakers

3 stirring rods

3 250-mL beakers

250-mL graduated cylinder

10-mL graduated cylinder

PROCEDURE A

Preparation

Solution A. This solution should be prepared at least 24 hours *before* use. In a 1-liter flask, dilute 22 mL of 37% CH_2O to 1.0 liter with distilled water. (This solution is 0.3M in CH_2O.)

Solution B. This solution must be prepared *within* 24 hours of use. In another 1-liter flask, dissolve 6.3 g of Na_2SO_3 and 21 g of $NaHSO_3$ in about 800 mL of distilled water. Dilute the solution to 1.0 liter with distilled water. (This solution is 0.05M in Na_2SO_3 and 0.2M in $NaHSO_3$.) (The shelf life of this solution may be prolonged to several days by the addition of 0.37 g of disodium ethylenediaminetetraacetate dihydrate [2].)

Place three 600-mL beakers in a row on the display table, and put a stirring rod in each. Set a 250-mL beaker behind each of the 600-mL beakers. Using a 250-mL graduated cylinder, pour 200 mL of solution A into each of the 600-mL beakers. Rinse the cylinder with distilled water and use it to pour 200 mL of solution B into each of the 250-mL beakers. Using the 10-mL graduated cylinder, add 2–3 mL of phenolphthalein indicator solution to the first 250-mL beaker, 5 mL of 2M $MgCl_2$ solution to the second, and 2–3 mL of thymolphthalein indicator solution to the third.

† 4,4′-bis(4-amino-1-naphthylazo)-2,2′-stilbene disulfonic acid is available from Kodak Laboratory and Research Products, 343 State Street, Rochester, New York, 14650; telephone: 1-800-225-5352.

Presentation

Quickly pour the contents of each 250-mL beaker into the 600-mL beaker in front of it, and stir the mixtures. After 20–30 seconds, the mixture in the first beaker will become red, the second white, and the third blue.

PROCEDURE B

Preparation

Prepare solutions A and B as directed in Procedure A.

Place three 600-mL beakers in a row on the display table, and put a stirring rod in each. Set a 250-mL beaker behind each of the 600-mL beakers. With a 250-mL graduated cylinder, pour 200 mL of solution A into each of the 600-mL beakers. Rinse the graduated cylinder with distilled water and use it to pour 200 mL of solution B into each of the 250-mL beakers. Using the 10-mL graduated cylinder, add 2–3 mL of *m*-nitrophenol indicator solution to the first 250-mL beaker, 2–3 mL of phenol red indicator solution to the second, and 1–2 mL of 4,4′-bis(4-amino-1-naphthylazo)-2,2′-stilbene disulfonic acid indicator solution to the third. Stir each mixture.

Presentation

Quickly pour the contents of each 250-mL beaker into the 600-mL beaker in front of it, and stir the mixtures. After 20–30 seconds, the mixture in the first beaker will change from colorless to yellow, the second from yellow to red, and the third from red to blue.

HAZARDS

Formaldehyde vapors are irritating to the eyes and respiratory system. It is toxic by ingestion.

m-Nitrophenol is toxic by ingestion.

The toxicological properties of 4,4′-bis(4-amino-1-naphthylazo)-2,2′-stilbene disulfonic acid are not known, and it should be handled with caution.

DISPOSAL

The waste solutions should be flushed down the drain with water.

DISCUSSION

The reaction between sulfite and formaldehyde in the presence of bisulfite was first described by Wagner [*3*]. The net reaction is represented by

$$H_2O + SO_3^{2-} + H{-}\overset{\overset{\large O}{\|}}{C}{-}H \longrightarrow \begin{matrix} H & & OH \\ & C & \\ H & & SO_3^- \end{matrix} + OH^- \qquad (1)$$

Because the reaction produces hydroxide ions, it causes a change in the pH of the solution, and an indicator changes color when the pH crosses its pK value. However, the pH change is gradual at first, because the initial mixture contains a buffer composed of bisulfite and sulfite ions. These ions are in equilibrium with each other.

$$HSO_3^- \rightleftarrows H^+ + SO_3^{2-} \qquad (2)$$

The value of the equilibrium constant for this ionization is 1.02×10^{-7} [*4*]. The initial mixture is acidic because the initial concentration of bisulfite ions (0.10M) is greater than that of sulfite ions (0.025M). The initial pH of the mixture can be estimated from the equilibrium constant and the initial concentrations of bisulfite ions and sulfite ions.

$$[H^+] = 1.02 \times 10^{-7} \frac{[HSO_3^-]}{[SO_3^{2-}]}$$

$$= 1.02 \times 10^{-7} \frac{0.10M}{0.025M}$$

$$= 4.1 \times 10^{-7}$$

This means that the initial pH of the solution is about 6.4.

The reaction of equation 1 consumes sulfite ions. As these ions are consumed, the equilibrium of equation 2 responds by shifting to the right, producing more sulfite ions. The hydrogen ions also produced in this shift neutralize the hydroxide ions formed by the first reaction. This maintains the pH of the solution near its initial value. However, nearly all the bisulfite ions will eventually disappear. When this happens, the pH of the solution will rise sharply, because the first reaction continues to produce hydroxide ions.

The pH of the final solution is about 10. This means that any substance that undergoes a change in appearance between pH 7 and pH 10 will cause the solution to change during the reaction. The chart lists several such substances.

Substance	Change	pH range of change
phenolphthalein	colorless to red-purple	8.0–9.6
thymolphthalein	colorless to blue	9.3–10.6
m-nitrophenol	colorless to yellow	6.7–8.6
phenol red	yellow to red	6.7–8.4
0.025M magnesium chloride	clear solution to white precipitate	9.0–10.0[a]

[a]The K_{sp} of $Mg(OH)_2$ (the white precipitate) is 5.66×10^{-12}. $Mg(OH)_2$ will begin to precipitate from a 0.025M Mg^{2+} solution when the concentration of hydroxide ions exceeds 1.5×10^{-5}M, that is, when the pH exceeds 9.2.

In this pH range, 4,4′-bis(4-amino-1-naphthylazo)-2,2′-stilbene disulfonic acid changes from red to blue, neutral red changes from red to amber, and bromothymol blue changes from yellow to blue. Combinations of these materials broaden the range of color changes that can be produced [*5*].

The sulfite-bisulfite solution must be prepared relatively soon before the presentation of the demonstration, because oxygen from the air oxidizes both sulfite and bisulfite ions [2]. The oxidation removes reactant from the solution, which will change the clock period. The oxidation also produces sulfuric acid, which changes the pH of the solution, and which can delay or even prevent the color change.

A detailed description of the mechanism of the reaction in this demonstration has been provided by Warnek [6]. The experimental rate law indicates that the reaction is first order with respect to formaldehyde, but it is zero order with respect to sulfite ions. Consequently, the rate-determining step of the reaction must be one which involves only formaldehyde. This strange observation is accounted for by the fact that formaldehyde exists in aqueous solution in its hydrated form, methylene glycol ($CH_2(OH)_2$) [7]. Only formaldehyde, not methylene glycol, reacts with bisulfite ions. Therefore, methylene glycol must undergo a dehydration before it can react with bisulfite ions.

$$CH_2(OH)_2 \longrightarrow H_2CO + H_2O$$

The dehydration is much slower than the addition of bisulfite to formaldehyde. Therefore, the dehydration of methylene glycol is the rate-determining step.

In aqueous formaldehyde solutions of concentration over about 4% w/w, the hydrated methylene glycol undergoes polymerization, forming polyoxymethylene glycols [8].

$$n\ CH_2(OH)_2 \rightleftarrows HO(CH_2O)_nOH + (n - 1)\ H_2O$$

When such concentrated solutions are diluted below a concentration of 4%, depolymerization occurs, which leaves methylene glycol as the main substance in solution. This depolymerization is slow, which accounts for the requirement that the formaldehyde solution be prepared at least 24 hours in advance.

The reaction in this demonstration can also be used in a laboratory investigation of kinetics [9]. Reactions with other soluble aldehydes, which do not hydrate as completely as formaldehyde, produce interesting results.

REFERENCES

1. R. L. Barrett, *J. Chem. Educ.* 32:78 (1955).
2. M. G. Burnett, *J. Chem. Educ.* 59:160 (1982).
3. C. Wagner. *Chem. Ber.* 62:2873 (1929).
4. R. C. Weast, Ed., *CRC Handbook of Chemistry and Physics,* 66th ed., p. D-163, CRC Press: Boca Raton, Florida (1985).
5. J. J. Fortman and J. A. Schreier, *J. Chem. Educ.* 68:324 (1991).
6. P. Warneck, *J. Chem. Educ.* 66:334 (1989).
7. M. Wadano, C. Trogus, and K. Hess, *Chem. Ber.* 67:174 (1934).
8. T. M. Gorrie, S. K. Raman, H. K. Rouette, and H. Zollinger, *Helv. Chim. Acta* 56:175 (1973).
9. T. Cassen, *J. Chem. Educ.* 53:197 (1976).

10.10

Luminol Chemiluminescent Clock Reactions

In a darkened room two colorless solutions are mixed together. After 20 seconds a bright blue glow appears, and considerable gas is evolved. The glow lasts approximately 2 seconds. When the room lights are turned on, the solution is a deep blue. Other mixtures and procedures are described in which the clock period and the glow duration may be varied. This is described as Demonstration 2.5 in Volume 1 of this series.

10.11

Two-Color Chemiluminescent Clock Reaction

In a darkened room, a colorless liquid is poured into a solution in a large beaker. The solution emits a red glow for a few seconds, foams, and then emits a blue glow for a few more seconds. This is described as Demonstration 2.6 in Volume 1 of this series.

10.12

Disproportionation of Acidified Sodium Thiosulfate

A small amount of a colorless liquid is added to a clear, colorless solution, and the liquids are mixed. After several seconds, the mixture becomes turbid, and the degree of turbidity gradually increases. The mixing is repeated with other liquids, and the time required for the turbidity to appear varies [*1, 2*].

MATERIALS

For preparation of stock solutions, see pages 14–15.

900 mL 0.20M sodium thiosulfate, $Na_2S_2O_3$ (stock solution)

300 mL distilled water

240 mL 2M sulfuric acid, H_2SO_4 (stock solution)

6 500-mL Erlenmeyer flasks

250-mL graduated cylinder

6 150-mL beakers

timing device capable of measuring seconds

PROCEDURE

Preparation

Place six 500-mL Erlenmeyer flasks in a row on the display table. Pour 200 mL of 0.20M $Na_2S_2O_3$ into each of the first two flasks, 150 mL into each of the middle two, and 100 mL into each of the last two. To the middle two flasks, add 50 mL of distilled water and swirl the flasks to mix the solution. To the last two flasks add 100 mL of distilled water and swirl the flasks. (The concentration of $Na_2S_2O_3$ in the first two flasks is 0.20M, in the second set it is 0.15M, and in the last pair it is 0.10M.)

Place a 150-mL beaker behind each of the flasks and pour 40 mL of 2M H_2SO_4 into each beaker.

Presentation

Quickly add the H_2SO_4 from one of the beakers to the first flask (one of the two with the 0.20M $Na_2S_2O_3$), swirl the flask to mix the solutions, and begin timing. The solution will remain clear for about 5–8 seconds, and then become turbid. Record the time required for the turbidity to appear. Repeat this with the second flask, and note the reproducibility of the time required for turbidity to develop.

Repeat the procedure described in the previous paragraph with the remaining two pairs of flasks. Note that the time for turbidity to appear lengthens as the initial concentration of $Na_2S_2O_3$ decreases.

DISPOSAL

The solutions should be flushed down the drain with water.

DISCUSSION

The yellowish white turbidity that develops in these solutions is due to colloidal sulfur produced by the disproportionation (self-oxidation/reduction) of thiosulfate ions in acidic aqueous solution [*3*].

$$S_2O_3^{2-} + H^+ \longrightarrow S + HSO_3^- \qquad (1)$$

Elemental sulfur is not soluble in the aqueous medium of the reaction, and it forms a solid in suspension. As the reaction proceeds, the particles of solid sulfur increase and cause the mixture to become turbid. When the production of sulfur is relatively rapid, the onset of turbidity is sudden. However, when it is slow, the onset is gradual, and the time at which it occurs is difficult to identify with precision. (This reaction is used under conditions that produce a gradual development of turbidity in Demonstration 9.41, Color of the Sunset: The Tyndall Effect, in Volume 3 of this series.) Qualitatively, the results of this demonstration indicate that, when the initial concentration of thiosulfate is decreased, the time required for turbidity to appear (the clock period) increases.

The disproportionation reaction is more complicated than equation 1 would indicate. The product of the reaction is a complex mixture which contains sulfur dioxide, hydrogen sulfide, several forms of colloidal sulfur, and a variety of polythionate ions [*4*]. The sulfur formed by the reaction occurs in the form of S_8, S_6, and some polymeric chains. The polythionate ions formed in the reaction include $S_3O_6^{2-}$ and $S_5O_6^{2-}$.

A sequence of reactions to account for the formation of these products has been suggested [*4*]. In the acidic medium of the reaction, the basic thiosulfate ions react with hydrogen ions forming hydrogen thiosulfate ions (equation 2).

$$^-S{-}SO_3^- + H^+ \rightleftarrows H{-}S{-}SO_3^- \qquad (2)$$

This reaction is an equilibrium, and equation 2 is written as the reverse of the ionization of hydrogen thiosulfate ion ($HS_2O_3^-$), a weak acid with K_a equal to 1.07×10^{-7} [*5*]. This ion is susceptible to attack by a thiosulfate ion, whose negatively charged terminal sulfur atom inserts between the outer sulfur and the hydrogen of $HS_2O_3^-$ (equation 3).

$$H{-}S{-}SO_3^- + {}^-S{-}SO_3^- \longrightarrow H{-}S{-}S{-}SO_3^- + SO_3^{2-} \qquad (3)$$

This increases the length of the chain of sulfur atoms. The process of sulfur-atom addition continues, producing ever longer chains of sulfur atoms (equation 4).

$$H{-}S_n{-}SO_3^- + {}^-S{-}SO_3^- \longrightarrow H{-}S_{n+1}{-}SO_3^- + SO_3^{2-} \qquad (4)$$

All the ions that contain chains of sulfur atoms ionize to some extent by losing a hydrogen ion (equation 5).

$$H{-}S_n{-}SO_3^- \rightleftarrows H^+ + {}^-S{-}S_{n-1}{-}SO_3^- \quad (5)$$

If the chain of sulfur atoms is long enough, the negatively charged sulfur atom at the end can wrap around and attach to a sulfur atom near the other end, forming a ring of elemental sulfur (equation 6).

$$^-S{-}S_x{-}S{-}S_y{-}SO_3^- \longrightarrow S_{x+2} + {}^-S_y{-}SO_3^- \quad (6)$$

The negatively charged sulfur atom of a thiosulfate ion may not always attach to the end sulfur in the chain. For example, if it attaches to the second sulfur, hydrogen sulfide ions and polythionate ions result (equation 7).

$$H{-}S{-}S_n{-}SO_3^- + {}^-S{-}SO_3^- \rightleftarrows H{-}S^- + {}^-O_3S{-}S{-}S_n{-}SO_3^- \quad (7)$$

This reaction is reversible because the hydrogen sulfide ion can attach to a sulfur atom in the polythionate ion and split it. (The production of HS^- by reaction 7 plays a key role in Demonstration 10.13, Precipitation of Arsenic(III) Sulfide.)

Sulfite ions are produced by many of the previous reactions. These sulfite ions are produced in an acidic solution. In this solution, they react with hydrogen ions nearly completely to form hydrogen sulfite ions (equation 8).

$$SO_3^{2-} + H^+ \rightleftarrows HSO_3^- \quad (8)$$

The hydrogen sulfite ions also react virtually completely with hydrogen ions, forming aqueous sulfur dioxide (equation 9).

$$HSO_3^- + H^+ \rightleftarrows SO_2(aq) + H_2O \quad (9)$$

This is the origin of the sulfur dioxide product.

Sodium thiosulfate is used in acidic solutions in several of the previous demonstrations. However, the concentrations of acid and of thiosulfate in those solutions are much lower than in this demonstration. Under those conditions the disproportionation of thiosulfate is slower than the other reactions of thiosulfate that occur in those demonstrations. Therefore, the disproportionation does not interfere with the other clock reactions.

REFERENCES

1. H. Garber and S. B. Arenson, *J. Chem. Educ.* 17:514 (1940).
2. D. H. Beach, *Chem 13 News* No. 102:16 (February 1979).
3. H. Bassett and R. G. Durant, *J. Chem. Soc.* 1927:1401 (1927).
4. R. E. Davis, *J. Am. Chem. Soc.* 80:3565 (1958).
5. R. C. Weast, Ed., *CRC Handbook of Chemistry and Physics,* 66th ed., p. D-163, CRC Press: Boca Raton, Florida (1985).

10.13

Precipitation of Arsenic(III) Sulfide

Two colorless solutions are mixed, and for a period of about 30 seconds, the mixture remains colorless. Then a yellow-orange precipitate appears [*1–4*].

MATERIALS

50 g sodium thiosulfate pentahydrate, $Na_2S_2O_3 \cdot 5H_2O$

ca. 200 mL distilled water

1.3 g sodium metaarsenite, $NaAsO_2$

200 mL 2M acetic acid, $HC_2H_3O_2$ (To prepare 1.0 liter of stock solution, carefully pour 115 mL of glacial [17.5M] acetic acid into 600 mL of distilled water and dilute the resulting solution to 1.0 liter.)

250-mL beaker, with graduations

150-mL beaker, with graduations

600-mL beaker

stirring rod

timing device capable of measuring seconds (optional)

PROCEDURE

Preparation

In a graduated 250-mL beaker, dissolve 50 g of $Na_2S_2O_3 \cdot 5H_2O$ in 50 mL of distilled water. Using the graduations on the beaker as a guide, dilute this solution to a volume of 100 mL. (This solution is 2M in $Na_2S_2O_3$.)

In a graduated 150-mL beaker, dissolve 1.3 g of $NaAsO_2$ in 75 mL of distilled water and dilute the resulting solution to 100 mL with distilled water. Pour the $NaAsO_2$ solution into the 600-mL beaker. Add 200 mL of 2M acetic acid to the beaker and stir the mixture. Leave the stirring rod in the 600-mL beaker. (This solution is 0.03M in $NaAsO_2$ and 1.3M in $HC_2H_3O_2$.)

Presentation

While stirring the solution in the 600-mL beaker, pour the $Na_2S_2O_3$ solution into it and start timing. Stir the mixture thoroughly. (The $Na_2S_2O_3$ solution is quite dense, and thorough stirring is required to make the mixture homogeneous. Immediately upon mixing, the solution is 1.0M in acetic acid, 0.50M in $Na_2S_2O_3$, and 0.025M in

$NaAsO_2$.) The mixture will remain clear for about 20 seconds, and then over a span of several seconds, a bright yellow precipitate will form in the mixture.

HAZARDS

Sodium metaarsenite is extremely poisonous by inhalation or ingestion. Dust irritates eyes and mucous membranes. Ingestion can cause severe stomach irritation and affect the liver, kidneys, and heart. Arsenic compounds are suspected carcinogens.

Glacial acetic acid can irritate the skin, and its vapors are irritating to the eyes and respiratory system. It should be handled only in a well-ventilated area.

DISPOSAL

The precipitated yellow arsenic sulfide should be filtered from the liquid. The recovered solid should be allowed to dry. Consult local authorities to locate an approved toxic-waste disposal site for the dry material which contains arsenic sulfide.

The filtrate should be flushed down the drain with plenty of water.

DISCUSSION

In this demonstration, two colorless solutions, one containing sodium metaarsenite and acetic acid and the other containing sodium thiosulfate, are mixed. After a short time a bright yellow precipitate of arsenic(III) sulfide (As_2S_3) forms suddenly throughout the mixture. The bright yellow As_2S_3 has been used for thousands of years as a pigment known as orpiment yellow and king's yellow [*5*]. The preparation of arsenic(III) sulfide by the reaction used in this demonstration was first reported in 1889 [*6*]. However, the clock effect of this reaction was first described in 1922, when the reaction was suggested as an alternative to the Landolt iodine clock reaction (Demonstration 10.1) [*1*].

In addition to arsenic(III) sulfide the reaction in this demonstration also produces a variety of polythionate ions, mainly pentathionate ions ($S_5O_6^{2-}$) [*7*]. The mixture of products makes a single chemical equation an inadequate representation of what is happening in the mixture. However, the reactions here seem to be related to the disproportionation of thiosulfate in acidic solution (Demonstration 10.12). In that reaction, a solution of sodium thiosulfate is mixed with sulfuric acid. After a short period a pale yellow precipitate of sulfur forms throughout the mixture. When a thiosulfate solution is acidified in the presence of arsenite, instead of a sulfur precipitate, one of arsenic(III) sulfide is produced. As the initial amount of arsenite in the mixture is reduced, the resulting precipitate contains less arsenic(III) sulfide and more sulfur. An arsenite to thiosulfate ratio of as little as 1:20 will inhibit the formation of significant amounts of elemental sulfur [*1*]. This suggests that competing reactions are involved, one producing sulfur and the other forming arsenic(III) sulfide.

A sequence of reactions has been suggested to account for the formation of the products in this demonstration [*4*]. In the acidic medium of the reaction, the basic thiosulfate ions react with hydrogen ions forming hydrogen thiosulfate ions (equation 1).

$$^{-}S{-}SO_3^{-} + H^{+} \rightleftarrows H{-}S{-}SO_3^{-} \quad (1)$$

This reaction is an equilibrium, and equation 1 is written as the reverse of the ionization of hydrogen thiosulfate ion ($HS_2O_3^-$), a weak acid with K_a equal to 1.07×10^{-7} [*8*]. This hydrogen thiosulfate ion is susceptible to attack by a thiosulfate ion, whose negatively charged terminal sulfur atom inserts between the outer sulfur and the hydrogen of $HS_2O_3^-$ (equation 2).

$$H—S—SO_3^- + {}^-S—SO_3^- \longrightarrow H—S—S—SO_3^- + SO_3^{2-} \qquad (2)$$

This increases the length of the chain of sulfur atoms. The process of sulfur-atom addition continues, producing longer chains of sulfur atoms (equations 3 and 4).

$$H—S_2—SO_3^- + {}^-S—SO_3^- \longrightarrow H—S_3—SO_3^- + SO_3^{2-} \qquad (3)$$

$$H—S_3—SO_3^- + {}^-S—SO_3^- \longrightarrow H—S_4—SO_3^- + SO_3^{2-} \qquad (4)$$

The negatively charged sulfur atom of a thiosulfate ion may not always attach to the end sulfur in the chain. For example, if it attaches to the second sulfur, hydrogen sulfide ions and polythionate ions result (equation 5).

$$H—S—S_n—SO_3^- + {}^-S—SO_3^- \rightleftarrows H—S^- + {}^-O_3S—S—S_n—SO_3^- \qquad (5)$$

This reaction is reversible because the hydrogen sulfide ion can attach to a sulfur atom in the pentathionate ion and split it. In the absence of arsenite, this is what happens, and the sequence of reactions continues until elemental sulfur rings and chains form from the $H—S_n—SO_3^-$ ions (see Discussion in Demonstration 10.12). However, in the presence of arsenite ions, the equilibrium of reaction 5 is driven to the right by the reaction of HS^- with AsO_2^- to form the insoluble arsenic(III) sulfide.

The reason for the time interval before the sudden appearance of As_2S_3 has not been definitively identified. A possible explanation is that a time interval is necessary before the concentration of arsenic(III) sulfide builds up to that of a saturated solution. The rapid precipitation of a rather large amount of arsenic(III) sulfide suggests that the solution first becomes quite supersaturated before precipitation begins. The supersaturation may occur because of the lack of sufficient nucleation sites for the formation of crystals. The solubility of arsenic(III) sulfide is only 0.00005 g/100 mL of H_2O (about 2×10^{-6} moles/liter) at 18°C [*9*].

REFERENCES

1. G. S. Forbes, H. W. Estill, and O. J. Walker, *J. Amer. Chem. Soc.* 44:97 (1922).
2. P. S. Chen, *Entertaining and Educational Chemical Demonstrations,* p. 30, Chemical Elements Publishing Co.: Camarillo, California (1974).
3. P. S. Bailey, C. A. Bailey, J. Anderson, P. G. Koski, and C. Rechsteiner, *J. Chem. Educ.* 52:524 (1975).
4. K. W. Watkins, *J. Chem. Educ.* 64:255 (1987).
5. M. Windholz, Ed., *The Merck Index,* 10th ed., p. 829, Merck and Co.: Rahway, New Jersey (1983).
6. G. Vortmann, *Chem. Ber.* 22:2307 (1889).
7. A. Kurtenacker and K. Matejka, *Z. anorg. Chem.* 229:19 (1936).
8. R. C. Weast, Ed., *CRC Handbook of Chemistry and Physics,* 66th ed., p. D-163, CRC Press: Boca Raton, Florida (1985).
9. R. C. Weast, Ed., *CRC Handbook of Chemistry and Physics,* 66th ed., p. B-75, CRC Press: Boca Raton, Florida (1985).

10.14

Bromate Oxidation of Manganese

Two colorless solutions are mixed, and for a period of about 20 seconds, the mixture remains colorless. Then, over a span of several seconds, the mixture becomes purple.

MATERIALS

40 mL 0.25M potassium bromate, $KBrO_3$ (To prepare 1 liter of solution, dissolve 42 g of $KBrO_3$ in 800 mL of distilled water and dilute the resulting solution to 1.0 liter.)

80 mL 6.0M sulfuric acid, H_2SO_4 (To prepare 1 liter of solution, set a 2-liter beaker containing 500 mL of distilled water in a pan of ice, and slowly pour 330 mL of concentrated [18M] H_2SO_4 into the beaker. After the resulting mixture has cooled, dilute it to 1.0 liter with distilled water.)

43 g sodium pyrophosphate decahydrate, $Na_4P_2O_7 \cdot 10H_2O$

80 mL 0.10M manganese(II) sulfate, $MnSO_4$ (To prepare 1 liter of solution, dissolve 22 g of $MnSO_4 \cdot 4H_2O$ in 600 mL of distilled water and dilute the resulting solution to 1.0 liter.)

ca. 200 mL distilled water

2 600-mL graduated beakers

magnetic stirrer with stir bar

10-mL graduated cylinder

stirring rod

timing device capable of measuring seconds (optional)

PROCEDURE [*1*]

Preparation

Pour 40 mL of 0.25M $KBrO_3$ into one of the 600-mL beakers, place a stir bar in the beaker, and set the beaker on the magnetic stirrer.

In the other 600-mL graduated beaker, combine 8 mL of 6.0M H_2SO_4 and 125 mL of distilled water. Add 43 g of $Na_4P_2O_7 \cdot 10H_2O$ to the beaker and stir the mixture until all the solid has dissolved. Dilute the resulting solution to 200 mL with distilled water, using the graduations on the beaker as a guide. Add 72 mL of 6.0M H_2SO_4 and 80 mL of 0.10M $MnSO_4$ to the beaker.

Presentation

Turn on the stirrer. Pour the 360 mL of solution into the 40 mL of solution in the beaker on the stirrer. (The initial concentrations of this mixture are 1.2M H_2SO_4, 0.25M $Na_4P_2O_7$, 0.025M $KBrO_3$, and 0.020M $MnSO_4$.) The mixture remains colorless for about 20 seconds and then, over a span of several seconds, the mixture turns purple.

HAZARDS

Concentrated sulfuric acid is both a strong acid and a powerful dehydrating agent; it must be handled with great care. The dilution of sulfuric acid is a highly exothermic process and releases sufficient heat to cause burns. Therefore, when diluting concentrated sulfuric acid, always add the acid to water, slowly and with stirring. The receiving container should be immersed in crushed ice if the concentration of the resulting solution will be 6M or more.

Potassium bromate is toxic by ingestion. It can cause vomiting, diarrhea, and kidney damage.

DISPOSAL

The solutions should be flushed down the drain with water.

DISCUSSION

The reaction used in this demonstration is the oxidation by bromate ions of nearly colorless $Mn^{2+}(aq)$ to highly colored pyrophosphate complexes of manganese(III). The overall reaction may be represented by the equation

$$4\ Mn^{2+}(aq) + 12\ H_4P_2O_7(aq) + BrO_3^-(aq) \longrightarrow 4\ Mn(H_2P_2O_7)_3^{3-}(aq) + HBrO(aq) + 19\ H^+(aq) + 2\ H_2O(l)$$

The reaction has been used as a colorimetric method for the determination of manganese [2]. The appearance of purple in this demonstration is not so sudden as is the appearance of color in the Landolt iodine clock reaction. Here the solution remains colorless for about 20 seconds, and then the purple appears over a span of several seconds.

For this reaction, the clock period (the time between mixing and the appearance of color) is independent of the initial manganese(II) and pyrophosphate concentrations, provided, of course, that there is enough to produce a detectable amount of $Mn(H_2P_2O_7)_3^{3-}$. However, the clock period does depend on the initial concentrations of sulfuric acid and of bromate ions. An increase in the initial concentration of acid produces a shorter clock period. An increase in the initial bromate ion concentration also yields a shorter clock period, although the effect is much smaller than that produced by increased acid concentration. A relatively high concentration of sulfuric acid (1.2M) is used in the demonstration to produce a conveniently short clock period. When the clock period is short, the purple color develops reasonably quickly, producing an acceptable clock effect. As the clock period lengthens, the purple color takes longer to develop, and the clock effect is less striking. A 12-fold excess of sodium pyrophosphate over manganese sulfate is used to complex the manganese(III) virtually completely and produce a deep color.

Precisely how the reactions in this demonstration produce a clock effect has not been determined. However, the reduction of bromate ions in the presence of manganese(II) also occurs in the Belousov-Zhabotinsky oscillating reactions described in chapter 7 in Volume 2 of this series. Bromide ions have been recognized to play a significant role in these oscillating reactions. Bromide ions may be involved in this clock reaction as well. If bromide ions are present in the initial mixture, the clock period is lengthened. This suggests that the purple color does not appear until all the bromide ions have been consumed by some reaction in the mixture. Bromide ions may be produced by reactions between bromate ions and manganese(II) that occur in the initial mixture [3].

The pyrophosphate solution is prepared by dissolving tetrasodium pyrophosphate in a solution of sulfuric acid. This is done because the tetrasodium salt is not as soluble as the trisodium salt ($Na_3HP_2O_7$).

REFERENCES

1. R. L. Rich and R. M. Noyes, *J. Chem. Educ.* 67:606 (1990).
2. F. D. Snell and C. T. Snell, *Colorimetric Methods of Analysis,* 3d ed., Vol. 2, p. 396, Van Nostrand: New York (1949).
3. C. M. Singh, H. C. Mishra, and R. N. Upadhyay, *J. Indian Chem. Soc.* 57:835 (1979).

10.15

Periodate-Thiosulfate Reaction

A clear, colorless liquid is poured into another clear, colorless liquid, and the mixture turns red. The red mixture is poured into a third clear, colorless liquid, and this mixture becomes deeper red. After several seconds, the red fades, leaving a cloudy white mixture. After several seconds more, the mixture turns blue [*1*].

MATERIALS

30 mg 2,2′,4,4′,4″-pentamethoxytriphenylmethanol, $C_{24}H_{26}O_6$†

60 mg thymolphthalein, $C_{28}H_{30}O_4$

100 mL acetone, CH_3COCH_3

15 g sodium thiosulfate pentahydrate, $Na_2S_2O_3 \cdot 5H_2O$

ca. 550 mL distilled water

ca. 5 mL 1.0M hydrochloric acid, HCl (To prepare 1.0 liter of stock solution, carefully pour 83 mL of concentrated [12M] HCl into about 700 mL of distilled water and dilute the cooled solution to 1.0 liter.)

0.08 g cadmium nitrate tetrahydrate, $Cd(NO_3)_2 \cdot 4H_2O$

5.4 g sodium periodate (sodium metaperiodate), $NaIO_4$

250-mL beaker

400-mL beaker

250-mL graduated cylinder

stirring rod

600-mL beaker

1-liter beaker

PROCEDURE

Preparation

In the 250-mL beaker, dissolve 30 mg of pentamethoxytriphenylmethanol and 60 mg of thymolphthalein in 100 mL of acetone. In the 400-mL beaker, dissolve 15 g of $Na_2S_2O_3 \cdot 5H_2O$ in 150 mL of distilled water. Pour the acetone solution all at once into this $Na_2S_2O_3$ solution and stir the mixture. The mixture will be slightly turbid. If

† Available as catalog no. 25,807-5 from Aldrich Chemical Company, 940 West Saint Paul Avenue, Milwaukee, Wisconsin 53233; telephone: 1-800-558-9160.

it is blue, add 1.0M HCl drop by drop while stirring the mixture, until the blue just disappears. (This solution is 0.24M in $Na_2S_2O_3$.)

In the 600-mL beaker, combine 4.5 mL of 1.0M HCl and 120 mL of distilled water. In this mixture dissolve 0.08 g of $Cd(NO_3)_2 \cdot 4H_2O$. (This solution is 0.002M in $Cd(NO_3)_2$.)

In the 1-liter beaker, dissolve 5.4 g of $NaIO_4$ in 250 mL of distilled water. (This solution is 0.10M in $NaIO_4$.)

Arrange the 400-mL, 600-mL, and 1-liter beakers in a row.

Presentation

Pour the contents of the 400-mL beaker into the 600-mL beaker. The mixture will immediately turn red. Pour the red mixture from the 600-mL beaker into the 1-liter beaker. This mixture will immediately become deeper red. After several seconds, the red will fade, leaving a white precipitate in the mixture. After several more seconds the mixture will turn blue (and remain turbid).

HAZARDS

Cadmium salts are highly toxic by ingestion and inhalation. They are also carcinogenic.

Concentrated hydrochloric acid can irritate the skin. Its vapors are extremely irritating to the eyes and respiratory system. It should be handled only in a well-ventilated area.

Sodium periodate is a strong oxidizing agent, and fires may result if it contacts combustible materials. It is irritating to the skin and is toxic if ingested.

Acetone is flammable and should be handled away from open flames and sparks. It is toxic by ingestion and by inhalation of its vapors.

DISPOSAL

Cadmium wastes should be converted to the highly insoluble sulfide, and this should be buried in a landfill approved for the disposal of toxic heavy metals. To form the sulfide of cadmium, add 1 g of sodium sulfide ($Na_2S \cdot 9H_2O$) or 1 mL of ammonium sulfide ($(NH_4)_2S$) to the solution remaining after the demonstration. Stir the mixture occasionally for 1 hour. Add aqueous ammonia until the mixture is just basic to litmus. Filter off the insoluble materials and allow them to dry. Consult local authorities to locate an approved toxic-waste disposal site for the dry materials, which contain cadmium sulfide. To destroy excess sulfide in the filtrate solution, add 50 mL of 5% sodium hypochlorite solution (liquid laundry bleach) and stir occasionally for 1 hour. Flush this solution down the drain with plenty of water.

DISCUSSION

The mixture in this demonstration undergoes a pH change from below 3 to about 11. As the pH of the mixture changes, so do the colors of the two pH indicators in the mixture. In addition, as the solution becomes less acidic, a white precipitate forms.

The indicators used in this demonstration are 2,2′,4,4′,4″-pentamethoxytriphenylmethanol and thymolphthalein. Molecular 2,2′,4,4′,4″-pentamethoxytriphenylmethanol indicator is colorless. In aqueous solution it ionizes to form a red cation as the pH falls through the range of 3.2–1.5 [*1*].

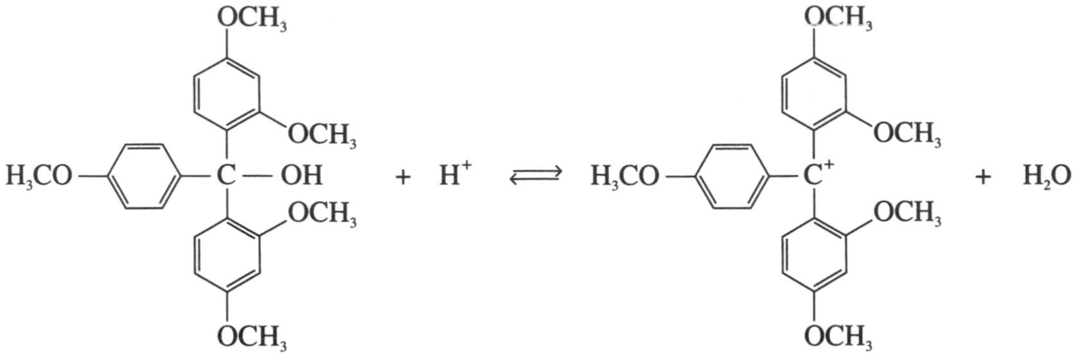

OCH_3 OCH_3 H_3CO C—OH OCH_3 OCH_3 + H^+ ⇌ H_3CO C^+ OCH_3 OCH_3 OCH_3 OCH_3 + H_2O

colorless red

The hydrochloric acid used in the procedure is sufficiently acidic to shift this equilibrium to the right and turn the solution red. Then, as the pH gradually increases, the equilibrium shifts to the left and the red disappears. The pH of the mixture increases and becomes sufficiently high to change the thymolphthalein from its colorless form to its blue form, which occurs in the pH range of 9.3–10.5 [*2*]. At pH values between the color changes of the two indicators, the mixture is white, because it contains a suspension of an insoluble cadmium compound.

The reaction mixture contains an oxidizing agent, sodium periodate, and a reducing agent, sodium thiosulfate. The overall reaction between them produces a pH change in the mixture. There are probably several reactions occurring in the mixture, and a number were suggested in the original report of this demonstration [*1*]. These include the oxidation of thiosulfate ions by periodate ions to produce tetrathionate ions,

$$IO_4^-(aq) + 2S_2O_3^{2-}(aq) + 2\,H^+(aq) \longrightarrow IO_3^-(aq) + S_4O_6^{2-}(aq) + H_2O(l) \quad (1)$$

the oxidation of thiosulfate ions to sulfite ions,

$$2\,IO_4^-(aq) + S_2O_3^{2-}(aq) + H_2O(l) \longrightarrow 2\,IO_3^-(aq) + 2\,SO_3^{2-}(aq) + 2\,H^+(aq) \quad (2)$$

and the oxidation of sulfite ions to sulfate ions.

$$IO_4^-(aq) + SO_3^{2-}(aq) \longrightarrow IO_3^-(aq) + SO_4^{2-}(aq) \quad (3)$$

Other reactions are possible, as well, making it quite difficult to formulate a comprehensive description of the process which produces the clock effect.

The identity of this cadmium compound is not known, but a likely candidate is a basic cadmium periodate ($Cd_2(H_4IO_6)(OH)_3$). The solubility product for this compound is $[Cd^{2+}]^2[H_4IO_6][OH^-]^3 = 1 \times 10^{-42}$ [*3*]. In the reaction mixture, the initial concentration of cadmium ions is 4.3×10^{-4}M, that of hydrochloric acid is 7.5×10^{-2} M, and that of periodate ions is 4.2×10^{-2}M. The periodate ions turn into $H_4IO_6^-$ through reaction with water molecules. If each periodate ion consumes two hydrogen ions as it reacts (equation 1), then the pH of the solution at which precipitation begins can be calculated. If x is the amount by which the concentration of iodate has decreased at some point in the reaction, then $0.042 - x$ is the concentration of iodate ions and

0.075 − 2x is the concentration of hydrogen ions at that point. These expressions can be substituted in the solubility product expression, resulting in the equation

$$(4.3 \times 10^{-4})^2 (0.042 - x) \left(\frac{1 \times 10^{-14}}{0.075 - 2x} \right)^3 = 2.5 \times 10^{-43}$$

The hydrogen ion concentration at which this equation holds is about 1.6×10^{-3}M. Therefore, the basic cadmium periodate salt will start to precipitate as the pH rises above 2.8.

REFERENCES

1. J. L. Lambert, M. J. Chejlava, G. T. Fina, and N. L. Luce, *J. Chem. Educ.* 60: 141 (1983).
2. M. Windholz, Ed., *The Merck Index,* 10th ed., Merck and Co.: Rahway, New Jersey (1983).
3. R. M. Smith and A. E. Martell, *Critical Stability Constants,* Vol. 4, p. 129, Plenum Press: New York (1976).

11

Electrochemistry: Batteries, Electrolytic Cells, and Plating

Rodney Schreiner, Bassam Z. Shakhashiri,
Ronald I. Perkins, Earle S. Scott,
and Larry E. Judge

Chemical reactions involve energy. For most chemical reactions, this energy takes the form of heat. Many reactions release heat, causing their vicinity to become warmer. A few reactions absorb heat, causing their surroundings to become cooler. Demonstrations dealing with reactions that release or absorb heat are described in "Thermochemistry," Chapter 1 of this series. Energy involved in some chemical reactions takes the form of light. Chemiluminescent reactions produce light, and they are described in Chapter 2, "Chemiluminescence." Some chemical reactions are produced or initiated by light energy, for example, the reaction between chlorine and hydrogen, shown in Demonstration 1.45; such reactions are called photochemical reactions. Another group of reactions involves electric energy. Some of these reactions produce electric energy and others are produced by it. This chapter contains demonstrations of both types of these reactions.

ELECTRIC CURRENT AND CONDUCTORS

When an electric conductor, such as a metal wire, is connected to two objects with different electric potentials (energies), an electric charge will flow through the conductor. The flow will be from higher potential to lower potential. This flow of electric charge is called an *electric current*. Electric charge is measured in coulombs, and an electric current is expressed in coulombs per second, or amperes (1 ampere = 1 coulomb/second). The change in energy that occurs when a current flows across a potential difference depends on the magnitude of the potential difference and on how much charge flows. A potential difference is expressed in joules per coulomb, or volts (1 volt = 1 joule/coulomb). The amount of energy released when a current flows through a conductor is equal to the potential difference (volts) driving the current multiplied by the amount of charge that flows (coulombs). This energy can produce heat, light, or chemical reactions.

Electric charge moves only when charged particles move, because charge is a property of particles. Therefore, when an electric current flows through a conductor, charged particles are moving within the conductor. In most metals, the charged particles that move are electrons. In the case of aqueous solutions of acids, bases, and salts, and in molten salts, the mobile charged particles are ions.

Electric charge is conserved in any process. Therefore, when charged particles move through a conductor from one object to another, either the objects themselves change total charge, or there is some other conductor connecting the two objects in which charges move in the opposite direction. In most electric devices, the latter is the case, and charge moves around a closed path, or *circuit*. In a flashlight, the circuit is relatively short, passing from one pole of the battery through the lamp back into the other pole of the battery, and then through the battery. For devices that are plugged into a household electric outlet, the path may travel many miles from the generators in the electric power plant and back again.

When a current flows through an electric conductor, whether metallic or ionic, it creates a magnetic field around the conductor, which provides a means for measuring the current [*1*]. When this magnetic field interacts with that of a permanent magnet fastened to a pointer, the pointer moves. (This is shown in Demonstration 11.1, where a current in a wire and in sulfuric acid rotates a compass needle.) The amount of pointer movement is a measure of the magnitude of the current. A device that employs this principle is called an *ammeter*. When installed in an electric circuit, an ammeter measures the current flowing through that circuit. Because an electric current is driven by a difference in electric potential (voltage), the same principle may be used to measure a difference in electric potential. When a standard conductor is connected between two points having different electric potentials, a current flows that is proportional to the potential difference. A device that makes use of this principle is called a *voltmeter*. The rotation of a magnet by the magnetic field produced by an electric current is also the basis of the operation of an electric motor. A magnetic field can also change the electric conductivity of some materials. This is the basis of the operation of transistors, which have made possible the miniaturization of electric devices such as radios and calculators.

When current flows through most conductors, some of the electric potential energy that drives the current is converted to heat. The heat is generated because the conductor resists the flow of current to some degree. This resistance is analogous to friction in a mechanical device. The coils of an electric range are resistors, electric conductors that offer resistance to the current that flows through them. The salt solution between the metal strips in a steam vaporizer is a resistor; the solution becomes hot and boils when an electric current flows through it. The disposable battery tester that comes with some batteries uses a liquid-crystal temperature-sensing strip over a resistor strip to indicate how much heat is released when the strip is connected to the terminals of the battery, thereby indicating the potency of the battery. There are some materials whose temperature does not change when an electric current flows through them. These materials have no resistance to electricity. They are called *superconductors*. Superconductors are important for applications that rely on the magnetic fields generated by electric current. Because none of the electric energy that flows through a superconductor is wasted as heat, more of the energy is available for use through magnetic effects, such as in motors.

CHEMICAL REACTIONS AND ELECTRIC CURRENT

When a metallic conductor is placed in contact with an ionic conductor and a current is passed through both of them, a change in the identity of the charge-carrying particles occurs at the place where the two conductors meet. Electrons must either enter or leave the metal conductor. These electrons leave or enter the ions in the ionic conductor. When this happens, a chemical reaction takes place. The metallic conductor at whose surface these chemical reactions take place is called an *electrode*. Electric charges are neither created nor destroyed in any chemical process. Therefore, when charge moves from a metallic conductor to an ionic conductor, there must be a corresponding location where the same amount of charge is moving from the ionic conductor to another metallic conductor. Thus, electrodes are always used in pairs: from one electrode electrons move into the ionic conductor; at the other electrode electrons are removed from the ionic conductor. At the electrode where electrons move into an ionic conductor, a chemical reduction occurs; this electrode is called the *cathode*. At the electrode where electrons are removed from the ionic conductor, an oxidation occurs; this electrode is called the *anode*.

The first person to observe electric energy obtained from a chemical reaction was Alessandro Volta [2]. In the 1790s he discovered that an electric current could be produced with two dissimilar metal disks, such as silver and zinc, separated by a blotting paper soaked in brine. When several of these "sandwiches" of disks and damp blotter were assembled into a stack, the current produced was increased. These assemblies, or batteries, of similar sandwiches are the origin of the term *battery,* which we apply to a chemical source of electric energy. Volta did not, apparently, notice any chemical change associated with the flow of this current. In the first decade of the 19th century, Humphry Davy used variations of Volta's batteries to produce chemical changes. He used the electric current extracted from these batteries to break down some compounds. He produced elemental potassium and sodium from their hydroxides. The process in which an electric current produces a chemical reaction is called *electrolysis*.

A chemical reaction can be used to produce an electric current. The production of an electric current takes place in a device called a *voltaic cell*. An electric current also can produce a chemical reaction. The device which does this is called an *electrolytic cell*. Voltaic cells and electrolytic cells have several common features. Both cells contain an electrically conductive liquid or gel, called the *electrolyte*. A salt (or acid or base) is dissolved in this liquid or gel. Both cells also contain two electrodes. When the electrodes of a voltaic cell are connected by a wire, a current flows through the wire. When the electrodes of the electrolytic cell are connected to a generator of electric current, a chemical reaction occurs in the cell.

An example of an oxidation-reduction reaction is the reaction that occurs when a bar of zinc is placed in a solution containing copper(II) ions. The zinc metal oxidizes to zinc ions in solution, and copper ions in the solution reduce to metallic copper.

$$Cu^{2+}(aq) + Zn(s) \longrightarrow Cu(s) + Zn^{2+}(aq)$$

This reaction occurs by the transfer of electrons from the zinc metal atoms to the copper ions. The process can be represented by the following two equations, called *half reactions* because each is only half of the process:

$$Zn(s) \longrightarrow Zn^{2+}(aq) + 2\,e^-$$

$$2\,e^- + Cu^{2+}(aq) \longrightarrow Cu(s)$$

If the aqueous copper ions and the zinc metal are physically separated, they cannot react. However, if they are connected by a wire, electron transfer can occur. When one end of a wire is attached to a zinc bar immersed in a salt solution (e.g., aqueous zinc sulfate, $ZnSO_4$) and the other end of the wire is immersed in a solution of copper(II) ions, electrons flow from the zinc metal to the copper ions. However, this flow of electrons is only momentary. Soon, the negative charge at the copper end of the wire builds up and repels other electrons flowing from the zinc. Some way must be provided to return negative charge to the zinc. This can be done by allowing ions to move between the solutions around the copper and the zinc. A gel containing ions (i.e., a *salt bridge*) will do this without allowing the copper ions to contact the zinc directly. Negative ions (anions) from the copper solution move through the salt bridge to the zinc solution, and cations from the zinc solution move to the copper solution. The salt bridge completes the flow of charge through the cell, that is, it completes the circuit. (Demonstration 11.13 shows that ions move in a gel when an electric current passes through it, and Demonstration 11.1 shows that current flowing through a solution produces a magnetic field, just as current in a wire does.) Electrons leave the zinc atoms (producing zinc ions) and flow through the wire to the solution containing copper(II) ions. The electrons combine with copper(II) ions forming copper metal. The net charge transferred from one solution to the other by the flow of electrons in the wire is offset by the flow of ions through the salt bridge.

ELECTRODE POTENTIALS

In a voltaic cell, a spontaneous oxidation-reduction causes electrons to flow from the anode through an electronic conductor to the cathode. The electrons flow from the anode to the cathode because these electrodes have different electric potentials. The potential of each electrode depends on the composition of the solution in which it is immersed. The potential of a single electrode cannot be measured directly; only the difference between the potentials of two electrodes can be measured. Therefore, an absolute electrode potential cannot be determined. However, a relative potential can be assigned to an electrode if some electrode is adopted as a standard.

The standard electrode that has been chosen is one involving ions that occur in all aqueous solutions, namely hydrogen ions. The standard hydrogen electrode, in which a platinum electrode is immersed in a 1.0M solution of hydrogen ions and surrounded by hydrogen gas at a pressure of 1.0 atmosphere, has been assigned a potential of exactly 0 volts. When this electrode is connected to a second electrode, the difference in the potentials of the two electrodes corresponds to the potential of the second electrode. If hydrogen ions are reduced to hydrogen gas at the hydrogen electrode, then electrons flow from the other electrode to the hydrogen electrode, and the half reaction at the hydrogen electrode is

$$2\,H^+(aq) + 2\,e^- \longrightarrow H_2(g)$$

This means that the other electrode has a higher potential than that of the hydrogen electrode. In this case the potential of the other electrode is positive, and this potential is equal to the cell potential. If hydrogen gas is oxidized to hydrogen ions at the hydrogen electrode, then electrons flow from the hydrogen electrode to the other electrode, and the half reaction at the hydrogen electrode is

$$H_2(g) \longrightarrow 2\,H^+(aq) + 2\,e^-$$

In this situation, the potential of the hydrogen electrode (i.e., 0 volts) is higher than that of the other electrode. Therefore, the other electrode has a negative potential, namely, the negative of the cell potential.

The potentials of many standard electrodes have been measured and their potentials are tabulated in reference books, such as the *CRC Handbook of Chemistry and*

Physics [3]. Because the potentials depend on such factors as solute concentration, gas pressure, and temperature, the tabulated values are generally reported as *standard reduction potentials*. These are the potentials of cathodes whose reactions occur when solute concentrations are exactly 1M, gas pressures are exactly 1 atmosphere, and the temperature is exactly 25°C. If the reaction that occurs at an electrode is an oxidation, then the electrode potential is an oxidation potential, which is the negative of the reduction potential. The cell potential is the sum of the two electrode potentials: a reduction potential and an oxidation potential.

COMMERCIAL VOLTAIC CELLS AND BATTERIES

Batteries have been used for over 150 years as a source of electric energy. One of the earliest commercial batteries is the gravity cell, or Daniell cell, which was developed by John Frederic Daniell in 1836 (see Demonstration 11.6) [2]. This cell uses the reaction between zinc and copper ions. It contains a 0.01M zinc sulfate solution floating on a much denser saturated copper sulfate solution. The zinc electrode is suspended in the zinc sulfate solution, and the copper electrode is immersed in the copper sulfate solution. No salt bridge is needed, because ions can move across the interface between the two solutions. This battery is not suited to mobile applications, because moving it will cause the solutions to mix. Mixing of the solutions reduces their ability to produce current. Other batteries have been designed to overcome this limitation.

The dry cell is one of the most common batteries today. It is called a dry cell, not because it does not contain water, but because it contains no fluids (see Demonstration 11.4) [4]. The water it does contain is mixed with ammonium chloride to form a paste. This paste is placed in a zinc cup. A graphite carbon rod surrounded by manganese dioxide is inserted in the paste. The zinc cup and the carbon rod are the electrodes. The zinc electrode is oxidized to zinc ions while current is drawn from the battery. The carbon rod carries electrons to the manganese dioxide, which is reduced. The dry cell is often called a zinc-carbon battery, after the materials of its electrodes. The dry cell is also called a Leclanché cell, after Georges Leclanché, the French engineer who demonstrated a forerunner around 1867. The original Leclanché cell consisted of a zinc electrode dipping into an ammonium chloride solution and a carbon electrode dipping into manganese dioxide. The "dry cell" was introduced in 1888 by Gassner, who substituted a moist paste of ammonium chloride for the solution, and formed the zinc electrode into a container for the cell.

Both the gravity cell and the dry cell are *primary cells*. A primary cell is one that produces electricity from the chemicals that are placed in the cell when it is made. (Demonstrations 11.4 through 11.6 show primary cells.) Once all the chemicals in a primary cell have reacted, the cell is discarded. A *secondary cell* is one that must be charged with energy from some other source before it can be used. This type of cell is usually rechargeable. To recharge the cell, the chemical reactions that produce electricity are reversed by passing a current through the cell in the opposite direction of the current produced by the cell. In a secondary cell, both the reactants and the products of the chemical reactions that produce electricity are insoluble in the liquid in the cell. This condition is required if the cell is to be rechargeable. If the products or reactants were not insoluble, they would not stay at the electrodes, and would not be available when the reaction is reversed.

The most common secondary cell is the automobile battery, or lead-acid cell (see

Demonstration 11.5). The lead-acid battery has electrodes composed of lead sulfate pressed into a lead grid. These electrodes are immersed in a fairly concentrated sulfuric acid solution. Lead sulfate is essentially insoluble in sulfuric acid. The cell is charged by passing an electric current through it. At one electrode, the lead sulfate is reduced to lead. At the other, it is oxidized to lead dioxide. When the battery discharges, the lead dioxide is reduced to lead sulfate, while lead at the other electrode is oxidized to lead sulfate. The charging process is the reverse of the discharging process. These processes can be repeated indefinitely, until mechanical effects due to impacts, temperature changes, and aging cause the lead sulfate and lead dioxide to flake from the electrodes. Because the reactants and products of the reactions in the lead-acid battery are all solids, the battery is able to withstand agitation and is useful in transportation.

A primary cell in which the reactants are supplied continuously from outside the cell is called a fuel cell. A fuel cell developed for applications in space exploration is the hydrogen-oxygen fuel cell (see Demonstration 11.7). In this cell hydrogen gas is passed over one electrode, and oxygen gas over the other. The electrolyte around the electrodes is an aqueous alkali, such as sodium hydroxide. At one electrode, hydrogen is oxidized to water. At the other, oxygen is reduced to water.

Voltaic cells, or batteries, use chemical reactions to generate an electric current. An electric current can also be used to cause a chemical reaction that would otherwise not occur. The charging of a secondary battery is a process in which an electric current is used to produce a reaction. Many oxidation-reduction reactions can be produced by passing an electric current through an electrochemical cell. This process is called *electrolysis,* and the cell in which it occurs is called an *electrolytic cell.* When a current is driven through an aqueous electrolyte, reactions occur at the electrodes that would not occur if the current were not imposed. In some solutions, the water solvent itself is oxidized to oxygen at the anode, and reduced to hydrogen at the cathode (Demonstrations 11.14 through 11.16). In other solutions, the ions of the electrolyte are oxidized or reduced (Demonstrations 11.17, 11.18, and 11.22).

ELECTROLYTIC CELLS

A common type of electrolytic cell is used for *electroplating,* the coating of some object with a metal reduced from its ions in the electrolyte. Demonstrations 11.23, 11.25 through 11.27, and 11.29 show several electroplating processes. Such electroplating is used for decorative purposes, such as silver plating of tableware. It is used as well to produce a durable and corrosion-resistant surface on utility items, for example, in chromium plating of plumbing fixtures. Electrolysis is also used as a technique for refining metals. Virtually all copper and aluminum metal has been purified by electrolysis.

In addition to being reduced from its compounds in electrolytic cells, aluminum metal is frequently made the anode in another type of cell. When aluminum is made the anode, it is oxidized to aluminum oxide in a process called *anodization* [*5*]. The aluminum oxide produced in the process adheres to the aluminum, producing a strong and corrosion-resistant surface. This anodized aluminum surface is also porous and accepts dyes, so the color of the surface can be changed (see Demonstration 11.31) [*6*]. As well as being corrosion-resistant, the oxide coating formed in anodization is a poor conductor of electricity. Because the anodized surface contains very small pores, the smallest ions, hydrogen ions, can get to the metal to accept electrons and be reduced to hydrogen gas. However, anions, which are much larger than hydrogen ions, cannot get near enough to the aluminum metal to release their electrons. Electrons flow from the electrode into the solution, but not in the other direction. Thus, an aluminum electrode passes current in only one direction. It can be used to convert, or rectify, alternating current to direct current (see Demonstration 11.20).

There is a relationship between the amount of charge passed through a solution and the amount of material deposited in an electrolytic cell. Michael Faraday formulated the laws of electrochemical stoichiometry in the early 19th century. These laws state that the number of moles of product formed in an electrolysis cell is proportional to the number of moles of electrons passed through the cell [*3*]. In other words, a mole of electrons (96,480 coulombs of charge) produces a fixed mass of a substance (e.g., 107.9 g of silver or 31.77 g of copper). A current that transfers 1 mole of electrons is called 1 *faraday* of electricity. An electrolytic cell can be used to measure the amount of current passed in a circuit. The amount of substance produced at an electrode in an electrolytic cell reveals the amount of current passed. When an electrolytic cell is used to determine the current passed through it, the cell is called a coulometer, because the total charge passed through the cell (in coulombs) is determined, as well. Demonstration 11.19 describes a couple of coulometers.

REFERENCES

1. F. W. Sears, M. W. Zemansky, and H. D. Young, *University Physics,* 6th ed., Addison-Wesley: Reading, Massachusetts (1982).
2. A. J. Ihde, *The Development of Modern Chemistry,* Dover Publications: New York (1984).
3. R. C. Weast, Ed., *CRC Handbook of Chemistry and Physics,* 66th ed., CRC Press: Boca Raton, Florida (1985).
4. D. Linden, Ed., *Handbook of Batteries and Fuel Cells,* McGraw-Hill Book Co.: New York (1984).
5. V. F. Henley, *Anodic Oxidation of Aluminium and Its Alloys,* Pergamon Press: New York (1982).
6. K. R. Van Horn, *Aluminum,* Vol. 3, *Fabrication and Finishing,* American Society for Metals: Metals Park, Ohio (1967).

11.1

Magnetic Field from a Conducting Solution

A wire is placed atop a magnetic compass on an overhead projector. When the ends of the wire are connected to the terminals of a battery, the compass needle turns perpendicular to the wire. The wire is replaced with a glass tube containing a colorless liquid with electrodes at each end. When the electrodes are connected to a dc power supply, the compass needle turns.

MATERIALS

overhead projector

25 mL 2M sulfuric acid, H_2SO_4 (To prepare 1 liter of solution, set a 2-liter beaker containing 500 mL of distilled water in a pan of ice water. While stirring the water, slowly pour 110 mL of concentrated [18M] H_2SO_4 into the beaker. The mixture will become very hot. If all the ice melts, add more. Once the mixture has cooled to room temperature, dilute it to 1.0 liter with distilled water.)

ca. 30-cm length of glass tubing, with outside diameter of 15 mm

glass-working torch

stand, ca. 50 cm tall, with clamp to hold glass tubing

2 15-cm carbon rods, ca. 8 mm in diameter (A suitable rod may be obtained from a welding supply shop or recovered by disassembling a discharged 1.5-volt 15-cm × 6.5-cm diameter dry cell [so-called #6 ignition battery].)

magnetic compass, ca. 5 cm in diameter, with transparent case

10-cm copper wire (insulated or bare), ca. 12 gauge

1.5-volt battery, with clip leads (e.g., D-cell in battery holder)

12-volt power supply, capable of delivering 1 ampere (e.g., automotive battery charger)

PROCEDURE

Preparation

Bend the 30-cm length of glass tubing into a U shape. The length of the horizontal section of the tube should be equal to the diameter of the magnetic compass and each vertical arm should be about 10 cm long.

Clamp the U-shaped tube to the stand and fill it with 2M H_2SO_4. Insert one of the carbon electrodes in each arm of the tube.

Presentation

Place the transparent magnetic compass on the overhead projector. Lay the copper wire over the compass and align it so that it is parallel with the compass needle. Clip one of the leads from the battery to one end of the wire. Touch the other lead to the other end of the wire. When contact is made, the compass needle will rotate until it is perpendicular to the wire. Remove the lead from the wire, and the needle will return to its prior position. Unclip the battery lead from one end of the wire and reattach it to the other end of the wire. Touch the second lead to the opposite end of the wire. This time the compass needle will rotate in the opposite direction to become perpendicular to the wire. Disconnect the battery, and the compass needle will return to its original position. Remove the wire from the projector.

Set the stand holding the tube of 2M H_2SO_4 on the overhead projector. Align the horizontal section of the tube so that it is parallel with the needle immediately over the compass. The bottom of the tube should be touching the top of the compass. Connect one lead from the 12-volt power supply to one of the electrodes in the tube. With the power supply turned off, connect the other lead to the other electrode. Turn on the power supply. The compass needle will immediately turn until it is perpendicular to the tube. Bubbles of gas will appear at each electrode. Turn off the power supply. The compass needle will return to its original position. Reverse the connections of the power supply to the electrodes. Turn on the power supply, and the compass needle will rotate in the opposite direction to become perpendicular to the tube. Again bubbles form at the electrodes. Turn off the power supply, and the needle will return to its original position.

HAZARDS

Because sulfuric acid is a strong acid and a powerful dehydrating agent, it must be handled with great care. Spills should be neutralized with an appropriate agent, such as sodium bicarbonate ($NaHCO_3$), and then wiped up.

DISPOSAL

The 2M sulfuric acid should be neutralized by the addition of sodium bicarbonate ($NaHCO_3$) until fizzing stops, and the neutralized mixture should be flushed down the drain with water. Rinse the carbon rods with distilled water to remove any sulfuric acid.

DISCUSSION

This demonstration shows one of the physical effects of the passage of an electric current, namely, a magnetic field. The flow of electric current produces a magnetic field, whether the current flows through a metallic conductor in the form of electrons or through an electrolyte solution in the form of ions. The magnetic field is detected in this demonstration with a magnetic compass. The magnetic compass contains a magnetized needle mounted at its center on a pivot. When the needle is placed in a magnetic field, it aligns itself parallel with the field. In the absence of other fields, the earth's magnetic field causes the needle to align itself in a north-south direction.

The connection between electric and magnetic phenomena was observed in 1819 by the Danish scientist Hans Christian Oersted [1]. He saw the same effect shown in this demonstration, namely, that a compass needle moved when an electric current flowed through a nearby wire.

A moving electric charge generates a magnetic field. This magnetic field will interact with any other magnetic field. All atoms contain moving charges, namely, the electrons that surround the nucleus. All atoms are affected by nearby magnetic fields, but most substances show only very small effects from a nearby magnetic field. However, there are a few substances, called ferromagnetic materials, for which the magnetic effect is very large. Among these materials is iron. In ferromagnetic materials, there are regions in which the magnetic fields of individual atoms are aligned [2]. These aligned fields accentuate the magnetic effects experienced by the individual atoms. In some ferromagnetic materials, the magnetic fields of the individual regions can be permanently aligned. These materials can be turned into "permanent magnets." The needle of a compass is a permanent magnet.

When a compass is placed in a magnetic field, the needle aligns itself with the field. Because the earth has a weak magnetic field oriented generally along its axis of rotation, a compass needle usually points north. It points north unless the compass is placed in a magnetic field stronger than that of the earth. In this demonstration, the compass is placed near a wire running in the north-south direction. When an electric current passes through the wire, the compass needle rotates out of the north-south alignment. This indicates that the magnetic field created by the flowing electric current is stronger than that of the earth. It also shows that the magnetic field created by a flowing current is perpendicular to the direction of current flow. The direction in which the compass needle turns also depends on the direction of current flow in the wire.

The compass needle deflects when a voltage is applied between electrodes in a nearby solution. This indicates that electric charges are moving in the solution. These moving charges are ions. Positive hydrogen ions and negative sulfate ions move through the solution of sulfuric acid, carrying the current.

The electric conductivity of an electrolyte solution is not as great as that of a metal. Therefore, the voltage applied between the electrodes must be greater than that applied to the wire, in order to produce a similar electric current in the two conductors. In spite of the higher voltage, the current in the solution is likely to be only a tenth of that in the wire. The weaker current in the solution will produce a weaker magnetic field, so the compass needle may not rotate as far or as quickly as it does near the conducting wire. Furthermore, the diameter of the solution conductor is greater than that of the wire conductor. This causes the magnetic field produced by the current in the solution to be more diffuse than that near the wire. This, too, will contribute to a less dramatic rotation of the needle. Therefore, it is necessary to place the tube of conducting solution as close to the compass needle as possible.

When electric current flows through a solution, two types of conduction occur. In the solution, the movement of ions conducts the electric current; sulfate anions move in one direction and hydrogen ions move in the opposite direction. In the wires connected to the electrodes and in the electrodes themselves, the current is conducted by moving electrons. At the surface of the electrodes, the current changes from electron-carried to ion-carried. This transformation can occur only if electrons are added to or removed from the ions. Such addition and removal of electrons from ions result in chemical transformation. The complete processes occurring at each electrode are quite complicated. However, the net effect at the electrode where electrons enter the solution is the reduction of hydrogen ions in the solution.

$$2\,H^+(aq) + 2\,e^- \longrightarrow H_2(g)$$

The bubbles produced at this electrode are hydrogen gas. At the other electrode, where electrons leave the solution, the net transformation is the oxidation of water to oxygen gas.

$$2\,H_2O(l) \longrightarrow O_2(g) + 4\,H^+(aq) + 4\,e^-$$

The overall chemical transformation occurring as current flows through the sulfuric acid solution is the decomposition of water to its elements, hydrogen and oxygen.

$$2\,H_2O(l) \longrightarrow 2\,H_2(g) + O_2(g)$$

REFERENCES

1. G. Shortley and D. Williams, *Elements of Physics*, 4th ed., Prentice-Hall: Englewood Cliffs, New Jersey (1965).
2. P. W. Atkins, *Physical Chemistry*, W. H. Freeman: San Francisco (1978).

11.2

An Activity Series: Zinc, Copper, and Silver Half Cells

A zinc strip is dipped into a beaker containing a colorless solution, and a copper strip is dipped into a blue solution. The two strips are connected to the terminals of a voltmeter, and the meter registers 0 volts. When a U-tube containing a gel is inverted and one arm is inserted in each beaker, the voltmeter indicates a potential of 1.1 volts.†

MATERIALS

200 mL distilled water

3 g powdered agar

15 g potassium nitrate, KNO_3

500 mL 1.0M copper nitrate, $Cu(NO_3)_2$ (To prepare 1 liter of solution, dissolve 242 g of $Cu(NO_3)_2 \cdot 3H_2O$ in 600 mL of distilled water and dilute the resulting solution to 1.0 liter.)

500 mL 1.0M zinc nitrate, $Zn(NO_3)_2$ (To prepare 1 liter of solution, dissolve 297 g of $Zn(NO_3)_2 \cdot 6H_2O$ in 600 mL of distilled water and dilute the resulting solution to 1.0 liter.)

250 mL 1.0M silver nitrate, $AgNO_3$ (To prepare 250 mL of solution, dissolve 42 g of $AgNO_3$ in 150 mL of distilled water and dilute the resulting solution to 250 mL.)

2 zinc strips, ca. 25 mm × 100 mm × 0.25 mm thick

2 copper strips, ca. 25 mm × 100 mm × 0.25 mm thick

silver strip, ca. 25 mm × 100 mm × 0.25 mm thick (A 10-cm silver wire may be substituted.)

400-mL beaker

hot plate

stirring rod

3 U-shaped drying tubes, ca. 10 cm tall

600-mL beaker

piece of plastic food wrap, ca. 15 cm square

rubber band

5 300-mL tall-form beakers, or 5 400-mL low-form beakers

† This demonstration was developed by Professor Kenneth Watkins of Colorado State University while he was a visiting professor at the University of Wisconsin–Madison in 1981.

voltmeter capable of reading 0–2 volts, with clip leads (The voltmeter should have a display large enough to be easily visible to the observers. A large digital voltmeter or an overhead projection meter is suitable.)

1.5-volt battery-operated wall clock, without battery (optional)

2 30-cm wire leads, with an alligator clip on each end (optional)

PROCEDURE

Preparation

Construct three salt bridges as described in this paragraph. Heat 150 mL of distilled water to boiling in a 400-mL beaker. Add 3 g of agar and stir the mixture as it boils until a uniform suspension forms (5–10 minutes). Remove the beaker from the heat, add 15 g of KNO_3, and stir the mixture until the KNO_3 dissolves. Pour the warm liquid into the three U-shaped drying tubes, filling them *completely*. Flush any excess liquid down the drain. Set the tubes upright inside the 600-mL beaker. Allow the tubes to stand undisturbed for 6–8 hours, during which time the agar will set to a firm gel. (In less than 6 hours, the agar gel will not be firm enough; in more than 8 hours, the gel may begin to dry and crack.) On one salt bridge label one arm "Zn" and the other "Cu"; on the second salt bridge label one arm "Zn" and the other "Ag"; and on the last salt bridge, label one arm "Cu" and the other "Ag." (These labels will help to avoid contaminating the solutions, if the salt bridges are used in more than one presentation of the demonstration.) Pour 50 mL of distilled water into the 600-mL beaker and place the gel-filled drying tubes, the salt bridges, in the beaker with open ends up. Seal the beaker with plastic wrap and stretch a rubber band around the beaker to secure the wrap. (The salt bridges can be stored indefinitely if the beaker is airtight.)

Pour 250 mL of 1.0M $Cu(NO_3)_2$ into each of two of the 300-mL tall-form beakers, 250 mL of 1.0M $Zn(NO_3)_2$ into each of two more 300-mL beakers, and 250 mL of 1.0M $AgNO_3$ into the remaining 300-mL beaker.

Clip one of the 30-cm wire leads to the positive terminal of the 1.5-volt wall clock, and clip the other lead to the negative terminal.

Presentation

Place one of the zinc strips in one of the beakers of $Cu(NO_3)_2$ solution and one of the copper strips in one of the beakers of $Zn(NO_3)_2$ solution. Allow them to rest for about 15 seconds. Remove the zinc strip and note that, where it was immersed in the $Cu(NO_3)_2$ solution, it has acquired a brown coating. Lean the strip against the outside of the beaker so the brown coating is visible. Remove the copper strip from the $Zn(NO_3)_2$ solution and note that it is unchanged.

Set the remaining beaker of $Cu(NO_3)_2$ solution next to the remaining beaker of $Zn(NO_3)_2$ solution. Insert the remaining copper strip in the beaker of $Cu(NO_3)_2$ solution and the remaining zinc strip in the beaker of $Zn(NO_3)_2$ solution. Clip the positive (red) lead of the voltmeter to the copper strip and the common (black) lead to the zinc strip. Connect the solutions with the "Zn-Cu" salt bridge by inserting the appropriately labelled arm of the salt bridge in each beaker. Record the reading of the voltmeter. Reverse the leads on the metal strips and note that the sign of the voltmeter reading changes, but its magnitude does not. Disconnect the voltmeter and remove the salt bridge from the beakers.

Insert the silver strip in the beaker of $AgNO_3$ solution. Set this beaker next to the second beaker of $Cu(NO_3)_2$ solution. Connect the positive lead of the voltmeter to the silver strip and the common lead to the copper strip. Connect the solutions with the "Cu-Ag" salt bridge by inserting the appropriately labelled arm of the salt bridge in each beaker. Record the reading of the voltmeter. Disconnect the voltmeter and remove the salt bridge.

Set the $AgNO_3$ beaker next to the second $Zn(NO_3)_2$ beaker. Connect the positive lead of the voltmeter to the silver strip and the common lead to the zinc strip. Connect the solutions with the "Zn-Ag" salt bridge by inserting the appropriately labelled arm of the salt bridge in each beaker. Record the reading of the voltmeter. Note that the measured potential of the zinc-silver cell is the sum of the measured potentials of the zinc-copper cell and the copper-silver cells.

Connect the positive lead from the battery-operated wall clock to the silver strip and the negative lead to the zinc strip. The clock will run for at least several hours when connected.

HAZARDS

Copper compounds can be toxic if taken internally, and dust from copper compounds can irritate mucous membranes.

Silver nitrate is irritating to the eyes and mucous membranes. If taken internally, silver nitrate can be toxic. Silver nitrate will stain the skin; these stains can be bleached by rinsing with an aqueous solution of sodium thiosulfate ($Na_2S_2O_3$) followed by water.

DISPOSAL

The solutions in the 300-mL beakers may be retained for repeated presentations of this demonstration. To dispose of them, flush them down the drain with water.

The salt bridges may be kept indefinitely sealed in a beaker with water. If the salt bridge should dry out or become discolored, the U-tube can be cleaned by immersing it in water and heating the water to boiling. The gel will soften. The softened gel can be flushed down the drain with water.

The brown coating on the first zinc strip can be removed by buffing the strip with emery cloth or sandpaper.

DISCUSSION

Three standard half cells are combined in different combinations to produce three different electrochemical cells. The potentials of these three electrochemical cells are measured. The measurements show that the third cell potential is the sum of the previous two. The potential of one of the cells is close to 1.5 volts, and this cell can drive a 1.5-volt clock. The half cells are the copper–copper(II) ion half cell, the zinc–zinc(II) ion half cell, and the silver–silver(I) ion half cell. The concentrations of the ions are 1M, and therefore the measured cell potentials are close to the standard values.

When zinc metal is placed in a solution containing copper ions, a spontaneous oxidation-reduction reaction takes place. Metallic copper is deposited on the surface of the zinc strip, and zinc ions go into solution.

$$Zn(s) + Cu^{2+}(aq) \longrightarrow Zn^{2+}(aq) + Cu(s)$$

In this reaction each zinc atom loses two electrons and each copper ion gains two electrons. This process transforms the zinc metal into zinc ions and copper ions into copper metal. The transformation of zinc to zinc ions can be represented by the equation

$$Zn(s) \longrightarrow Zn^{2+}(aq) + 2\ e^-$$

This equation is called a half reaction because it is only half of the process occurring in the reaction. The half reaction for the copper ions is

$$Cu^{2+}(aq) + 2\ e^- \longrightarrow Cu(s)$$

Neither of these half reactions can occur by itself without some other reaction also occurring. However, they can occur in separate locations.

The solution in which zinc becomes zinc ions does not need to be the same solution in which the copper ions are converted to copper metal. The solution containing copper ions can be placed in one beaker, and the zinc strip can be immersed in a solution without copper ions in a different beaker. However, this alone is not sufficient for the reaction to occur. There must be a way for the electrons released by the zinc atoms to reach the copper ions. This is accomplished by connecting the zinc metal to the copper solution with a wire. The wire conducts electrons from the zinc to the copper ions. Even with the wire, the reaction does not proceed very far. Also needed is some means of conducting charge back to the solution in contact with the zinc electrode to counterbalance the charge transferred through the wire. This is accomplished by connecting the two solutions with a "salt bridge." The salt bridge is a gel containing ions that can migrate from one end to the other. This migration of ions balances the charge transferred through the wire. When the salt bridge is placed in contact with the two solutions, the reaction between zinc metal and copper ions proceeds. Zinc metal dissolves in one solution, and the copper ions in the other solution are converted to copper atoms where the wire contacts the solution.

When the copper strip is immersed in copper nitrate solution and the zinc strip is immersed in zinc nitrate, the two metal strips are at different electric potentials. If the two solutions are in contact, for example, through a salt bridge, this potential difference can be measured with a voltmeter.

The potential difference arises because copper has a greater reduction potential than zinc. When a piece of metal is immersed in a solution, the metal has a tendency to produce more ions. This also produces an excess of electrons in the metal. This process continues until an equilibrium is established between the charged metal and the ions in the solution [*1*]. When the equilibrium is established, the metal has a net negative charge and the solution has a net positive charge. This creates a difference in electric potential between the metal and the bulk of the solution. It is not possible to measure this potential directly, because to do so requires placing an electrically conductive probe in contact with the solution. Any conductive probe in contact with the solution will establish its own equilibrium with the solution, and thus have an electrode potential of its own. For this reason, it is possible to measure only differences in electric potential between conductive materials in contact with solutions.

When zinc metal is placed in a solution of zinc nitrate, an equilibrium between the metal and ions in solution is established. This equilibrium leaves the zinc metal with a slight negative charge and the solution with a positive charge. A similar situation arises when copper metal is placed in a copper nitrate solution. The charge that builds up on the zinc strip is greater than that on the copper strip. Therefore, there is a difference in the potential of the zinc strip and that of the copper strip.

When the voltmeter is connected to the zinc strip in the zinc nitrate solution and the copper strip in the copper nitrate solution, it registers no potential difference. This is a result of the way in which a voltmeter operates [*2*]. In order to measure a potential difference, a voltmeter draws a tiny current through a calibrated resistance. However, no current can flow between the electrodes unless current can also flow between the solutions. Therefore, a salt bridge is placed between the solutions to carry electric current in the form of ions between the solutions. When the two solutions are connected by the salt bridge, the voltmeter registers a potential difference between the two metal electrodes.

In a voltaic cell the reactants are physically separated, and electrons must travel through the external circuit (wires and voltmeter) from the zinc electrode to the copper electrode. As electrons leave the zinc electrode, zinc ions enter the solution. The electrons move from the zinc electrode to a region of lower potential, namely through the external circuit to the copper electrode. At the interface between the copper electrode and the copper nitrate solution, electrons combine with Cu^{2+} ions, forming copper atoms. The copper atoms accumulate as a copper plating on the copper electrode.

The salt bridge, which contains mobile K^+ and NO_3^- ions in a gel, completes the circuit (path of electric charge). As Cu^{2+} ions are converted to copper atoms in the copper nitrate solution, an excess of NO_3^- ions builds up in the solution. As zinc atoms are converted to Zn^{2+} in the zinc nitrate solution, an excess of cations develops in this solution. The salt bridge provides the means to counteract this charge build up. Excess NO_3^- ions from the copper nitrate solution enter one end of the bridge, and NO_3^- ions migrate out into the zinc nitrate solution. Simultaneously, excess cations from the zinc nitrate solution, namely Zn^{2+}, enter the salt bridge, and potassium ions migrate into the copper nitrate solution.

In the copper-zinc cell, the potential was positive when the positive voltmeter lead was connected to the copper strip. Therefore, the voltmeter registers a positive potential when its positive lead is connected to the strip where the metal is reduced, that is, the cathode. In the silver-copper cell, the voltmeter registers a positive reading when its positive lead is connected to the silver strip. Therefore, in the silver-copper cell, the silver is reduced and the copper oxidized. The overall reaction in this cell is

$$Cu(s) + 2\,Ag^+(aq) \longrightarrow Cu^{2+}(aq) + 2\,Ag(s)$$

In the silver-zinc cell, the voltmeter reading is positive when its positive lead is connected to the silver strip. Therefore, its cell reaction is

$$Zn(s) + 2\,Ag^+(aq) \longrightarrow Zn^{2+}(aq) + 2\,Ag(s)$$

The cell reactions reveal that zinc is the most active of these three metals, that is, the most likely to form ions and reduce the ions of the other two metals. Silver is the least active of these three. The three metals can be arranged in a series based on their activities: Zn > Cu > Ag. Furthermore, the measured potential of the zinc-silver cell is the sum of the measured potentials of the zinc-copper cell and the copper-silver cell. In other words, the potential difference between zinc and silver is the sum of the difference between zinc and copper and the difference between copper and silver. Represented symbolically, this is

$$E_{Zn} - E_{Ag} = (E_{Zn} - E_{Cu}) + (E_{Cu} - E_{Ag})$$

where E is the potential of the electrode. This algebraic relationship indicates that a numerical value can be assigned to each of these potentials. However, these values cannot be determined solely by measurement, because all that can be measured is the difference between electrode potentials; an absolute potential cannot be measured. Therefore, a reference point must first be established. This is done by assigning an arbitrary potential to some electrode. By convention, the value of 0 volts is assigned to the electrode at which hydrogen ions are reduced to hydrogen gas.

$$2\,H^+(aq) + 2\,e^- \longrightarrow H_2(g)$$

The selection of this electrode as the standard is reasonable, for the hydrogen ion is the most common ion, present in *all* aqueous solutions.

The measured potential of the zinc-silver cell is close to that of a 1.5-volt battery. However, not all 1.5-volt battery-operated devices could be driven by this zinc-silver cell. Some such devices draw more current than this cell can provide. The flow of charge through the salt bridge is limited by the mobility of the potassium and nitrate ions in the gel. This mobility is not great enough to carry a large current. A battery-operated wall clock is designed to run for over a year on a single 1.5-volt battery. It does not draw much current, and the zinc-silver cell in this demonstration can provide enough to operate the clock.

REFERENCES

1. P. W. Atkins, *Physical Chemistry,* p. 327, W. H. Freeman: San Francisco (1978).
2. F. W. Sears, M. W. Zemansky, and H. D. Young, *College Physics,* 4th ed., p. 460, Addison-Wesley: Reading, Massachusetts (1977).

11.3

The "Standard" Orange Electrode

Two holes are cut in an orange, a graphite rod is inserted in one hole, and a gel-filled tube is inserted in the other. The other end of the gel-filled tube is inserted, in turn, in each of several beakers containing solutions and strips of metal. A voltmeter is connected between the metal strip in the beaker and the graphite rod in the orange, and the voltage is recorded for each beaker. The recorded voltages can be used to predict the voltages of cells made from combinations of the beakers.†

MATERIALS

100 mL 1.0M zinc nitrate, $Zn(NO_3)_2$ (To prepare 100 mL of solution, dissolve 30 g of $Zn(NO_3)_2 \cdot 6H_2O$ in 60 mL of distilled water and dilute the resulting solution to 100 mL.)

100 mL 1.0M copper(II) nitrate, $Cu(NO_3)_2$ (To prepare 100 mL of solution, dissolve 24 g of $Cu(NO_3)_2 \cdot 3H_2O$ in 60 mL of distilled water and dilute the resulting solution to 100 mL.)

100 mL 1.0M lead(II) nitrate, $Pb(NO_3)_2$ (To prepare 100 mL of solution, dissolve 33 g of $Pb(NO_3)_2$ in 60 mL of distilled water and dilute the resulting solution to 100 mL.)

ca. 25 mL 1M sodium chloride, NaCl (To prepare 100 mL of solution, dissolve 5.8 g of NaCl in 60 mL of distilled water and dilute the resulting solution to 100 mL.)

large orange

25 g sodium sulfide nonahydrate, $Na_2S \cdot 9H_2O$ (See Disposal section for use.)

3 250-mL beakers, with labels

U-shaped drying tube, 10 cm tall

2 cotton balls

knife to cut orange

10-cm graphite rod, ca. 2.5 mm in diameter

zinc metal strip, 2 cm × 10 cm × ca. 0.8 mm thick

voltmeter that reads 0–2 volts, with clip leads

copper metal strip, 2 cm × 10 cm × ca. 0.8 mm thick

lead metal strip, 2 cm × 10 cm × ca. 0.8 mm thick

† This demonstration was developed by Professor Joseph Conrad of the University of Wisconsin–River Falls.

PROCEDURE

Preparation

Label three 250-mL beakers with "1.0M $Zn(NO_3)_2$," "1.0M $Cu(NO_3)_2$," and "1.0M $Pb(NO_3)_2$." Pour 100 mL of the appropriate solution into each beaker.

Hold the U-shaped drying tube upright and fill it completely with 1M NaCl solution. Pack a wad of cotton tightly into the opening of each arm of the tube to make a plug about 1 cm long. Invert the tube momentarily to be sure the liquid remains in the tube. If the tube leaks, wad the cotton more tightly.

Presentation

Before cutting into the orange, squeeze it repeatedly to break some of its internal membranes and free some of the juice inside the orange. Place the orange on the table and cut two 1-cm holes through its skin. Insert the graphite rod in one hole.

Set the beaker of 1.0M $Zn(NO_3)_2$ next to the orange. Invert the NaCl-filled U-tube, and insert one arm in the remaining hole in the orange and the other arm in the beaker of $Zn(NO_3)_2$. Be sure the cotton plugs in the arms of the U-tube are making contact with the juice in the orange and the solution in the beaker. Insert the zinc strip in the $Zn(NO_3)_2$ solution. Attach the reference (common, black) lead of the voltmeter to the graphite rod in the orange and the positive (red) lead to the zinc strip in the beaker. Record the voltage registered by the meter. (If the voltmeter being used cannot read negative voltages directly, and the reading goes below 0, reverse the leads on the graphite rod and metal strip. Record this reading as negative.)

Repeat the procedure of the previous paragraph with the remaining two beakers of solution and their appropriate metal strips. Record the voltmeter reading for each.

Set the beaker of 1.0M $Zn(NO_3)_2$ beside the beaker of 1.0M $Pb(NO_3)_2$. Place the zinc strip in the zinc solution and the lead strip in the lead solution. Invert the NaCl-filled U-tube, and insert one arm in each beaker. Connect the leads of the voltmeter to the metal strips and record its reading. Repeat this with the other two combinations of metals and solutions (zinc and lead, copper and lead).

HAZARDS

Lead compounds are harmful when taken internally. The effects of exposure to small concentrations can be cumulative, causing loss of appetite and anemia.

Copper compounds can be toxic if taken internally, and dust from copper compounds can irritate mucous membranes.

DISPOSAL

The lead nitrate waste should be converted to lead sulfide by adding 25 g of sodium sulfide nonahydrate ($Na_2S \cdot 9H_2O$) to the solution. The lead sulfide precipitate should be collected and buried in a landfill designed for heavy metals. (Local regulations on the disposal of hazardous wastes should be consulted.) The solution should be flushed down the drain with large volumes of water.

The other solutions should be flushed down the drain with water.

The leftover orange should not be eaten. Discard it in a solid-waste receptacle.

DISCUSSION

This demonstration uses a most unconventional electrode, an orange, as a reference electrode. Its unconventionality emphasizes the arbitrary and relative nature of the values of electrode potentials.

The electrode potential of an electrically conductive solid in contact with an electrically conductive solution arises from a difference in electron energy between the bulk of the solid and the bulk of the solution. It is not possible to measure this difference directly. The measurement of an electrode potential requires placing another electrically conductive probe in contact with the solution. Another conductive probe in contact with the solution will have its own electrode potential. Therefore, a measurement of energy difference between electrodes is actually a measurement of the difference in differences. The energy (potential) of one electrode differs from that of the bulk of the solution by a certain amount, say, $\Delta E_1 = E_{electrode\ 1} - E_{solution}$. The potential of another electrode differs from the bulk of the solution by another amount, $\Delta E_2 = E_{electrode\ 2} - E_{solution}$. The measured potential between the two electrodes is $\Delta E = \Delta E_1 - \Delta E_2$. Thus, all measurements of electrode potentials are, by their nature, relative. When an electrode potential is reported, it makes sense only if the electrode used as a reference is also reported.

The potential of an electrode depends on both the nature of the solid conductor and the composition of the solution with which it is in contact. Because of this, a reference electrode is usually used with its own standard solution. The solid conductor of the reference electrode is not in contact with the solution containing the electrode whose potential is to be measured. Instead, the reference electrode solution is brought into electrical contact with the solution around the electrode being measured. This is accomplished by joining the two solutions with an intermediate solution, a "salt bridge."

For convenience sake, a small number of reference electrodes have been chosen. The fundamental reference electrode is the standard hydrogen electrode. This reference is fundamental because its potential has been assigned the value of 0 volts. Other reference electrodes are used. A common reference electrode is the standard calomel electrode (SCE). This demonstration uses an uncommon reference, namely, a graphite rod inserted in an orange.

The voltage of each of three metal electrodes in corresponding 1M ion solutions is measured relative to the graphite rod in the orange. The measured values correspond to the difference in potential between the orange cell and the metal cells. If the potential of the "standard orange electrode" is assigned a value of 0 volts, then the measured values can be assigned to the metal electrodes. (These metal cell potentials will be called E^*, the potential relative to the standard orange electrode [SOE].) These E^* values can be used to predict the measurement that would be obtained if the metal electrodes were to be measured versus each other. These predicted values can be compared with the actual measured values to illustrate the utility of the reference electrode.

First, the E^* values can be tabulated. (These are typical values from the demonstration. There is considerable variation among oranges and, therefore, among the SOE potentials.)

$E^*(Zn^{2+}/Zn)$	-0.90 volt	$Zn^{2+} + 2\,e^- \longrightarrow Zn$
$E^*(Pb^{2+}/Pb)$	-0.50 volt	$Pb^{2+} + 2\,e^- \longrightarrow Pb$
$E^*(Cu^{2+}/Cu)$	-0.10 volt	$Cu^{2+} + 2\,e^- \longrightarrow Cu$

Then, the potential of one metal cell versus another can be calculated. For example, for the cell

$$Pb + Cu^{2+} \longrightarrow Pb^{2+} + Cu$$

the potentials referred to the SOE would give

$$\begin{aligned} E^*(\text{cell}) &= E^*(Cu^{2+}/Cu) - E^*(Pb^{2+}/Pb) \\ &= (-0.10) - (-0.50) \\ &= 0.40 \text{ volt} \end{aligned}$$

This corresponds to what is found when the cell potential is measured. Since E*(cell) is positive, the reaction for the cell would be spontaneous as written.

Because an orange is a complicated and variable mixture of substances, the measurements obtained will vary from orange to orange. However, for a particular orange, there is reasonable correlation between the measured electrode potentials of the metal cells and the published values relative to the standard hydrogen electrode. The measured cell potentials can be converted to values relative to the hydrogen electrode. This is done be taking the difference between one of the measured values, for example, for the Cu^{2+}-Cu cell, and the standard value versus the hydrogen electrode, and subtracting this value from the other values measured versus the orange. However, the larger the absolute value of the cell potential, the greater the deviation from accepted values. Thus, the measured potential versus the orange for the zinc cell is likely to be significantly out of line with the standard value.

Precisely what reactions account for the potential of the orange itself are difficult to determine. The juice of an orange contains, in addition to several inorganic salts, quite a few organic substances. Some of these are capable of undergoing electrochemical reactions, for instance, citric acid, ascorbic acid, and NADH (nicotinamide adenine dinucleotide hydride). Furthermore, because the orange contains so many substances, the reactions may depend on the potential of the cell connected to it. Cells with higher potentials may cause reactions within the orange that are different from cells with lower potentials. Because of this, the orange does not make a very reproducible cell, and is not a good standard cell.

Other fruit can be used in place of the orange. Grapefruits, lemons, limes, tangerines, and other citrus fruits are obvious substitutes. As long as the fruit is sufficiently juicy to thoroughly wet the electrode inserted in it, the fruit can serve as a reference cell. Tomatoes can be used; apples are less suitable. Cucumbers may work, and when they're pickled, they certainly will.

11.4

Constructing a Dry Cell

Black powder is sprinkled onto a damp felt pad, the pad is wrapped around a carbon rod, and the pad is wrapped, in turn, with zinc foil. When the carbon rod and the zinc foil are connected to the terminals of a battery-operated clock, the clock runs [*1*].

MATERIALS

100 mL 4M ammonium chloride, NH_4Cl (To prepare, dissolve 20 g of NH_4Cl in 85 mL of distilled water.)

4 g powdered manganese dioxide, MnO_2

1.5-volt battery-operated wall clock, without battery

2 20-cm wire leads, 24 gauge, with alligator clips on both ends

gloves, plastic or rubber

felt pad, 12 cm × 5 cm × ca. 5 mm thick

15-cm carbon rod, ca. 8 mm in diameter (A suitable rod may be obtained from a welding supply shop or recovered by disassembling a discharged 1.5-volt 15-cm × 6.5-cm diameter dry cell [so-called #6 ignition battery].)

zinc metal foil, 12 cm × 12 cm × ca. 0.25 mm thick

ca. 30 cm string

voltmeter, with probes

PROCEDURE

Preparation

Clip one lead to each of the battery terminals of the battery-operated wall clock.

Presentation

Wearing gloves, moisten the felt pad by immersing it in the saturated NH_4Cl solution and squeezing it. Sprinkle 4 g of MnO_2 in a layer onto the pad. Wrap the pad around the carbon rod with the MnO_2 between the carbon and the pad. Wrap the zinc foil tightly around the pad, making sure the zinc does not touch the carbon rod (Figure 1). Tie the assembly together with string. Press one of the probes of the voltmeter against the zinc and the other probe against the carbon rod. Record the voltage. Attach the lead from the positive terminal of the wall clock to the carbon rod, and clip the other lead to the zinc foil. The clock will start and run for at least half an hour.

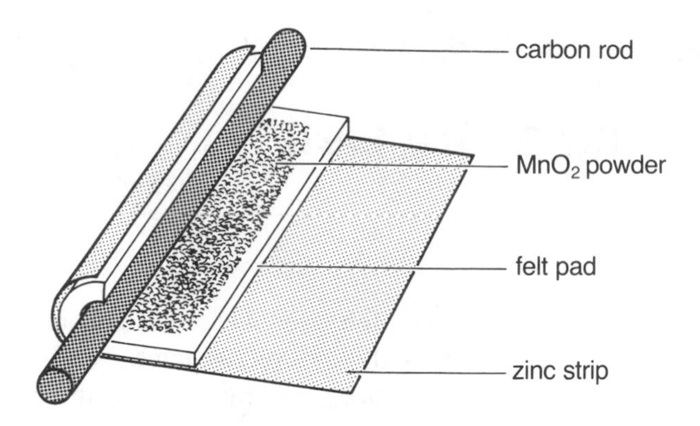

Figure 1. Assembly of dry cell.

HAZARDS

Manganese dioxide is a strong oxidizing agent. The felt pad should not be allowed to dry out in contact with manganese dioxide (see the Disposal section).

Ammonium chloride is toxic when taken internally.

DISPOSAL

The zinc foil can be rinsed, air dried, and reused in repeated presentations of this demonstration until it becomes too corroded to bend without breaking. It should then be discarded in a solid-waste receptacle.

The felt pad should be rinsed under running water to remove the manganese dioxide and ammonium chloride. It can be air dried for reuse or discarded in a solid-waste receptacle.

The ammonium chloride solution should be flushed down the drain with water.

DISCUSSION

This demonstration illustrates the construction of a common dry cell. As it shows, the term *dry cell* is something of a misnomer. The cell is not dry in the sense of being free of water, but rather it contains no fluids. The water it does contain is absorbed into a felt pad that keeps it from flowing. The dry cell is often called a zinc-carbon battery, after the materials of its electrodes. This name, too, is misleading, because while the zinc electrode is active, being consumed during discharge of the battery, the carbon electrode is inert. The dry cell is also called a Leclanché cell, after Georges Leclanché, the French engineer who demonstrated a forerunner around 1867 [*2*]. The original Leclanché cell consisted of a zinc electrode dipping into an ammonium chloride solution and a carbon electrode dipping into manganese dioxide. The "dry cell" was introduced in 1888 by Gassner, who substituted a moist paste of ammonium chloride for the solution, and formed the zinc electrode into a container for the cell. The dry cell is still used because it is easily fabricated from relatively inexpensive materials. Its efficiency is constantly being improved, even over the past 20 years.

In this demonstration, a dry cell is constructed from a carbon rod surrounded by manganese dioxide, which is wrapped in turn with a pad moistened with ammonium chloride solution and then with a zinc strip. The zinc and carbon electrodes are at different potentials. Initially, the cell produces about 1.5 volts, and electrons flow through the external circuit from the zinc to the carbon. The reaction that produces electrons is the oxidation of zinc metal [*3*].

$$Zn \longrightarrow Zn^{2+} + 2\,e^-$$

At the carbon electrode, manganese dioxide is reduced. The actual process involved at the carbon electrode has been a subject of debate. Kozawa and Powers postulated a single solid-phase, proton-electron mechanism, in which the manganese dioxide crystal lattice accepts the protons and electrons [*4*]. Upon continued discharge the Mn^{3+} and OH^- concentrations increase, suggesting the following cathode reaction:

$$MnO_2(s) + H_2O(l) + e^- \longrightarrow MnO(OH)(s) + OH^-(aq)$$

This is the prevalent reaction at a slow discharge rate, but other discharge reactions have been found and are dependent upon electrolyte concentration, discharge temperature, and type of manganese dioxide used. An overall cell reaction is represented by

$$Zn(s) + 2\,MnO_2(s) + 2\,H_2O(l) \longrightarrow Zn^{2+}(aq) + 2\,MnO(OH)(s) + 2\,OH^-(aq)$$

The hydroxide ions produced at the carbon electrode migrate toward the zinc electrode, carrying negative charge from the carbon electrode back to the zinc electrode. The zinc ions produced at the zinc electrode migrate in the opposite direction. Where OH^- and Zn^{2+} ions meet, a complex of zinc and ammonia forms from the ammonium chloride.

$$Zn^{2+}(aq) + 4\,NH_4^+(aq) + 4\,OH^-(aq) \longrightarrow [Zn(NH_3)_4]^{2+}(aq) + 4\,H_2O(l)$$

The construction of a typical commercial dry cell is illustrated in Figure 2. A zinc can serves as anode and container for the other components. The inner surface of the zinc can is coated with a damp paste of ammonium chloride, zinc chloride, and starch (to keep the paste stiff). This paste is electrically conductive. The zinc can is filled with a mixture of manganese dioxide and powdered graphite. The powdered graphite is used

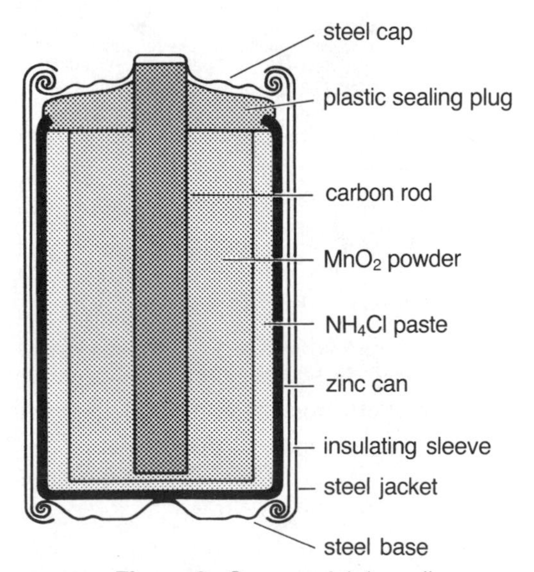

Figure 2. Commercial dry cell.

to increase the electric conductivity of the mixture. The ratio of manganese dioxide to carbon black typically varies from 3:1 to 10:1. However, in cells for high-current demand, a ratio of 1:1 is common. General purpose batteries use manganese dioxide obtained from natural manganese dioxide ore, whereas heavy duty batteries use manganese dioxide produced by the electrolysis of hot manganese sulfate solutions, which is more uniform [*5*]. At the center of the dry cell is a graphite rod, which serves as the cathode. Zinc chloride is included in the paste to serve as an electrolyte and to prevent the accumulation of ammonia gas inside the cell. Ammonia gas is produced by the reaction of ammonium chloride with the hydroxide ions, which are produced by the reduction of manganese dioxide. Ammonia reacts with zinc ions and chloride ions in the paste, forming the tetraammine zinc(II) ion ($[Zn(NH_3)_4]^{2+}$) and insoluble diamminedichlorozinc(II) ($Zn(NH_3)_2Cl_2$) [*6*]. The accumulation of ammonia gas inside the cell would cause the cell contents to expand and perhaps cause the cell to leak. To further minimize the risk of leakage, the zinc can is usually clad with a steel jacket.

REFERENCES

1. A. Joseph, P. F. Brandwein, E. Morholt, H. Pollack, and J. F. Castka, *A Sourcebook for the Physical Sciences,* p. 258, Harcourt, Brace, and World: New York (1961).
2. A. J. Ihde, *The Development of Modern Chemistry,* p. 469, Dover Publications: New York (1984).
3. J. E. Huheey, *Inorganic Chemistry,* 2d ed., p. 314, Harper and Row: New York (1978).
4. A. Kozawa and R. A. Powers, *J. Chem. Educ.* 49:587 (1972).
5. D. Linden, Ed., *Handbook of Batteries and Fuel Cells,* p. 5-4, McGraw-Hill Book Co.: New York (1984).
6. C. A. Vincent, F. Bonino, M. Lazzari, and B. Scrosati, *Modern Batteries: An Introduction to Electrochemical Power Sources,* Edward Arnold: London (1984).

11.5

The Lead Storage Battery

Two strips of lead fastened together with a porous pad between them are immersed in a colorless solution and attached to a source of direct current for 5 minutes. When an electric clock is substituted for the direct-current source, the clock runs for over an hour [*1*].

MATERIALS

300 mL 3M sulfuric acid, H_2SO_4 (To prepare 1 liter of solution, set a 2-liter beaker containing 500 mL of distilled water in a pan of ice water. While stirring the water, slowly pour 170 mL of concentrated [18M] H_2SO_4 into the beaker. The mixture will become very hot. If all the ice melts, add more. Once the mixture has cooled to room temperature, dilute it to 1.0 liter with distilled water.)

25 g sodium sulfide nonahydrate, $Na_2S \cdot 9H_2O$ (see Disposal section for use.)

1.5-volt battery-operated wall clock

or

dc motor that operates on 1.5 volts (Suitable motors are available from hobby shops and electronic parts stores.)

index card, 4 inches × 6 inches

tape or glue

ring stand, with clamp

fiberglass window screening, 20 cm square (available from hardware stores)

2 lead strips, 5 cm × 15 cm × 1 mm thick (Lead sheets of approximately this thickness, which can be cut with heavy-duty scissors, are available from hardware stores and plumbing supply shops.)

2 rubber bands

400-mL beaker

2 30-cm insulated wires, 18 gauge, with alligator clips on both ends

dc power supply capable of supplying 5 volts at 200 milliamperes (Suitable power supplies are available from electronic parts stores and hobby shops. An ac to dc adapter for electronic devices can be used.)

PROCEDURE

Preparation

If a motor is to be used in place of the battery-operated clock, cut a propeller about 15 cm in diameter from the index card, insert the motor shaft through its center, and secure it in place with tape or glue. Clamp the motor to the stand so the propeller can turn freely when the motor's shaft rotates (Figure 1).

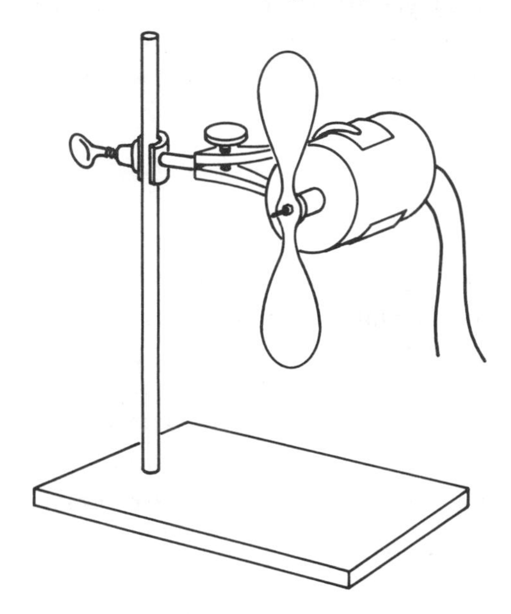

Figure 1. Direct-current motor mounted on stand.

Presentation

Fold the fiberglass screen into a pad about 5 cm × 10 cm. Sandwich the pad between the two strips of lead so that the strips do not touch each other (Figure 2). Use rubber bands to hold the assembly together. Bend the ends of the lead strips apart so clips can be attached to them.

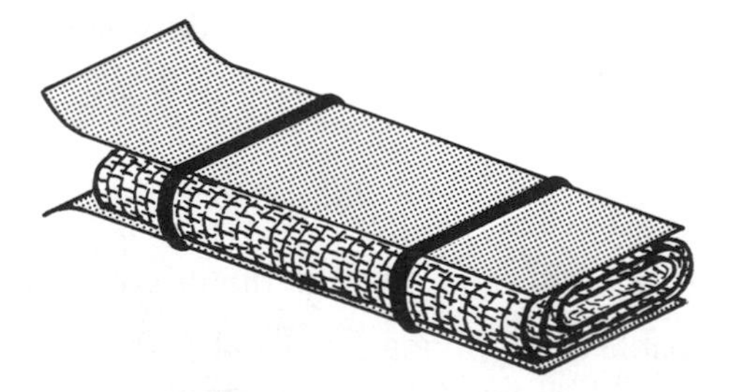

Figure 2. Fiberglass screen sandwiched between lead strips.

Construct the lead storage cell by pouring 300 mL of 3M H_2SO_4 into the 400-mL beaker and placing the sandwiched lead strips in the beaker. Clip a wire to the top of each lead strip. Connect the free end of one wire to one of the terminals of the power supply and connect the other wire to the other terminal. Bubbles will form at both lead strips, and one of the strips will become dark brown where it is immersed in the H_2SO_4.

After the cell has been connected to the power supply for 5 minutes, turn off the power supply, and disconnect the wires from it. The clock's battery holder has a connection marked with plus sign (+) and another marked with a minus sign (−). Connect the wire from the dark brown lead strip to the plus connection on the clock and the wire from the other lead strip to the minus connection. The clock will run for several hours. If the motor with propeller is used, connect one wire to each of the motor's terminals, and the motor will spin the propeller for over a minute.

HAZARDS

Lead compounds are harmful when taken internally. The effects of exposure to small concentrations can be cumulative, causing loss of appetite and anemia.

Because sulfuric acid is a strong acid and a powerful dehydrating agent, it must be handled with great care. Spills of concentrated and dilute solutions of sulfuric acid should be neutralized with an appropriate agent, such as sodium bicarbonate ($NaHCO_3$), and then wiped up.

DISPOSAL

The electrolyte solution should be neutralized by adding sodium bicarbonate to it until no more fizzing occurs. Then, convert the lead salts to lead sulfide by adding 25 g of sodium sulfide nonahydrate ($Na_2S \cdot 9H_2O$) in a basic solution to the neutralized electrolyte solution. The lead sulfide precipitate should be collected and buried in a landfill designed for heavy metals. (Local regulations on the disposal of hazardous wastes should be consulted.) The solution should be flushed down the drain with large volumes of water.

DISCUSSION

This demonstration illustrates the components of the lead storage battery and how they can be assembled into a functioning cell. However, the cell constructed here is rudimentary. It is not fabricated in the way commercial lead storage batteries are manufactured.

The lead storage battery is the most common battery in the market today. Its sales currently represent 60% of all battery sales worldwide. It was first developed by Raymond Gaston Plante in 1860. His early cell consisted of two lead strips and intermediate layers of cloth in a 10% sulfuric acid solution. This first cell, similar to the one constructed in this demonstration, had only a small capacity, because the amount of lead(IV) oxide adhering to the surface was dependent on the porosity of the lead plate. The corrosion of the lead plate after repetitive cycling or "roughening" causes the capacity of the cell to increase [*2*]. Commercial lead storage batteries do not behave this way. Commercial batteries are built with lead grids, instead of smooth plates, which are coated with a lead(II) oxide and sulfuric acid paste during construction. This promotes the adhesion of the lead(IV) oxide to the lead grid, and yields a cell capable of producing much more current before it totally discharges.

The initial charging of the cell in this demonstration produces hydrogen gas and lead(IV) oxide ($PbO_2(s)$) by the following reactions:

anode: $Pb(s) + 2\ H_2O(l) \longrightarrow PbO_2(s) + 4\ H^+(aq) + 4\ e^-$

cathode: $2\ H^+(aq) + 2\ e^- \longrightarrow H_2(g)$

overall: $Pb(s) + 2\ H_2O(l) \longrightarrow PbO_2(s) + 2\ H_2(g)$

A red-brown deposit forms on the anode; this is the lead(IV) oxide (PbO_2). Lead(II) sulfate ($PbSO_4$), an intermediate formed from lead metal, is oxidized to lead(IV) oxide.

$$Pb(s) + SO_4^{2-}(aq) \longrightarrow PbSO_4(s) + 2\ e^-$$

As the lead metal is oxidized, some of the lead(II) sulfate falls off the electrode. This is the white precipitate that can be found in the cell after charging.

Once charged, the cell discharges by the following electrode reactions:

$$Pb(s) + SO_4^{2-}(aq) \longrightarrow PbSO_4(s) + 2\ e^-$$

$$PbO_2(s) + 4\ H^+(aq) + SO_4^{2-}(aq) + 2\ e^- \longrightarrow PbSO_4(s) + 2\ H_2O(l)$$

The potential developed by the cell is approximately 2 volts. Commercial lead storage batteries are composed of several of these cells connected in series; 6-volt motorcycle batteries contain three such cells, and 12-volt automobile batteries contain six.

The charging reactions for the commercial cell and for the demonstration cell (after its initial charging) are:

$$PbSO_4(s) + 2\ H_2O(l) \longrightarrow PbO_2(s) + 4\ H^+(aq) + SO_4^{2-}(aq) + 2\ e^-$$

$$PbSO_4(s) + 2\ e^- \longrightarrow Pb(s) + SO_4^{2-}(aq)$$

This is merely the reverse of the discharge reaction. During recharging of the cell in the demonstration, bubbles of gas form at the electrodes. This occurs because the voltage applied to the cell is greater than that needed to charge the cell, namely, 2 volts. The overvoltage causes the electrolysis of water in the cell, and hydrogen gas is produced at the cathode and oxygen gas at the anode. This can happen when a commercial lead storage battery is charged. The generation of explosive hydrogen gas mandates that lead storage batteries be charged in well-ventilated areas away from sparks and flames.

REFERENCES

1. A. Joseph, P. Brandwein, E. Morholt, H. Pollack, and J. Castka, *A Sourcebook for the Physical Sciences,* p. 258, Harcourt, Brace and World, Inc.: New York (1961).
2. D. Linden, Ed., *Handbook of Batteries and Fuel Cells,* p. 14-2, McGraw-Hill Book Co.: New York (1984).

11.6

A Gravity Cell

A beaker contains a colorless solution floating on a blue solution. A copper disk is suspended in the blue solution, and a zinc disk is suspended in the colorless solution. When the copper and zinc disks are connected to a small electric motor, the motor runs. The motor can be kept running for over an hour [*1*].

MATERIALS

copper sheet, 10 cm × 30 cm × ca. 0.25 mm thick

zinc sheet, 10 cm × 30 cm × ca. 0.8 mm thick

ca. 500 mL 0.01M zinc sulfate, $ZnSO_4$ (To prepare 1 liter of solution, dissolve 2.9 g of $ZnSO_4 \cdot 7H_2O$ in 600 mL of distilled water and dilute the resulting solution to 1.0 liter.)

ca. 500 mL 1.0M copper sulfate, $CuSO_4$ (To prepare 1 liter of solution, dissolve 250 g of $CuSO_4 \cdot 5H_2O$ in 600 mL of distilled water and dilute the resulting solution to 1.0 liter.)

heavy-duty sheet-metal shears (tin snips)

2-liter beaker

dc motor which operates on 0.5 volt, with clip leads 30 cm long (Suitable motors are available from hobby shops and electronic parts stores.)

stand, with clamp for motor

piece of card stock, 5 cm × 30 cm (e.g., from a manila folder)

tape or glue

500-mL long-stemmed separatory funnel

or

long-stemmed filter funnel

wad of cotton or glass wool, ca. 1 cm in diameter

600-mL beaker

PROCEDURE

Preparation

Use sheet-metal shears to cut an electrode of the shape indicated in Figure 1 from the 0.25-mm thick copper sheet. Use the shears to cut an electrode of the same shape from the 0.8-mm thick zinc sheet. Bend the rectangular tab on each electrode so it is

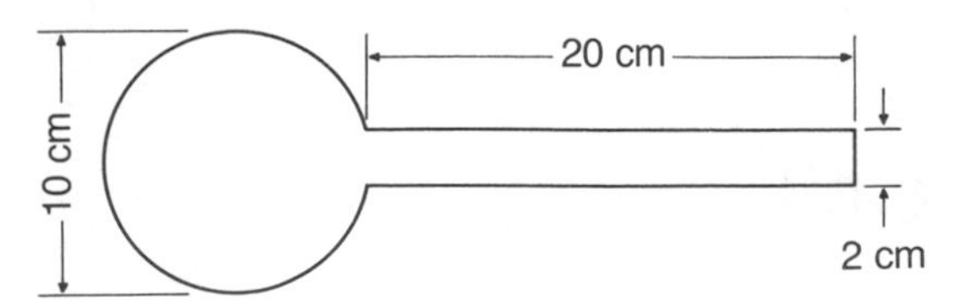

Figure 1. Shape of electrodes for gravity cell.

perpendicular to the electrode disk. Position the copper electrode in the 2-liter beaker so the disk is parallel with and 2 cm above the bottom of the beaker. Bend the tab on the copper electrode over the rim of the beaker to hold the electrode at this level (Figure 2). Position the zinc electrode in the beaker with its disk parallel with and about 4 cm above the copper disk. Situate the tab of the zinc electrode opposite the copper tab, and bend it over the rim of the beaker to hold the zinc electrode in place at this level. Loosening the tabs as little as possible, remove both electrodes from the beaker.

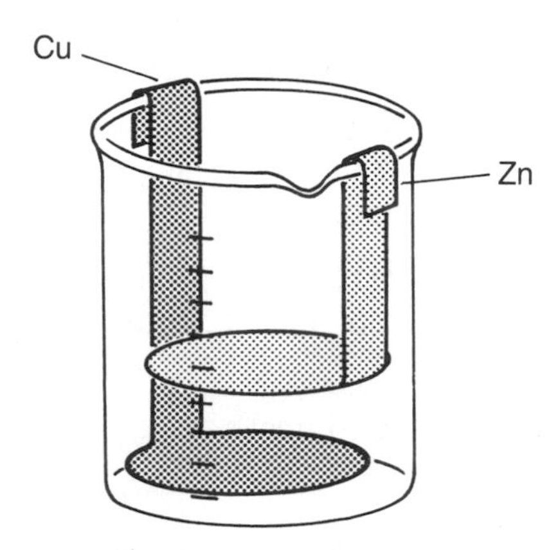

Figure 2. Electrodes in beaker.

Cut a propeller from the piece of card stock, insert the motor shaft through its center, and secure it in place with tape or glue. Clamp the motor to the stand so the propeller can turn freely when the motor's shaft rotates (Figure 3).

Fill the 500-mL long-stemmed separatory funnel with 1.0M $CuSO_4$ solution. If a long-stemmed filter funnel is to be used, stuff a small plug of cotton or glass wool in the top of the funnel's stem (this will prevent air bubbles from entering the stem when liquid is poured into the funnel), and pour 500-mL of 1.0M $CuSO_4$ into a 600-mL beaker.

Presentation

Position first the copper electrode and then the zinc electrode in the 2-liter beaker. Be sure the electrodes do not touch. Pour just enough 0.01M $ZnSO_4$ solution into the beaker to cover the copper electrode about 2 cm (about 500 mL).

Insert the tip of the long-stemmed funnel along the side of the beaker to its bottom. Open the stopcock slightly to allow the $CuSO_4$ solution to drain slowly into the beaker. (If a long-stemmed filter funnel is being used, slowly pour the $CuSO_4$ solution from the 600-mL beaker into the funnel.) The $CuSO_4$ solution must flow slowly to create as little turbulence and mixing of the solutions as possible. The blue $CuSO_4$ solution will

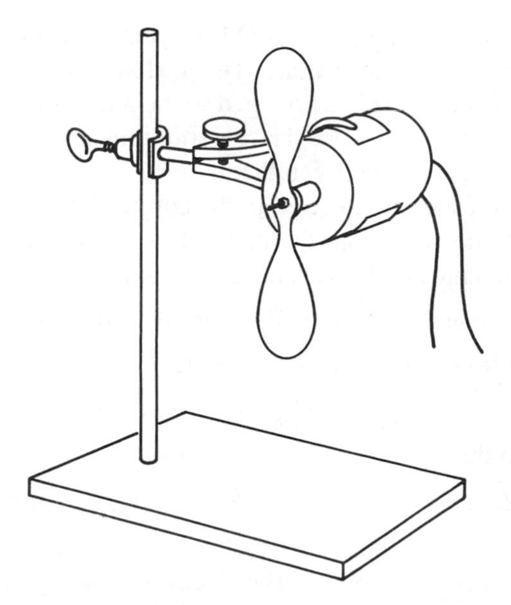

Figure 3. Direct-current motor mounted on stand.

slowly cover the bottom of the beaker and lift the colorless $ZnSO_4$ solution. Add enough blue $CuSO_4$ solution to raise the boundary between it and the colorless $ZnSO_4$ solution to halfway between the two electrode disks.

Clip one of the leads from the motor to the tab of the copper electrode, and clip the other lead to the zinc tab. If the motor does not start immediately, give the propeller a nudge to overcome its inertia. The motor will run for over an hour.

HAZARDS

Copper compounds can be toxic if taken internally, and dust from copper compounds can irritate mucous membranes.

DISPOSAL

The electrodes should be removed from the cell and rinsed with water. The rinse should be flushed down the drain. The mixture in the beaker should be stirred and flushed down the drain with water.

The electrodes may be reused in subsequent presentations of the demonstration. The zinc electrode will eventually disintegrate, at which time the fragments should be discarded in a solid-waste receptacle.

DISCUSSION

This demonstration shows the construction and functioning of a voltaic cell in which the anode and cathode cells are separated by the difference in the densities of their solutions, that is, by gravity. For this reason, the cell is called a gravity cell. It is also called a Daniell cell, after John Frederic Daniell, who introduced it in 1836 [2].

The cathode is copper metal suspended in a 1M solution of copper sulfate. The anode is zinc metal suspended in 0.01M zinc sulfate. The 0.01M zinc sulfate is much less dense than the 1M copper sulfate. The cell is arranged so the zinc sulfate solution floats on top of the copper sulfate solution. Because the two solutions are in contact, no salt bridge or other ion conductor is needed between them.

When the zinc electrode is connected to the copper electrode by means of a wire, electrons flow through the wire from the more easily oxidized zinc to the less easily oxidized copper. At the zinc anode, zinc atoms are oxidized to zinc ions.

$$Zn(s) \longrightarrow Zn^{2+}(aq) + 2\,e^-$$

At the cathode, copper ions are reduced to copper.

$$Cu^{2+}(aq) + 2\,e^- \longrightarrow Cu(s)$$

The overall reaction in the cell is

$$Zn(s) + Cu^{2+}(aq) \longrightarrow Zn^{2+}(aq) + Cu(s)$$

This is the same reaction that occurs when metallic zinc is placed in aqueous copper(II) sulfate and becomes coated with copper. The electrons flowing through the wire drive a motor in this demonstration. As current is drawn from the cell, ions drift across the boundary between the lighter zinc sulfate solution and the denser copper sulfate solution beneath.

The Daniell, or gravity, cell was one of the first batteries that produced a fairly stable current. It was widely used to supply power for telegraphy, electroplating, and research in the early years of electrochemistry.

REFERENCES

1. J. F. Skinner, *J. Chem. Educ.* 54:619 (1977).
2. A. J. Ihde, *The Development of Modern Chemistry,* p. 469, Dover Publications: New York (1984).

11.7

Electricity from a Fuel Cell

A damp sheet of paper is sandwiched between two pieces of wire gauze. This sandwich is mounted between two transparent plastic squares having two tubes connected to each. A tank of hydrogen gas is connected to a tube on one of the plastic squares and a tank of oxygen is connected to the other square. When the terminals of an electric motor are connected to the two pieces of wire gauze, the motor runs.†

MATERIALS

ca. 50 mL 3M hydrochloric acid, HCl (To prepare 1 liter of solution, slowly pour 250 mL of concentrated [12M] HCl into 600 mL of distilled water and dilute the resulting, cooled solution to 1.0 liter.)

ca. 500 mL distilled water

0.5 g hydrogen hexachloroplatinate(IV) hydrate (chloroplatinic acid), $H_2PtCl_6 \cdot xH_2O$

or

30 mL concentrated (12M) hydrochloric acid, HCl

10 mL concentrated (16M) nitric acid, HNO_3

0.2 g platinum metal (e.g., wire)

50 mL distilled water

100-mL beaker

30-mL porcelain evaporating dish

watch glass to cover evaporating dish

hot plate

cylinder of hydrogen gas, with pressure regulator

cylinder of oxygen gas, with pressure regulator

ca. 50 mL 1M sodium hydroxide, NaOH (To prepare 1 liter of stock solution, dissolve 40 g of NaOH in 600 mL of distilled water, and dilute the resulting, cooled solution to 1.0 liter.)

2 sheets of ¼-inch acrylic, 12 cm square

drill, with ⅜-inch and ¼-inch bits

4 5-cm lengths of acrylic tubing, with outside diameter of ⅜ inch

acrylic cement

† We wish to thank Ronald R. Esman of Abilene High School, Abilene, Texas, for providing us with a description of this demonstration, which he developed.

sheet of rubber, 8 cm × 16 cm × ca. 2 mm thick (e.g., piece of inner tube)

2 pieces of nickel screen, 8 cm × 12 cm, 50–100 mesh

scissors or tin snips to cut nickel screen

2 10-cm petri dishes

2 gas washing bottles

4 0.5-m lengths of plastic or rubber tubing, with outside diameter of ⅜ inch

index card, 4 inches × 6 inches

tape or glue

dc motor that operates on 1 volt, with clip leads (Suitable motors are available from hobby shops and electronic parts stores.)

stand, with clamp for motor

2 pieces of 8-cm round filter paper

4 3⁄16-inch × 1-inch bolts, with wing nuts

dropper

stand, with clamp for fuel cell

voltmeter, with clip leads

PROCEDURE

Preparation

Drill matching holes through the two 12-cm-square sheets of ¼-inch acrylic: Align the squares on top of each other and drill two ⅜-inch holes about 3 cm apart near the centers of both sheets (Figure 1). Drill four ¼-inch holes through both sheets, one hole near each corner.

Using acrylic cement, glue one end of the four 5-cm pieces of ⅜-inch acrylic tubing to each of the center holes in the squares (Figure 2). These form the entry and exit ports for the fuel gases.

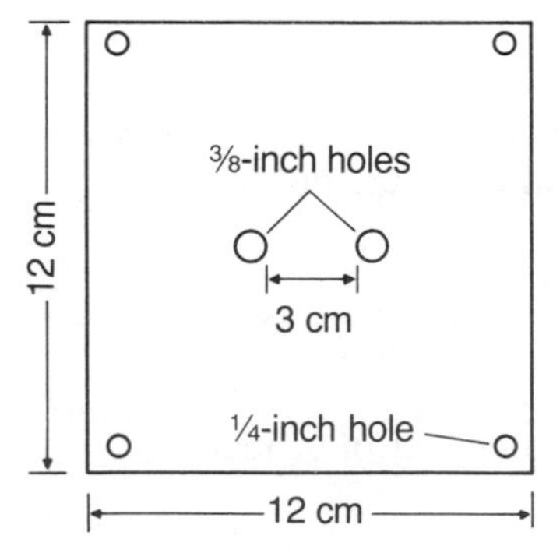

Figure 1. Location of holes in acrylic square.

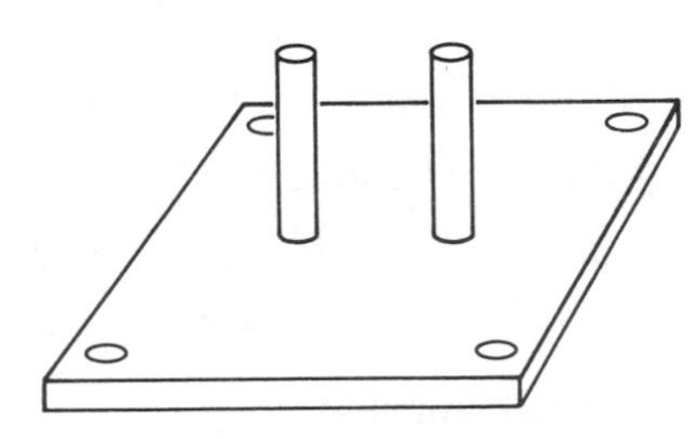

Figure 2. Gas ports attached to acrylic square.

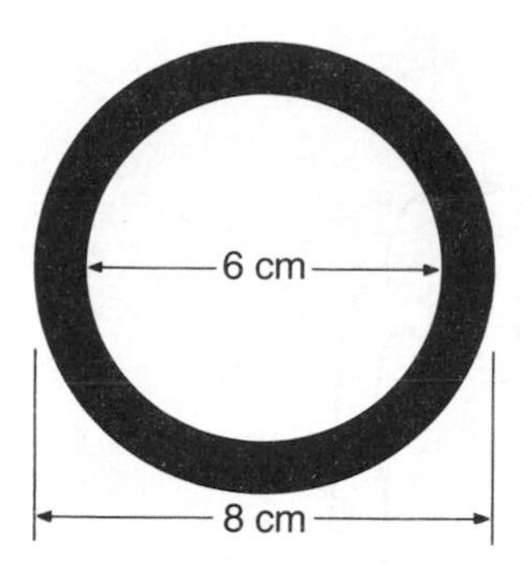

Figure 3. Gasket cut from rubber sheet.

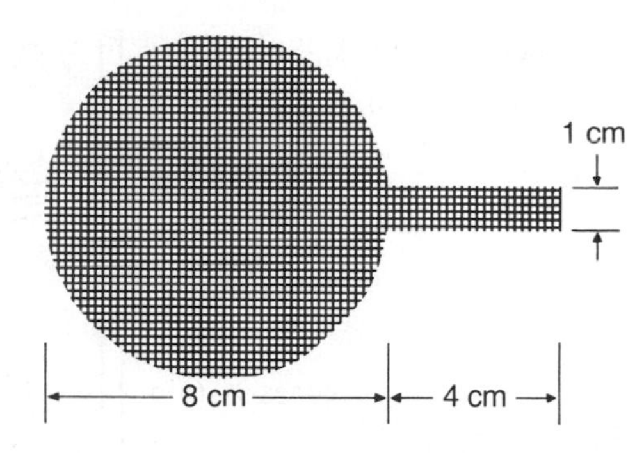

Figure 4. Electrode cut from nickel screen.

From the rubber sheet cut two disks 8 cm in diameter. Cut a 6-cm disk from the center of each of these, leaving a circular ring (Figure 3). These will be used as gaskets.

From the nickel screen, cut two circular electrodes with a diameter of 8 cm and having a 4-cm × 1-cm tab at the circumference (Figure 4). These electrodes must be cleaned and platinized (coated with a spongy black form of metallic platinum) before use. Clean the electrodes by soaking them in 3M HCl for several minutes and then rinsing them with distilled water. The electrodes are platinized by soaking them in a solution containing hexachloroplatinate(IV) ions. This solution is prepared by dissolving 0.5 g hydrogen hexachloroplatinate(IV) hydrate in 50 mL of distilled water. If $H_2PtCl_6 \cdot xH_2O$ is unavailable, the solution may be prepared from platinum metal as described in the following paragraph.

In a fume hood, prepare the hexachloroplatinate(IV) solution from platinum metal as follows. Combine 30 mL of concentrated (12M) HCl and 10 mL of concentrated (16M) HNO_3 in a 100-mL beaker. Place 0.2 g of platinum metal in the evaporating dish, and pour enough acid mixture from the beaker into the disk to cover the wire. Cover the dish with the watch glass, and set the evaporating dish on the hot plate. Set the hot plate to a low setting to warm the dish. Continue to warm the dish until all the platinum has dissolved. (This can take several hours.) It may be necessary to add more acid mixture as it evaporates. Once the platinum has dissolved, uncover the dish and continue heating it until all the liquid has evaporated. Cool the dish. Dissolve the orange residue in the cooled dish in 50 mL of distilled water.

Place each of the screen electrodes in a petri dish and pour enough of the hexachloroplatinate(IV) solution into each dish to cover the circular portion of the electrodes. Soak the electrodes for several hours. While the electrodes are soaking, nickel in the electrode will reduce platinum from the solution, forming a spongy black coating of platinum on the screen. Swirl the dishes and turn the electrodes periodically while soaking them to promote formation of a uniform coating of platinum black. Rinse the coated electrodes gently with distilled water and store them immersed in distilled water. They may be stored indefinitely in this condition. The hexachloroplatinate(IV) solution can be stored in a sealed container for future use.

Fill each of the gas washing bottles half full with water and close them. Use a piece of the 0.5-m plastic or rubber tubing to attach the outlet of the hydrogen cylinder to the inlet (the opening connected to the central tube) of one of the gas washing bottles. Attach another piece of tubing to the outlet of the bottle. In a similar fashion, attach the oxygen cylinder to the other gas washing bottle.

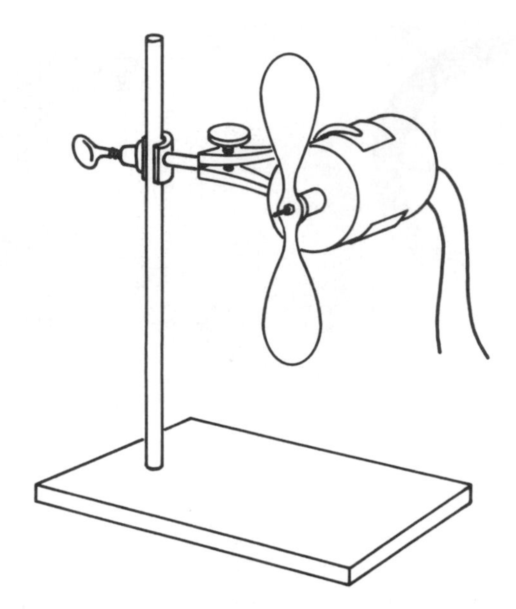

Figure 5. Direct-current motor mounted on stand.

Cut a propeller about 15 cm in diameter from the index card, insert the motor shaft through its center, and secure it in place with tape or glue. Clamp the motor to the stand so the propeller can turn freely when the motor's shaft rotates (Figure 5).

Presentation

Assemble the fuel cell. (Figure 6 shows the assembled cell, Figure 7 shows it in cross section.) Holding one of the acrylic squares with its tubes pointing down, center a rubber gasket ring on the square. Align one of the platinized electrodes atop the gasket. Be sure the tab protrudes beyond the edge of the acrylic square. Align both pieces of filter paper on the electrode. Cover the filter paper with the remaining electrode. Situate the tab on the second electrode so it, too, protrudes beyond the edge of the acrylic square, and so it does not touch the first electrode. Place the remaining gasket on top of the second electrode. Set the remaining acrylic square on top of the gasket and align the holes in the top square with those in the bottom square. Insert the four 1-inch bolts through the four corner holes in the acrylic squares. Attach the wing nuts to the bolts and tighten them to hold the assembly together. Be sure the two electrodes do not touch each other or the bolts holding the cell together.

Thoroughly moisten the filter paper with 1M NaOH solution by dropping it through the fuel ports. Clamp the cell assembly vertically to the stand. Attach the tube from the hydrogen gas washing bottle to the upper fuel port on one side of the cell. Attach the tube from the oxygen gas washing bottle to the lower gas port on the opposite side of the cell.

The cell is now ready to operate. Attach one of the wire leads from the voltmeter to each of the electrodes and note the voltage reading. Open the needle valve on the hydrogen regulator to produce a slow bubbling in the washing bottle. Note the change in the potential of the cell as detected by the voltmeter. Now, slowly open the valve on the oxygen regulator to produce a gentle bubbling in its washing bottle. Again, note the effect on the potential of the cell. A few minutes are required for the potential to become stable. After the potential has stabilized, turn off the hydrogen flow and note what

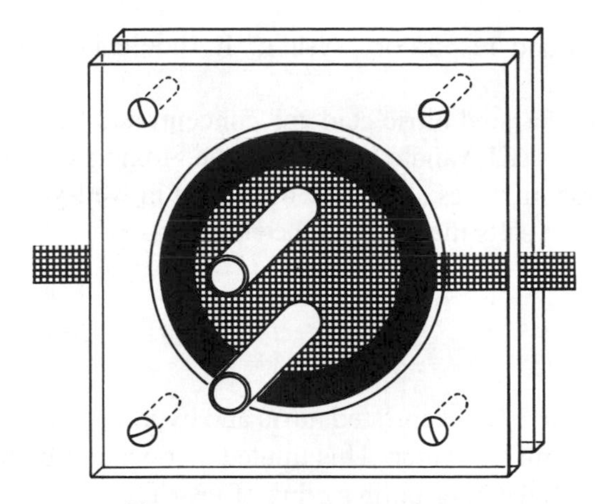

Figure 6. Assembled fuel cell.

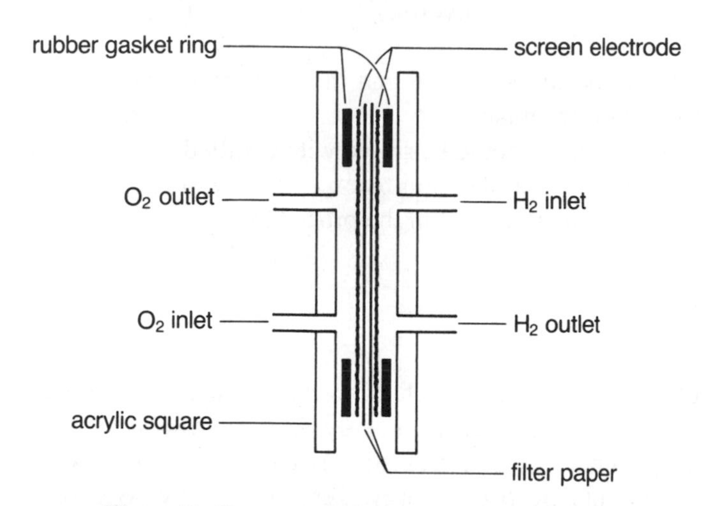

Figure 7. Cross-section of assembled fuel cell.

happens to the cell potential. Increase the hydrogen flow again to about what it was before.

Attach the wires from the electric propeller motor to the electrodes. The propeller will turn. While the motor is running, vary the flow rate of the hydrogen and note the effect on the speed of the motor. The cell can be operated continuously for several hours. However, if the filter paper should dry, the cell will cease to function.

HAZARDS

Hydrogen is extremely flammable. Therefore, this demonstration should be presented away from open flames and sparks.

Concentrated nitric acid is both a strong acid and a powerful oxidizing agent. Contact with combustible materials can cause fires. Contact with the skin can result in severe burns. The vapor irritates the respiratory system, eyes, and other mucous membranes, and therefore concentrated nitric acid should be handled only in a well-ventilated area.

Concentrated hydrochloric acid can irritate the skin. Its vapors are extremely irritating to the eyes and respiratory system. It should be handled only in a well-ventilated area.

Mixtures of concentrated nitric acid and concentrated hydrochloric acid (aqua regia) release chlorine gas (Cl_2) and other reactive and toxic gases (NO_2, NOCl, NO_2Cl, etc.). Therefore, these mixtures should be used only in well-ventilated areas.

Complex platinum salts may irritate the skin.

DISPOSAL

The excess mixture of concentrated nitric and hydrochloric acids should be diluted by pouring it into 500 mL of water. This diluted acid should be neutralized by adding sodium bicarbonate ($NaHCO_3$) until fizzing stops. The resulting mixture should be flushed down the drain with water.

The remaining 1M sodium hydroxide solution should be flushed down the drain with water.

The cell should be disassembled and the filter paper rinsed with water and discarded in a solid-waste receptacle.

The platinized screens should be rinsed with distilled water and stored in water for repeated use. Eventually, the platinum black coating will become ineffective and will need to be regenerated as described in the procedure.

DISCUSSION

A fuel cell is a voltaic cell which converts the chemical energy of a fuel and its oxidizing agent directly into electrical energy. In other words, the fuel and its oxidizer could be combined under other circumstances to generate heat, which could be converted mechanically into electrical energy. Certainly, the hydrogen used in this demonstration would react with oxygen in a highly exothermic reaction. The fuel cell produces electrical energy directly, without the production of heat and the energy loss associated with mechanical conversions. Mechanical electric generators have efficiencies between 40% and 50%, whereas the efficiency of a fuel cell can be close to 100%.

Electrochemical reactions are heterogeneous. They involve electron exchange reactions which occur at the interface between two or three phases, such as the interface between a metal electrode and a solution. In the hydrogen-oxygen fuel cell used in this demonstration, three phases are involved: the solid platinum surface, the liquid solution of sodium hydroxide, and a gas (hydrogen at one electrode and oxygen at the other).

At the hydrogen electrode, hydrogen gas flows over the platinum surface. Hydrogen gas is chemisorbed onto the platinum surface; that is, it forms chemical bonds with atoms of platinum at the electrode's surface [*1*]. When the hydrogen is chemisorbed, it dissociates into hydrogen atoms. Some hydrogen atoms lose an electron to the metal and combine with hydroxide ions from the electrolyte, forming water [*2*]. Hydrogen gas is oxidized to water; therefore this electrode is the anode. Having gained electrons in this way, the metal electrode becomes negatively charged. Thus, the reactions at the anode are

$$H_2(g) + Pt(s) \rightleftarrows 2\ H \cdot Pt(\text{chemisorbed})$$

$$2\ H \cdot Pt + 2\ OH^-(aq) \rightleftarrows 2\ H_2O(l) + Pt(s) + 2\ e^-$$

Oxygen gas is reduced at the cathode. Oxygen molecules strike the platinum surface and are weakly bound. The adsorbed oxygen molecules dissociate into oxygen atoms. These oxygen atoms capture electrons from the metal, producing chemisorbed oxide ions. These react with water molecules from the solution at the interface and produce hydroxide ions in the solution. The electrode, having lost some electronic charge to the electrolyte phase, becomes positively charged. The reactions occurring at the cathode are

$$O_2(g) + Pt(s) \rightleftarrows 2\ O{\cdot}Pt(\text{chemisorbed})$$
$$2\ O{\cdot}Pt + 4\ e^- \rightleftarrows 2\ O^{2-}{\cdot}Pt$$
$$O^{2-}{\cdot}Pt + H_2O(l) \rightleftarrows 2\ OH^-(aq) + Pt(s)$$

When the two electrodes are connected by a conductor of electrons, such as the leads from the motor, electrons flow from the negative electrode (H_2) through the circuit of the electrical device to the positive electrode (O_2). In order for electron current to flow continuously the hydroxide ions produced at the cathode must migrate toward the anode. The cell reactions are

cathode: $$O_2(g) + 2\ H_2O(l) + 4\ e^- \rightleftarrows 4\ OH^-(aq)$$
anode: $$2\ H_2(g) + 4\ OH^-(aq) \rightleftarrows 4\ H_2O(l) + 4\ e^-$$
overall: $$2\ H_2(g) + O_2(g) \rightleftarrows 2\ H_2O(l)$$

Water is the product of the reaction of hydrogen with oxygen. The intermediate hydroxide ions travel through the sodium hydroxide solution from the cathode to the anode, and the intermediate electrons travel from the anode to the cathode through the external circuit.

The standard cell potential (E°) of a hydrogen-oxygen fuel cell is 1.23 volts. This value is the sum of the standard reduction potentials in basic solution [*3*]:

$$2\ H_2O(l) + 2\ e^- \rightleftarrows 2\ OH^-(aq) + H_2(g) \qquad E^\circ = -0.83 \text{ volt}$$
$$O_2(g) + 2\ H_2O(l) + 4\ e \rightleftarrows 4\ OH^-(aq) \qquad E^\circ = 0.40 \text{ volt}$$

The fuel cell in this demonstration develops a maximum potential of about 1 volt.

REFERENCES

1. J. O'M. Bockris and S. Srinivasar, *Fuel Cells: Their Electrochemistry,* McGraw-Hill Book Co.: New York (1969).
2. J. Weissbart, *J. Chem. Ed.* 38:267 (1961).
3. R. C. Weast, Ed., *CRC Handbook of Chemistry and Physics,* 66th ed., pp. D-152, D-153, CRC Press: Boca Raton, Florida (1985).

11.8

A Zinc-Acid Cell

A miniature lamp is connected to a carbon rod and a zinc strip. When the rod and strip are immersed in hydrochloric acid, the lamp lights and bubbles form at both electrodes. The lamp is replaced by an ammeter and the current is measured. A voltmeter is substituted for the ammeter to measure the potential of the cell [*1, 2*].

MATERIALS

300 mL 3M hydrochloric acid, HCl (To prepare 1 liter of solution, slowly pour 250 mL of concentrated [12M] HCl into 600 mL of distilled water, wait for the mixture to cool to room temperature, and dilute it to 1.0 liter.)

2 40-cm insulated wires, 18–24 gauge

wire strippers

2 alligator clips

soldering iron and solder

3-volt flashlight lamp, with socket

400-mL beaker

12-cm carbon rod, ca. 7 mm in diameter

zinc strip, 12 cm × 5 cm × 0.5 mm thick

ammeter, with clip leads

voltmeter, with clip leads

clock or watch

PROCEDURE

Preparation

Assemble the apparatus shown in the figure. Strip 0.5 cm of the insulation from each end of both pieces of wire. Attach an alligator clip to one end of each of the insulated wires. Solder the free end of one of these wires to one of the terminals of the lamp socket, and solder the other wire to the remaining terminal. Insert the bulb in the socket.

Pour 300 mL of 3M HCl into the 400-mL beaker.

Presentation

Place the carbon rod in the 400-mL beaker and attach one of the alligator clips from the lamp socket to the carbon rod. Attach the other clip to one end of the zinc strip. Dip the zinc strip into the acid solution. The lamp glows and bubbles of gas form at both the zinc strip and the carbon rod. Remove the zinc strip from the acid. Disconnect the lamp from the carbon and zinc.

Clip one of the leads from the ammeter to the carbon rod and the other to the zinc strip. Insert the zinc strip in the acid. Raise and lower the zinc strip in the acid, thereby changing the surface area of zinc immersed. The ammeter will register a current, whose value depends on how much of the zinc surface is immersed in the acid. (If the ammeter has a needle and the needle deflects to below 0, reverse its connections to the carbon rod and zinc strip.) Remove the zinc strip from the acid and detach the ammeter.

Clip the leads from the voltmeter to the carbon rod and zinc strip. Immerse the zinc strip in the acid, and raise and lower it to vary the surface area of zinc immersed. The voltmeter will register a potential, whose value is virtually independent of the immersed area. Remove the zinc from the acid and disconnect the voltmeter.

Reconnect the lamp. Place the zinc strip back in the acid and record the time. Allow the lamp to burn continuously and record the time at which its glow is no longer visible (about 20 minutes). Use the ammeter and voltmeter to measure the current and potential after the lamp has ceased to glow. The current and the voltage will be lower.

HAZARDS

Concentrated hydrochloric acid can irritate the skin. Its vapors are extremely irritating to the eyes and respiratory system. It should be handled only in a well-ventilated area.

DISPOSAL

The hydrochloric acid remaining after this experiment should be neutralized by the addition of sodium bicarbonate ($NaHCO_3$) until fizzing stops, and the neutralized mixture should be flushed down the drain with water.

DISCUSSION

When the zinc strip is dipped into hydrochloric acid, the filament of the miniature bulb begins to glow, and continues to glow as long as zinc metal and hydrochloric acid remain. Bubbles appear at both the zinc strip and carbon electrode. These bubbles contain hydrogen gas.

The lamp glows whenever sufficient electron current flows through the filament to heat it to incandescence. Electrons are supplied by zinc atoms, which readily undergo oxidation.

$$Zn(s) \longrightarrow Zn^{2+}(aq) + 2\,e^-$$

These electrons travel through the external circuit (wires and bulb filament) to the carbon electrode, where hydrogen ions are reduced.

$$2\,e^- + 2\,H^+(aq) \longrightarrow H_2(g)$$

The net reaction is

$$Zn(s) + 2\,H^+(aq) \longrightarrow Zn^{2+}(aq) + H_2(g)$$

The bubbles produced at the carbon electrode (cathode) result from the production of hydrogen gas. The Zn^{2+} ions formed by the oxidation of zinc in the zinc anode disperse throughout the hydrochloric acid solution.

The bubbles that appear at the zinc strip are produced by the direct reaction of hydrochloric acid with zinc. These bubbles appear whenever an active metal such as zinc is placed in acid. This reaction would occur without the external circuit which connects the zinc strip to the carbon electrode. The reaction occurs at the surface of the zinc strip and is the same reaction as that which produces the electric current.

$$Zn(s) + 2\,H^+(aq) \longrightarrow Zn^{2+}(aq) + H_2(g)$$

This side reaction could be eliminated by separating the oxidation and reduction processes. The cell could be arranged with the carbon electrode immersed in hydrochloric acid solution, the zinc electrode immersed in a salt solution, and the two solutions connected by a salt bridge. However, the salt bridge would increase the internal resistance of the cell, and much less current would be produced. The salt bridge would increase the cell resistance because the ions it contains can move only relatively slowly through its gelatinous medium. The current produced by such an arrangement would be insufficient to light the lamp. The current is measured with an ammeter. In addition to the internal cell resistance, the resistance of the bulb filament and the wires also affect the magnitude of the current.

The voltage of the cell steadily decreases as it operates. This happens because, as it operates, the concentrations of the products of its overall reaction increase and its condition approaches equilibrium. The voltage of an electrochemical cell as a function of the concentration of reactants and products is given by the Nernst equation. For the oxidation-reduction reaction represented by

$$aA + bB \longrightarrow cC + dD$$

the Nernst equation is

$$E = E^\circ - \frac{2.303\,RT}{nF} \log \frac{a_C^c\, a_D^d}{a_A^a\, a_B^b}$$

In this equation E is the cell voltage, E° is the standard cell potential, n equals the number of moles of electrons transferred in the balanced equation, R is the gas constant

(8.314 joule/mole·Kelvin), T is the absolute temperature, and F is the Faraday constant (9.648×10^4 coulomb/mole). At 25°C (298 K) the value of 2.303RT/F is 0.0591 volt·mole. The ratio in the log term is the ratio of the activities of the products to reactants, and each activity is raised to the power of the corresponding coefficient in the balanced chemical equation. Activities are approximately proportional to the concentration of the substance.

For the reaction in this demonstration, the Nernst equation has the form

$$E = 0.76 - \frac{0.059}{2} \log \frac{[Zn^{2+}]}{[H^+]^2}$$

The activities of Zn^{2+} and H^+ have been replaced by their molar concentrations, to which they are proportional. The activities of pure substances, such as solid zinc and hydrogen gas, are equal to 1.0, and are not shown explicitly in the equation.

The initial cell potential of about 1.8 volts is a result of the low initial concentration of Zn^{2+} and, less significantly, of the high initial hydrogen ion concentration. As the cell produces current, the ratio of $[Zn^{2+}]$ to $[H^+]^2$ increases, and the cell voltage (E) decreases. If the cell operates long enough, the reaction comes to equilibrium, and E falls to 0.

The Copper-Magnesium Cell, Demonstration 11.10, is similar to this one, replacing the zinc with magnesium and the carbon rod with copper.

REFERENCES

1. F. G. Villarreal, *J. Chem. Ed.* 34:A481 (1957).
2. D. R. Martin, *J. Chem. Ed.* 25:495 (1948).

11.9

A Zinc-Iodine Cell

A colorless liquid is placed in a beaker. A porous ceramic cup is placed in the beaker and filled with the same liquid. A metal strip is inserted in the beaker, and a carbon rod is inserted in the cup. When a black solid is placed inside the cup, a voltmeter attached to the carbon rod and metal strip registers a change in potential. When a motor is connected to the metal strip and the carbon rod, the motor runs.

MATERIALS

ca. 225 mL 0.9M potassium iodide, KI (To prepare 1 liter of solution, dissolve 150 g of KI in 800 mL of distilled water and dilute the resulting solution to 1.0 liter.)

10 g iodine, I_2

250-mL beaker

2 stands, each with clamp

100-mL porous ceramic cup (e.g., Central Scientific catalog no. 79284)

index card, 4 inches × 6 inches

tape or glue

dc motor that operates on 1.25 volts, with clip leads (Suitable motors are available from hobby shops and electronic parts stores.)

zinc strip, 3 cm × 15 cm × ca. 0.5 mm thick

15-cm carbon rod, ca. 5 mm in diameter (Suitable carbon rods may be obtained by disassembling a hobby battery or from a welding supply shop.)

voltmeter capable of reading 1.5 volts, with clip leads

gloves, plastic or rubber

PROCEDURE [*1*]

Preparation

Place the beaker on the base of one of the stands. Clamp the porous ceramic cup to a stand so it can be lowered at least halfway into the beaker by adjusting the position of the clamp.

Cut a propeller about 15 cm in diameter from the index card, insert the motor shaft through its center, and secure it in place with tape or glue. Clamp the motor to the other stand so the propeller can turn freely when the motor's shaft rotates (Figure 1).

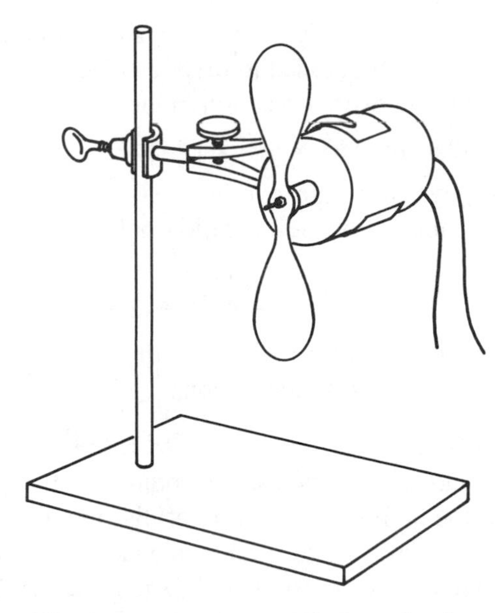

Direct-current motor mounted on stand.

Presentation

Pour 150 mL of 0.9M KI solution into the 250-mL beaker. Lower the clamp holding the porous cup so the cup dips as deeply as possible into the solution in the beaker. Pour 0.9M KI solution into the cup until the level of solution inside the cup is the same as it is outside (about 75 mL). Insert the zinc strip in the solution in the beaker. Insert the carbon rod in the solution in the porous cup.

Connect one lead of the voltmeter to the zinc strip and the other to the carbon rod. Note the voltmeter reading. Wearing gloves, add 10 g of iodine to the porous cup. Use the carbon rod to stir the mixture. Note the change in the voltmeter reading.

Disconnect the voltmeter from the zinc strip and carbon rod. Attach one lead from the motor to the zinc strip and the other to the carbon rod. The motor will run.

HAZARDS

Iodine is corrosive and can cause burns to the skin and mucous membranes.

DISPOSAL

The solutions should be flushed down the drain with water.

DISCUSSION

The oxidation of zinc by elemental iodine is a highly exothermic reaction. The direct reaction between these elements is quite vigorous and makes an impressive chemical demonstration in its own right (see Demonstration 1.19 in Volume 1 of this series). When the reactants are suitably arranged in an electrochemical cell, the energy released by the reaction can be harnessed to drive an electric motor.

In this demonstration, the overall reaction in the electrochemical cell is

$$Zn(s) + I_3^-(aq) \longrightarrow Zn^{2+}(aq) + 3\ I^-(aq)$$

The reaction is divided physically into two half cells, in which the oxidation and reduction processes are separated. In the oxidation half cell, zinc metal is oxidized to zinc ions.

$$Zn(s) \longrightarrow Zn^{2+} + 2\ e^-$$

The elemental zinc in this reaction is provided by the electrode itself. In the reduction half cell, triiodide ions are reduced to iodide ions.

$$I_3^-(aq) + 2\ e^- \longrightarrow 3\ I^-(aq)$$

The electrolyte in both half cells is potassium iodide. In the reduction half cell, the potassium iodide serves the additional function of dissolving iodine. Elemental iodine is not very soluble in water, but is quite soluble in aqueous iodide solutions. Elemental iodine combines with iodide ions to form triiodide ions, which increases the solubility of iodine in aqueous media.

$$I_2(s) + I^-(aq) \longrightarrow I_3^-(aq)$$

The two half cells are separated by a porous ceramic cup. Because the cup is porous, ions from the electrolyte in the two half cells can diffuse through the cup. The movement of ions balances the movement of charge through the external circuit (i.e., through the motor). The entire surface of the cup is porous. If only a portion were porous, the resistance to the flow of ions between the half cells would be greater and the amount of current that could be drawn from the cell would be reduced.

The standard potential for the reduction of zinc ions to zinc is −0.762 volt, and that for the reduction of triiodide ions is 0.536 volt [2]. Therefore, the standard potential for this cell is 1.298 volts. However, the actual potential of the cell used in this demonstration will be different from this value, because the concentrations of ions are not standard, namely, 1M. The initial concentration of zinc ions in the zinc half cell is 0, because no zinc salt is added to the solution. The initial concentration of triiodide ions in the iodine half cell is about 0.5M.

REFERENCES

1. *J. Chem. Educ.* 25:495 (1948).
2. R. C. Weast, Ed., *CRC Handbook of Chemistry and Physics,* 66th ed., p. D-155, CRC Press: Boca Raton, Florida (1985).

11.10

A Copper-Magnesium Cell

A flashlight bulb is connected with wires to a silvery metal strip and a copper metal strip. When the metal strips are immersed in 12M sulfuric acid, the bulb glows. When the strips are immersed in 1M hydrochloric acid, the bulb glows more brightly. The potentials between the electrodes and the currents flowing through the bulb are measured [*1*].

MATERIALS

150 mL 1M hydrochloric acid, HCl (To prepare 1 liter of solution, pour 80 mL of concentrated [12M] HCl into 700 mL of distilled water and dilute the resulting solution to 1.0 liter.)

150 mL 12M sulfuric acid, H_2SO_4 (To prepare 1 liter of solution, set a 2-liter beaker containing 300 mL of distilled water in a pan of crushed ice. Slowly pour 640 mL of concentrated [18M] H_2SO_4 into the beaker. The mixture will become very hot. If all the ice melts, add more. Once the mixture has cooled to room temperature, dilute it to 1.0 liter with distilled water.)

2 50-cm insulated wires, 18 gauge

wire strippers

soldering iron and solder

2.4-volt, 0.3-ampere flashlight lamp, with socket

2 alligator clips

2 250-mL beakers, with labels

copper strip, 5 cm × 15 cm × ca. 1 mm thick

magnesium strip, 5 cm × 15 cm × ca. 1 mm thick

voltmeter capable of reading 2 volts with 0.05-volt accuracy, with probes

ammeter capable of reading 1 ampere with 0.05-ampere accuracy, with probes

PROCEDURE

Preparation

Strip 0.5 cm of the insulation from each end of both pieces of wire. Solder one of the wires to one terminal of the lamp socket and the other wire to the other terminal (see figure). Attach an alligator clip to the free end of each wire. Seat the bulb in the socket.

Label one of the beakers "1M HCl" and the other "12M H_2SO_4." Pour 150 mL of 1M HCl and 150 mL of 12M H_2SO_4 into the appropriate beakers.

Presentation

Attach one of the alligator clips to one end of the copper strip and the other clip to the magnesium strip. Immerse the two metal strips in the beaker of 1M HCl. The lamp will glow and bubbles will appear at both metal strips. With the lamp still connected to the metal strips, attach the probes of the voltmeter to the two metal strips and record the voltage. Connect the ammeter in series with the lamp by removing the lamp's clip from the copper strip and attaching it to one probe of the ammeter, and attaching the other probe of the ammeter to the copper electrode. Record the value of the current through the circuit. Remove all the clips from the metal strips and rinse the strips.

Repeat the procedure of the previous paragraph using the beaker of 12M H_2SO_4 in place of the 1M HCl. The light will not glow as brightly in the 12M H_2SO_4, and the voltage and current readings will be lower.

HAZARDS

Because sulfuric acid is both a strong acid and a powerful dehydrating agent it must be handled with great care. The dilution of concentrated sulfuric acid is a highly exothermic process and releases sufficient heat to cause burns.

Hydrochloric acid can irritate the skin. Hydrochloric acid vapors are extremely irritating to the eyes and respiratory system. Therefore, it should be handled only in a well-ventilated area.

DISPOSAL

The waste solutions should be combined with 500 mL of water in a 2-liter beaker and neutralized by adding sodium bicarbonate ($NaHCO_3$) to them until fizzing stops. The mixture should be flushed down the drain with water.

DISCUSSION

This demonstration shows that electrical energy can be extracted from a magnesium strip dipped into acid. To extract this energy, one terminal of an electric motor is connected to the magnesium strip and the other terminal is connected to a copper strip dipped into the same solution. This extraction of electric energy may be surprising, because the oxidation and reduction processes do not appear to be separated in this cell. In most voltaic cells (batteries), the oxidation and reduction processes must be separated in order to force electric current through an external circuit. When magnesium is immersed in acid, it reacts directly with the acid; that is, both oxidation and reduction occur at the surface of the magnesium.

$$\text{oxidation:} \quad Mg(s) \longrightarrow Mg^{2+}(aq) + 2\,e^-$$

$$\text{reduction:} \quad 2\,H^+(aq) + 2\,e^- \longrightarrow H_2(g)$$

The piece of magnesium is both anode and cathode.

In this demonstration, a copper electrode is connected by a wire to the magnesium. This copper electrode serves as an auxiliary cathode. The reduction of hydrogen ions to hydrogen gas occurs there, as well as at the surface of the magnesium. Electrons flow through the wire from the magnesium to the copper. The electron flow allows the oxidation and reduction to occur more rapidly. The rate at which magnesium dissolves in acid is limited by the rate at which the magnesium ions can leave the magnesium metal and the rate at which the hydrogen ions can approach the magnesium. While the magnesium is dissolving, magnesium ions and hydrogen ions are moving in opposite directions at the surface of the magnesium metal. This creates a sort of bottleneck, which limits the rate of the reaction. When the magnesium is connected by a wire to the piece of copper, then the reduction of hydrogen ions can occur at the copper surface. This alleviates the bottleneck of ions moving in opposite directions at the magnesium surface.

The two electrodes are dipped in turn into 1M hydrochloric acid and in 12M sulfuric acid. The lamp glows more brightly with hydrochloric acid than with sulfuric acid, indicating that a greater current flows between electrodes with hydrochloric acid than with sulfuric acid. This is confirmed by the ammeter readings. Furthermore, the potential between the electrodes is greater with 1M hydrochloric acid than it is with 12M sulfuric acid. Because the external circuit and electrodes are the same in both cells, the differences can be ascribed to differences between 1M hydrochloric acid and 12M sulfuric acid. One difference that can affect the current is the conductivity of the solutions themselves. The conductivity of 12M sulfuric acid may be less than that of 1M hydrochloric acid. In dilute acids, transfer of hydrogen ions between water molecules produces highly conductive solutions. However, the 12M sulfuric acid is 72% sulfuric acid by weight and 32% by mole. Most of the water molecules will be strongly associated to ions from the acid, and not available to exchange hydrogen ions. Another factor affecting the current is the rate of the reactions at the electrodes. Magnesium may react faster with 1M hydrochloric acid than with 12M sulfuric acid, releasing electrons to the circuit at a higher rate, and creating a greater current.

REFERENCE

1. H. N. Alyea and F. B. Dutton, Ed., *Tested Demonstrations in Chemistry,* 6th ed., p. 17, Journal of Chemical Education: Easton, Pennsylvania (1965).

11.11

A Concentration Cell

Two copper strips are immersed in identical copper solutions in two beakers. A salt bridge connects the beakers, and a voltmeter is connected to the copper strips. When a sodium sulfide solution is added to one of the beakers, a precipitate forms in the beaker and the voltmeter reading increases.

MATERIALS

ca. 100 mL 0.20M potassium nitrate, KNO_3 (To prepare 1 liter of solution, dissolve 20 g of KNO_3 in 600 mL of distilled water and dilute the resulting solution to 1.0 liter.)

800 mL 0.10M copper sulfate, $CuSO_4$ (To prepare 1 liter of solution, dissolve 25 g of $CuSO_4 \cdot 5H_2O$ in 600 mL of distilled water and dilute the resulting solution to 1.0 liter.)

50 mL 1.0M sodium sulfide, Na_2S (To prepare 100 mL of solution, dissolve 24 g of $Na_2S \cdot 9H_2O$ in 60 mL of distilled water and dilute the resulting solution to 100 mL.)

U-shaped drying tube, ca. 10 cm tall

2 strips of filter paper, 1 cm × 4 cm

2 corks to fit arms of drying tube

2 600-mL beakers

50-mL buret, with stand

2 copper strips, 3 cm × 15 cm × ca. 1 mm thick

voltmeter capable of reading 0–1 volt, with clip leads

PROCEDURE

Preparation

Assemble the salt bridge as illustrated in the figure. Fill the U-tube completely with 0.20M KNO_3. Dampen the two strips of filter paper with the KNO_3 solution and drape one strip over the rim of each arm of the U-tube. Cork each arm of the U-tube so that the strips of filter paper are immersed in the solution inside the tube and protrude about 1 cm outside the tube.

Pour 400 mL of 0.10M $CuSO_4$ into each of the two 600-mL beakers. Fill the 50-mL buret with 1.0M Na_2S and mount it on the stand.

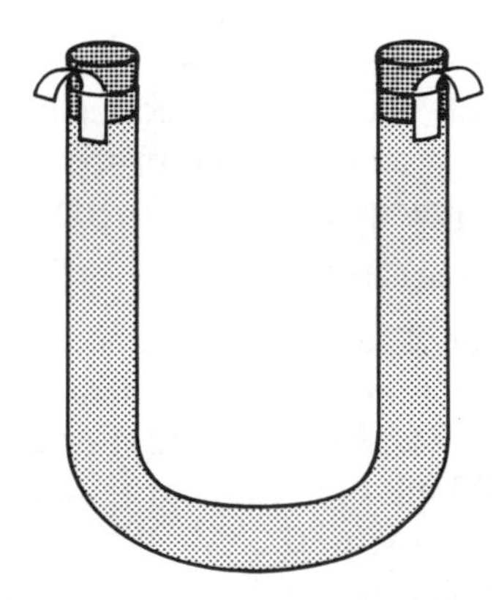

Presentation

Place the beakers of $CuSO_4$ solution side by side. Invert the U-tube filled with KNO_3 solution and insert one arm in each beaker. Place one of the copper strips in each beaker. Connect the leads of the voltmeter to the copper strips. Note the voltmeter reading (close to 0).

Position the buret over the beaker containing the copper strip to which the common (black) voltmeter lead is attached. Add a few drops of the Na_2S solution to this $CuSO_4$ solution. A precipitate will form. The voltmeter reading will increase from 0. Add a total of 20 mL of Na_2S solution to the beaker. Record the voltmeter reading. Continue to add Na_2S solution until 50 mL has been added. The maximum voltmeter reading will be about 0.6 volt.

HAZARDS

Copper compounds can be toxic if taken internally, and dust from copper compounds can irritate mucous membranes.

Sodium sulfide is a skin irritant. It liberates toxic hydrogen sulfide gas when it comes in contact with acids. Sodium sulfide is a flammable solid and should be handled away from flames.

DISPOSAL

The copper solution which is unchanged in the demonstration may be retained in a sealed container for repeated presentations of the demonstration. Alternatively, it may be flushed down the drain with water.

The copper solution to which sodium sulfide was added should be filtered. The solid (CuS) should be discarded in a receptacle for solid wastes. The filtrate should be flushed down the drain with water.

The buret should be rinsed with water and the rinse flushed down the drain.

DISCUSSION

This demonstration shows the effect of the concentration of the ions in an electrochemical cell on the cell potential. Initially, the two beakers contain copper sulfate solutions of identical concentrations. When the beakers are made the two compartments of an electrochemical cell with copper electrodes, the potential of the cell is 0. When sodium sulfide solution is added to one of the beakers, the voltmeter reading increases. At the same time, a precipitate of copper(II) sulfide forms.

$$Cu^{2+}(aq) + S^{2-}(aq) \longrightarrow CuS(s)$$

This reaction has the effect of reducing the concentration of copper(II) ions in the solution to which the Na_2S solution has been added. For copper(II) sulfide, the solubility product is 1.3×10^{-36} [*1*]. For example, after 20 mL of 1.0M sodium sulfide solution have been added to 400 mL of 0.1M copper sulfate solution, the concentration of copper(II) ions drops from 0.1M to 0.05M. When 40 mL of sodium sulfide have been added, the amount of copper ions and sulfide ions in the mixture is the same. At that point, the concentration of copper ions in solution is 1×10^{-18}M.

After sodium sulfide solution has been added to one of the beakers, the copper strips are no longer in contact with solutions having the same copper ion concentration. This causes the cell potential to increase from 0. This is an example of a *concentration cell,* one in which the two compartments of the cell are identical except for a difference in the concentration of a component. Because the copper ion concentration in one compartment is lower than in the other, and the oxidation reaction is favored where the copper ion concentration is low, the copper electrode is more likely to dissolve in that compartment.

$$Cu(s) \longrightarrow Cu^{2+}(aq) + 2\,e^-$$

This makes the electrode in the low-concentration compartment the anode. In the other compartment, where the copper ion concentration is higher, the ions are reduced at the electrode.

$$Cu^{2+}(aq) + 2\,e^- \longrightarrow Cu(s)$$

This makes this electrode the cathode.

The potential of a concentration cell is given by the Nernst equation.

$$E = E° - \frac{0.059}{n} \log \frac{[Cu^{2+}]_{anode}}{[Cu^{2+}]_{cathode}}$$

For a concentration cell, E° is 0, because the cell potential is zero when both compartments are identical. For the copper concentration cell, the number of electrons transferred per copper atom is 2, so $n = 2$. After 10 mL of sodium sulfide solution have been added to one of the beakers, $[Cu^{2+}]_{anode} = 0.05M$ and $[Cu^{2+}]_{cathode} = 0.10M$. Therefore, the cell potential is

$$E = -\frac{0.059}{2} \log \frac{0.05M}{0.1M}$$

The cell potential is 0.009 volt. When 40 mL of 1.0M sodium sulfide have been added, the concentration of copper ions in the anode compartment is 1×10^{-18}M and the cell potential is 0.5 volt.

Virtually any half-cell reaction can be used in a concentration cell. For example, this type of cell can be constructed using inert platinum electrodes and different concentrations of oxidizing agent in the half cells [*2*].

REFERENCES

1. R. C. Weast, Ed., *CRC Handbook of Chemistry and Physics,* 66th ed., p. B-222, CRC Press: Boca Raton, Florida (1985).
2. J. T. Stock and W. C. Purdy, *J. Chem. Educ.* 34:A169 (1957).

11.12

A Potentiometric Silver Series

Two beakers containing colorless liquids are connected with an inverted U-tube. A silver wire electrode dips into each solution, and each electrode is connected to a voltmeter. When silver nitrate solution is added to one of the beakers, the voltmeter reading decreases. This beaker is replaced with a different beaker. When a series of solutions is added to the new beaker, precipitates form and redissolve, and the voltmeter reading increases with each addition.†

MATERIALS

ca. 900 mL 0.20M potassium nitrate, KNO_3 (To prepare 1 liter of solution, dissolve 20 g of KNO_3 in 600 mL of distilled water and dilute the resulting solution to 1.0 liter.)

200 mL distilled water

ca. 610 mL 0.020M silver nitrate, $AgNO_3$ (To prepare 1 liter of solution, dissolve 3.4 g of $AgNO_3$ in 600 mL of distilled water and dilute the resulting solution to 1.0 liter.)

12 mL 1.0M potassium chromate, K_2CrO_4 (To prepare 1 liter of solution, dissolve 194 g of K_2CrO_4 in 600 mL of distilled water and dilute the resulting solution to 1.0 liter.)

24 mL 1.0M ammonia, NH_3 (To prepare 1 liter of solution, pour 67 mL of concentrated [15M] NH_3 into 600 mL of distilled water and dilute the resulting solution to 1.0 liter.)

16 mL 1.0M potassium chloride, KCl (To prepare 1 liter of solution, dissolve 75 g of KCl in 600 mL of distilled water and dilute the resulting solution to 1.0 liter.)

12 mL concentrated (15M) ammonia, NH_3

16 mL 1.0M potassium bromide, KBr (To prepare 1 liter of solution, dissolve 119 g of KBr in 600 mL of distilled water and dilute the resulting solution to 1.0 liter.)

24 mL 1.0M sodium thiosulfate, $Na_2S_2O_3$ (To prepare 1 liter of solution, dissolve 250 g of $Na_2S_2O_3 \cdot 5H_2O$ in 600 mL of distilled water and dilute the resulting solution to 1.0 liter.)

16 mL 1.0M potassium iodide, KI (To prepare 1 liter of solution, dissolve 166 g of KI in 600 mL of distilled water and dilute the resulting solution to 1.0 liter.)

† This demonstration was developed by Professor Jerry Bell of Simmons College, Boston, Massachusetts.

100 mL 3.0M sodium thiosulfate, $Na_2S_2O_3$ (To prepare 1 liter of solution, dissolve 750 g of $Na_2S_2O_3 \cdot 5H_2O$ in 600 mL of distilled water and dilute the resulting solution to 1.0 liter.)

25 mL concentrated (18M) sulfuric acid, H_2SO_4 (See Disposal section for use.)

100 mL concentrated (15M) aqueous ammonia, NH_3 (See Disposal section for use.)

20 g sodium sulfide, $Na_2S \cdot 9H_2O$ (See Disposal section for use.)

U-shaped drying tube, ca. 10 cm tall

2 strips of filter paper, 1 cm × 4 cm

2 corks to fit arms of drying tube

4 600-mL beakers

3 magnetic stirrers, with stir bars

single-edged razor blade

2 #5 solid rubber stoppers

2 15-cm silver wires, ca. 1 mm in diameter

stand, with 2 clamps and 2 clamp holders

500-mL graduated cylinder

voltmeter capable of reading 0–1000 millivolts, with clip leads

5 10-mL graduated cylinders

2 25-mL graduated cylinders

50-mL graduated cylinder

PROCEDURE

Preparation

Assemble the salt bridge as illustrated in Figure 1. Fill the U-tube completely with 0.20M KNO_3. Dampen the two strips of filter paper with the KNO_3 solution and drape

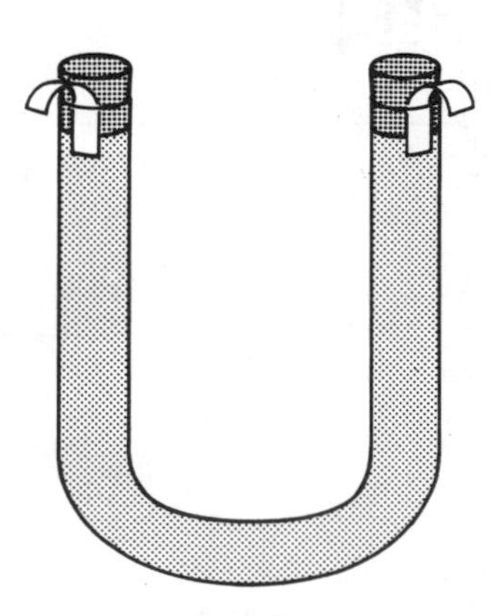

Figure 1.

one strip over the rim of each arm of the U-tube. Cork each arm of the U-tube so that the strips of filter paper are immersed in the solution inside the tube and protrude about 1 cm outside the tube.

Number the four 600-mL beakers 1 through 4. Place a magnetic stir bar in each beaker.

Assemble the electrode as shown in Figure 2. With the single-edged razor blade, cut a slit about 1 cm deep in the side of each of the rubber stoppers. Open each slit and lay one of the silver wires in each stopper. Leave about 2 cm of the wire extending from the top of the stopper and 10 cm from the bottom.

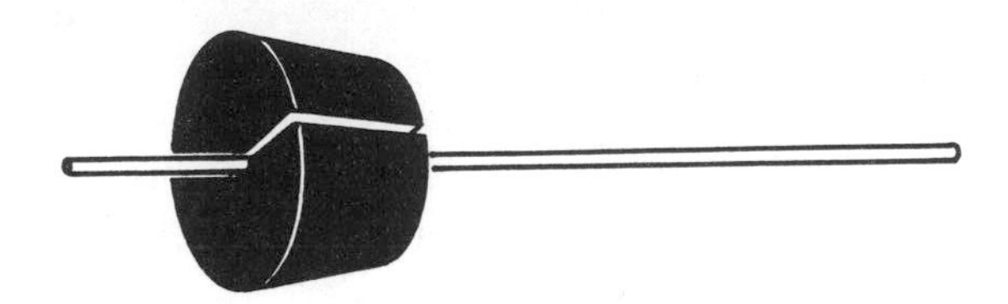

Figure 2. Stopper and silver wire electrode assembly.

Arrange the apparatus as illustrated in Figure 3. Place the three magnetic stirrers in a row. Set beakers 1 through 3 on each stirrer. Be sure beakers 1 and 2 are sufficiently close to each other that the inverted U-tube can be positioned with one arm in each. The arms of the inverted U-tube should reach at least halfway to the bottom of the beakers. Clamp the stoppers holding the silver wires to the stand so the wires extend to near the bottom of beakers 1 and 2. Be sure the wires themselves do not touch any metal parts of the clamps. Remove the U-tube and position it upright in beaker 4.

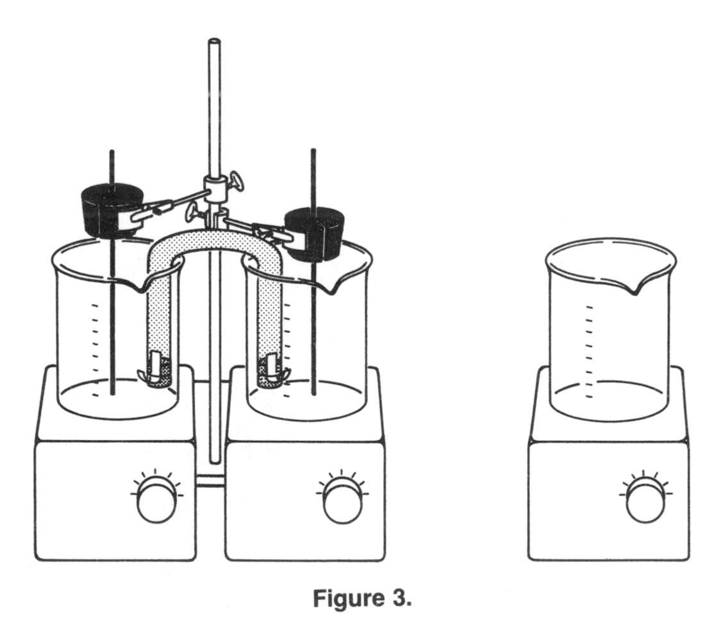

Figure 3.

Presentation

Pour 200 mL of 0.20M KNO_3 and 200 mL of distilled water into beaker 1. (This beaker then contains 400 mL of 0.10M KNO_3.) Pour 200 mL of 0.20M KNO_3 and

200 mL of 0.020M $AgNO_3$ into beaker 2. (This beaker then contains 400 mL of a solution which is both 0.10M in KNO_3 and 0.010M in $AgNO_3$.) Invert the KNO_3-filled U-tube and position it with one arm in beaker 1 and the other in beaker 2. Attach the negative or common voltmeter lead (usually black) to the end of the silver wire electrode of beaker 1. Connect the positive lead (usually red) to the end of the silver wire electrode of beaker 2. Start the stirrers.

Add several milliliters of $AgNO_3$ solution to beaker 1. The voltmeter reading will decrease. Record the reading. Add several more milliliters of 0.020M $AgNO_3$ to beaker 1. The voltmeter reading will decrease again. Record the reading. Note that the voltage difference between the electrodes decreases as the difference between the concentrations of silver ions in beakers 1 and 2 decreases. Reverse the voltmeter leads on the electrodes, moving the positive lead to the electrode in beaker 1 and the negative lead to the electrode in beaker 2. The voltmeter reading will change sign, but its absolute value will remain the same. Note that when the positive lead is connected to the electrode in the solution with the higher silver concentration, the voltmeter reading is positive.

Pour 200 mL of 0.020M $AgNO_3$ and 200 mL of 0.20M KNO_3 into beakers 3 and 4. Replace beaker 1 in the assembly with beaker 3. Connect the positive lead of the voltmeter to the electrode in beaker 2 and the negative lead to the electrode in beaker 3. The voltmeter reading is close to 0. Set beaker 4 on the third stirrer and start the stirrer. (All three beakers on the stirrers contain 400 mL of solution that is 0.010M in $AgNO_3$ and 0.10M in KNO_3.)

Pour 6 mL of 1.0M K_2CrO_4 into beaker 3. A red-brown precipitate forms in the beaker. The voltmeter reading is positive, indicating that the concentration of silver ions in beaker 2 is greater than that in beaker 3. Record the voltmeter reading. Pour 6 mL of 1.0M K_2CrO_4 into beaker 4. A red precipitate appears in beaker 4.

Pour 12 mL of 1.0M NH_3 into beaker 2. The voltmeter reading decreases but remains positive. The decrease indicates that the concentration of silver ions in beaker 2 has decreased. Because the reading is still positive, the concentration is still greater in beaker 2 than in beaker 3. Record the value of the voltmeter reading. Add 12 mL of 1.0M NH_3 to beaker 4. The red precipitate remains.

Pour 8 mL of 1.0M KCl into beaker 3. The precipitate mixture turns yellow but remains cloudy. The voltmeter reading increases, indicating that the concentration of silver ions in beaker 3 has decreased. Add 8 mL of 1.0M KCl to beaker 4. The solution becomes yellow and remains cloudy.

Pour 6 mL of 15M NH_3 into beaker 2. The solution remains clear, and the voltmeter reading decreases but remains positive. Add 6 mL of 15M NH_3 to beaker 4.

Pour 8 mL of 1.0M KBr into beaker 3. There is little change in the appearance of the solution, but the voltmeter reading increases. Add 8 mL of 1.0M KBr to beaker 4. The mixture remains yellow and cloudy.

Pour 12 mL of 1.0M $Na_2S_2O_3$ into beaker 2. The solution remains clear, but the voltmeter reading decreases and becomes negative. Add 12 mL of 1.0M $Na_2S_2O_3$ to beaker 4. The mixture becomes clear yellow; all the precipitate dissolves.

Pour 8 mL of 1.0M KI into beaker 3. The appearance of the mixture does not change, but the voltmeter reading becomes positive. Add 8 mL of 1.0M KI to beaker 4, and a precipitate forms.

Pour 50 mL of 3.0M $Na_2S_2O_3$ into beaker 2. The voltmeter reading becomes negative again. Add 50 mL of 3.0M $Na_2S_2O_3$ to beaker 4. All the precipitate dissolves, and the solution is clear yellow.

HAZARDS

Concentrated aqueous ammonia can irritate the skin, and its vapors are harmful to the eyes and mucous membranes. It should be handled only in a well-ventilated area.

Potassium chromate is harmful to the skin, eyes, and mucous membranes. The solid is corrosive. Chromates are suspected carcinogens.

Potassium bromide is toxic if taken internally.

Potassium nitrate is a strong oxidizing agent, and it may cause explosions when heated with combustible materials. Direct contact with the solid is irritating to the skin.

Silver nitrate is corrosive and contact with the solid can burn the skin. It is toxic if taken internally. Solutions of silver nitrate can stain the skin and clothing; these stains can be removed by rinsing with an aqueous solution of sodium thiosulfate ($Na_2S_2O_3$) followed by water.

DISPOSAL

The solutions from the four beakers should be combined in a 3- to 4-liter beaker and acidified by adding 25 mL of concentrated (18M) sulfuric acid (H_2SO_4). Allow this mixture to rest for several hours. Then add 100 mL of concentrated (15M) aqueous ammonia (NH_3) and stir the mixture for several minutes. Add 20 g of sodium sulfide ($Na_2S \cdot 9H_2O$) and stir the mixture again for several minutes. Filter the solid from the mixture and dispose of it in a landfill designed for chemical wastes. (Consult local regulations regarding the disposal of hazardous wastes.) The filtrate should be flushed down the drain with water.

DISCUSSION

In this demonstration, various substances are added to the compartments of a silver concentration cell. Each compartment starts with identical silver nitrate solutions in contact with silver electrodes. As each substance is added to one of the compartments, it reacts with the silver ions in the solution, reducing the concentration of silver ions in that compartment. Whenever the concentration of silver ions in one of the compartments decreases, the electrical potential of the cell also changes. Each substance is also added to a single beaker of silver nitrate solution. This allows a comparison to be made between the measured cell potentials and the reactivity of each substance with silver ions.

The cell used in this demonstration is a Ag-Ag^+ concentration cell. Each compartment of this cell contains silver metal in contact with a solution containing silver ions. In this case the silver ions are from silver nitrate. (Potassium nitrate is added to the solution to decrease the cell resistance and produce more stable voltage measurements.) The magnitude of the cell potential is a measure of the ratio of the silver ion concentrations in the two compartments. A quantitative relationship between cell potential and the concentration ratio is given by the Nernst equation.

$$E = 0.0592 \log \frac{[Ag^+]_a}{[Ag^+]_b}$$

This relationship shows that the greater the difference in silver ion concentrations in the two compartments, the greater the cell potential.

Initially the two compartments contain a 0.010M silver nitrate solution. Because the silver ion concentrations are the same, the cell potential is 0. Then, a solution of potassium chromate is added to one compartment. A precipitate of silver chromate forms, reducing the concentration of silver ions in that compartment. Because the silver ion concentrations in the two compartments are no longer the same, the cell potential is no longer 0. The silver ion concentration in one compartment is still 0.010M. The cell potential is measured. Therefore, the silver ion concentration in the compartment to which potassium chromate has been added can be calculated using the previous equation.

The measurement of cell potential after addition of potassium chromate provides an estimate of the solubility product of silver chromate. The initial number of silver ions in the solution is known (expressed in moles: 0.010M × 0.4 liter). The number that precipitated upon addition of chromate ions is the difference between the initial number and the number that remains (0.40 liter × calculated concentration). The number of chromate ions removed by precipitation of silver chromate is half the number of silver ions removed. The number of chromate ions remaining in solution is the difference between the number added and the number precipitated. Therefore, the concentrations of silver ions and chromate ions remaining in solution can be determined. These concentrations are related to the solubility product by

$$K_{sp} = [Ag^+]^2\,[CrO_4^{2-}]$$

Other substances that react with silver ions are added to the compartments of the cell. Aqueous ammonia does not cause a visible change in the silver nitrate solution, but the change in the cell potential reveals that ammonia does react with silver ions. The cell potential changes but does not change sign after the first addition of ammonia. This indicates that ammonia does not bind silver ions as strongly as chromate ions. This is why adding ammonia to the beaker containing silver chromate precipitate (beaker 4) does not dissolve the precipitate. Thiosulfate ions also produce no apparent change when added to a cell compartment. But in this case, the cell potential changes sign. This means that thiosulfate ions reduce the silver ion concentration more than any of the substances added to the other cell compartment. Therefore, when thiosulfate ions are added to beaker 4, all precipitates dissolve.

The reactions between silver ions and the added substances are represented by the following equations:

$$Ag^+ + 2\,NH_3 \rightleftarrows Ag(NH_3)_2^+$$

$$2\,Ag^+ + CrO_4^{2-} \rightleftarrows Ag_2CrO_4$$

$$Ag^+ + Cl^- \rightleftarrows AgCl$$

$$Ag^+ + Br^- \rightleftarrows AgBr$$

$$Ag^+ + 2\,S_2O_3^{2-} \rightleftarrows Ag(S_2O_3)_2^{3-}$$

$$Ag^+ + I^- \rightleftarrows AgI$$

When potassium chloride solution is added to the mixtures containing red-brown silver chromate, the mixture turns yellow, but remains cloudy. This happens because silver chloride binds the silver ions more strongly than silver chromate, as revealed by the cell potential measurement. When silver chromate is converted to silver chloride, the red-brown precipitate is destroyed, producing a yellow chromate ion solution. The silver chloride precipitate is white but appears yellow because the solution is colored by chromate ions.

11.13

Migration of Copper(II) and Dichromate Ions

A U-tube contains a brown gel at the bottom and colorless solution in each arm above the gel. An electrode is inserted in the solution in each arm and a current is passed through the tube. After about 40 minutes the solution in the arm with the negative electrode has turned blue, and the solution in the arm with the positive electrode has turned yellow [*1*] (Procedure A). Several drops of a brown solution are placed in a well cut into a colorless gel in a petri dish located on an overhead projector. Two electrodes connected to the terminals of a 9-volt battery are inserted in the gel on either side of the well. Within a few minutes, the gel between one electrode and the well has become blue, and the gel between the other electrode and the well has turned yellow (Procedure B).

MATERIALS FOR PROCEDURE A

2.5 g copper sulfate pentahydrate, $CuSO_4 \cdot 5H_2O$

2.5 g sodium dichromate dihydrate, $Na_2Cr_2O_7 \cdot 2H_2O$, or 5.0 g potassium dichromate, $K_2Cr_2O_7$

325 mL distilled water

ca. 30 mL 1M sulfuric acid, H_2SO_4 (To prepare 1 liter of solution, set a 2-liter beaker containing 500 mL of distilled water in a pan of ice water. While stirring the water, slowly pour 55 mL of concentrated [18M] H_2SO_4 into the beaker. The mixture will become very hot. If all the ice melts, add more. Once the mixture has cooled to room temperature, dilute it to 1.0 liter with distilled water.)

5 g agar

5 g sodium bisulfite, $NaHSO_3$ (See Disposal section for use.)

250 g sodium sulfide nonahydrate, $Na_2S \cdot 9H_2O$ (See Disposal section for use.)

250-mL beaker

glass stirring rod

hot plate

thermometer, $-10°C$ to $+110°C$

stand, with clamp

U-tube, 50 cm tall, with outside diameter of 2.5 cm

2 solid rubber stoppers to fit arms of U-tube (optional)

400-mL graduated beaker

piece of white poster board, 20 cm × 50 cm

ca. 30 cm adhesive tape

sharp knife

2 1-holed rubber stoppers to fit arms of U-tube

2 ca. 15-cm graphite rods, 8 mm in diameter (Electrodes of platinum foil may be substituted.)

labels for arms of U-tube

dc power supply capable of producing 1 ampere at 10 volts, with clip leads (An automotive battery charger is suitable.)

MATERIALS FOR PROCEDURE B

overhead projector

0.5 g sodium sulfate, Na_2SO_4

50 mL distilled water

0.5 g agar

10 mL 0.1M sodium sulfate, Na_2SO_4 (To prepare 1 liter of solution, dissolve 12 g of Na_2SO_4 in 600 mL of distilled water and dilute the resulting solution to 1.0 liter.)

10 mL 0.1M copper sulfate, $CuSO_4$ (To prepare 1 liter of solution, dissolve 25 g of $CuSO_4 \cdot 5H_2O$ in 600 mL of distilled water and dilute the resulting solution to 1.0 liter.)

10 mL 0.1M sodium dichromate, $Na_2Cr_2O_7$ (To prepare 1 liter of solution, dissolve 26 g of $Na_2Cr_2O_7$ in 600 mL of distilled water and dilute the resulting solution to 1.0 liter.)

150-mL beaker

hot plate

thermometer, −10°C to +110°C

glass stirring rod

petri dish

cork borer, #7 or larger

3 50-mL beakers

dropper, with bulb

2 ca. 8-cm platinum or Nichrome wires, 20 gauge

2 30-cm wire leads, with an alligator clip on each end

9-volt battery

PROCEDURE A

Preparation

Several hours before presenting the demonstration, prepare the agar gel. Place 2.5 g of $CuSO_4 \cdot 5H_2O$ and 2.5 g of $Na_2Cr_2O_7 \cdot 2H_2O$ in the 250-mL beaker, add 100 mL of distilled water, and stir the mixture. Add about 5 mL of 1M H_2SO_4 in 1-mL increments while stirring the mixture until all the solids dissolve. The solution will be clear brown. Heat the solution to about 90°C and add 5 g of agar. Stir the solution to dissolve the agar. Clamp the U-tube to the stand and pour the hot solution into the U-tube, filling it to a level several centimeters above the top of the bend (Figure 1). Allow the mixture to cool until the agar solidifies to a gel (about 2 hours). Once the agar has solidified, the U-tube may be stored for at least several weeks if both arms of the tube are sealed with solid rubber stoppers.

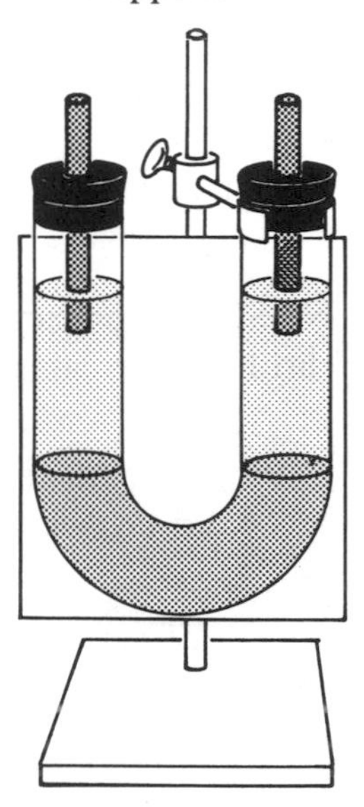

Figure 1.

Pour 25 mL of 1M H_2SO_4 into a 400-mL graduated beaker. Dilute the acid to 250 mL with distilled water. (The diluted H_2SO_4 is 0.1M.) After the agar has solidified, fill both arms of the U-tube to within several centimeters of the top with the diluted H_2SO_4. Tape the piece of poster board to the back of the U-tube.

Insert a sharp knife in the hole of the stoppers, and slit them open from the hole to the edge. Insert one of the graphite-rod electrodes through each hole. Seat the stoppers in the openings of the U-tube. Label one arm of the U-tube with a positive sign (+) and the other with a negative sign (−).

Presentation

With the power supply turned off, clip the cable from the positive terminal of the power supply to the electrode in the arm labelled positive. Clip the other cable to the electrode in the other arm. Turn on the power supply, and set the voltage to about 10 volts. After 45 minutes the solution around the negative electrode will be pale blue, and copper metal may have deposited on the electrode. The solution around the positive electrode will be yellow.

PROCEDURE B

Preparation

Dissolve 0.5 g Na_2SO_4 in 50 mL of distilled water in the 150-mL beaker. Heat the solution to about 90°C, add 0.5 g agar, and stir the mixture until the agar dissolves. Fill the petri dish with the hot solution. Allow the solution to cool until the agar solidifies (about 30 minutes). Use a cork borer to cut a round hole through the agar in the center of the dish, forming a well. (If the dish of agar is not to be used within several hours, it should be sealed in a plastic bag to keep it from drying out.)

Pour about 10 mL of 0.1M Na_2SO_4 into one of the 50-mL beakers, 10 mL of 0.1M $CuSO_4$ into the second beaker, and 10 mL of 0.1M $Na_2Cr_2O_7$ into the third.

Presentation

Place the three beakers and the petri dish on the stage of the overhead projector. Note the colors of the three solutions and their contents. Pour the blue $CuSO_4$ solution into the orange $Na_2Cr_2O_7$ solution. The mixture will be brown. Use the dropper to fill the well in the agar with the brown solution. Make a right-angle bend in each platinum wire about 1 cm from its end. Insert one of the platinum wires in the agar about 2 cm from the edge of the well (Figure 2). Insert the other platinum wire in the agar the same

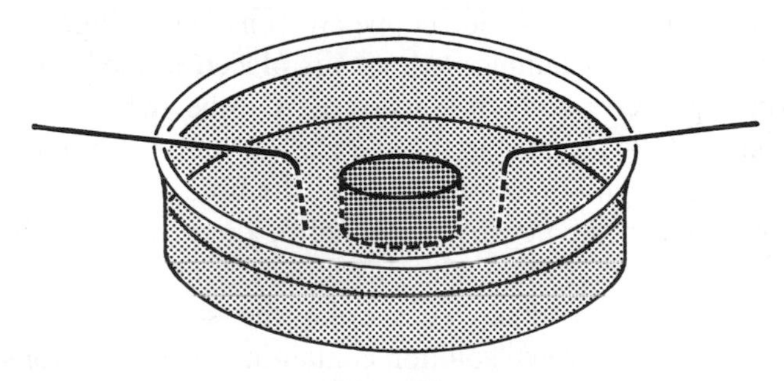

Figure 2.

distance from the opposite side of the well. Use the wire leads with alligator clips to connect the platinum wires to each of the terminals of a 9-volt battery. As soon as the connection is made, bubbling will occur at each platinum wire. Gradually, the agar at the edge of the well near the platinum wire attached to the positive (+) terminal of the battery will turn yellow. At the same time, the gel near the platinum wire attached to the negative (−) battery terminal will turn blue. Over a period of several minutes, the color will gradually spread into the agar and toward the platinum wires.

HAZARDS

The dust from dichromate salts irritates eyes and the respiratory tract. Frequent exposure can cause dermal ulceration, liver and kidney disease, and possibly cancer. Compounds of chromium(VI), such as dichromates, are suspected carcinogens.

Copper compounds can be toxic if taken internally, and dust from copper compounds can irritate mucous membranes.

The dc power supply is capable of delivering a serious shock. Do not touch any metal parts of the electrolysis apparatus while the power supply is turned on.

DISPOSAL

The agar mixture and dichromate solution should be disposed of as follows. Place 300 mL of water in a 1-liter beaker and add 5.0 g of sodium bisulfite ($NaHSO_3$). Using a Bunsen burner, gently heat the U-tube to redissolve the agar mixture. Once it has liquefied, pour the mixture into the beaker containing the sodium bisulfite. The yellow agar mixture will turn a dark green, indicating the reduction of chromium(VI) to chromium(III). Add 250 g $Na_2S \cdot 9H_2O$ to the green solution and stir the mixture periodically for an hour. Filter the solid chromium(III) sulfide from the liquid. Dispose of the solid in a landfill specified for heavy metal wastes. Flush the liquid down the drain with water.

DISCUSSION

This demonstration shows that, when an electric current flows through a solution, the ions in the solution move toward the electrodes. Positive ions move toward one electrode and negative ions move toward the other electrode. This is revealed by using ions having characteristic colors—blue copper(II) ions and orange dichromate ions. In Procedure A, a brown gel containing both ions is placed in the bottom of a U-tube. A colorless solution is poured into both arms of the U-tube, where it rests on top of the brown gel. An electrode is inserted in the colorless solution in each arm. When a current is passed through the solution in the U-tube, the blue copper(II) ions move toward the electrode connected to the negative terminal of the power supply. They color the solution around this electrode blue. The dichromate ions move toward the electrode connected to the positive terminal. They color the solution around this electrode yellow. In Procedure B, the brown solution containing copper(II) ions and dichromate ions is placed in a well in an agar gel in a petri dish. Electrodes connected to a 9-volt battery are inserted in the gel on either side of the well. Copper(II) ions move into the colorless gel toward the negative electrode. Dichromate ions move toward the positive electrode.

The terminology of electrochemistry was developed by Michael Faraday in 1833. He originated the terms *electrode, anode, ion, cathode, anion, cation, electrolyte,* and *electrolysis* [2]. By definition an ion is an atom or group of atoms bearing a net electric charge. An anion is a negative ion, and a cation is a positive ion. In this demonstration, the origin of the terms *cathode* and *anode* is illustrated. The blue color that develops around one of the electrodes indicates that the blue copper cations move toward this electrode. This electrode is called the cathode, because it attracts cations. Similarly, the yellow color around the other electrode indicates that the dichromate anions are attracted to this electrode. This electrode is called the anode.

Copper(II) ions and dichromate ions are used in this demonstration because they are differently colored ions that are compatible in solution. The colorless solution around the electrodes contains sulfuric acid as an electrolyte to conduct current. Rather than using a salt such as potassium sulfate, sulfuric acid is used as an electrolyte to prevent the formation of copper(II) hydroxide ($Cu(OH)_2$) precipitate in the cathodic arm. Were the acid not used, the potential applied to the electrodes would cause the reduction of water and the production of hydroxide ions. These hydroxide ions would precipitate the copper ions as copper(II) hydroxide. The copper(II) and dichromate ions are set in an agar gel to make the U-tube easier to fill without coloring the sulfuric acid solution in the arms of the tube.

REFERENCES

1. H. N. Alyea and F. B. Dutton, Eds., *Tested Demonstrations in Chemistry,* 6th ed., p. 76, Journal of Chemical Education: Easton, Pennsylvania (1965).
2. A. J. Ihde, *The Development of Modern Chemistry,* p. 136, Dover Publications: New York (1984).

11.14

Electrolysis of Water: Color Changes and Exploding Bubbles

A solution of sodium sulfate is electrolyzed to produce a mixture of hydrogen and oxygen gas. An indicator shows the pH change in the solution around each electrode (Procedures A and B). If the resulting solutions are mixed, the original color of the indicator returns [*1, 2*]. This can also be shown using an overhead projector (Procedure C) [*2*]. An alternative method uses the electrolysis of sodium sulfate to produce soap bubbles of gas which are exploded (Procedure D).

MATERIALS FOR PROCEDURE A

1.5 liters 1.0M sodium sulfate, Na_2SO_4 (To prepare 2.0 liters of solution, dissolve 284 g of anhydrous Na_2SO_4 in 1.7 liters of distilled water and dilute the resulting solution to 2.0 liters.)

either

60 mL 0.04% bromothymol blue indicator solution (To prepare 100 mL of solution, dissolve 0.04 g of the sodium salt of bromothymol blue (3′, 3-dibromothymolsulfonephthalein, sodium salt) in 100 mL of distilled water.)

or

60 mL 1% litmus indicator solution (To prepare 100 mL of solution, grind 2 g of litmus to a powder, boil it in 100 mL of distilled water for 5 minutes, cool the solution, and reconstitute it to 100 mL with distilled water.)

ca. 10 mL 0.1M sulfuric acid, H_2SO_4 (To prepare 1 liter of stock solution, pour 3.7 mL of concentrated [18M] H_2SO_4 into 500 mL of distilled water and dilute the resulting solution to 1.0 liter.)

ca. 10 mL 0.1M sodium hydroxide, NaOH (To prepare 1 liter of solution, dissolve 4.0 g of NaOH in 500 mL of distilled water and dilute the resulting solution to 1.0 liter.)

2 2-liter beakers

stirring rod

dropper

Hoffman electrolysis apparatus (e.g., Sargent-Welch no. S-29125-50)

dc power supply capable of delivering 2 amperes at 25 volts, with clip leads

2 labels to attach to Hoffman apparatus

piece of white poster board, slightly larger than Hoffman apparatus

ca. 50 cm adhesive tape

candle

matches

2 test tubes, 15 mm × 125 mm

wooden splint

MATERIALS FOR PROCEDURE B

500 mL 1.0M sodium sulfate, Na_2SO_4 (For preparation, see Materials for Procedure A.)

either

20 mL 0.04% bromothymol blue indicator solution (For preparation, see Materials for Procedure A.)

or

20 mL 1% litmus indicator solution (For preparation, see Materials for Procedure A.)

ca. 10 mL 0.1M sulfuric acid, H_2SO_4 (For preparation, see Materials for Procedure A.)

ca. 10 mL 0.1M sodium hydroxide, NaOH (For preparation, see Materials for Procedure A.)

2 1-liter beakers

stirring rod

dropper

U-tube, 30 cm tall, with outside diameter of 4 cm

ring stand, with clamp

white poster board, ca. 15 cm × 30 cm

ca. 20 cm adhesive tape

either

2 8-cm platinum wires, ca. 20 gauge

or

2 platinum electrodes, with leads sealed in glass (e.g., Sargent-Welch no. S-29125-30)

dc power supply capable of delivering 1 ampere at 20 volts, with clip leads

2 labels for U-tube arms

MATERIALS FOR PROCEDURE C

overhead projector

50 mL 1.0M sodium sulfate, Na_2SO_4 (For preparation, see Materials for Procedure A.)

either

2 mL 0.04% bromothymol blue indicator solution (For preparation, see Materials for Procedure A.)

or

2 mL 1% litmus indicator solution (For preparation, see Materials for Procedure A.)

ca. 1 mL 0.1M sulfuric acid, H_2SO_4 (For preparation, see Materials for Procedure A.)

ca. 1 mL 0.1M sodium hydroxide, NaOH (For preparation, see Materials for Procedure A.)

150-mL beaker

stirring rod

dropper

10-cm petri dish, ca. 3 cm deep

2 8-cm platinum wires, ca. 20 gauge

ca. 4 cm adhesive tape

dc power supply capable of delivering 1 ampere at 20 volts, with clip leads

MATERIALS FOR PROCEDURE D

ca. 250 mL 1.0M sodium sulfate, Na_2SO_4 (For preparation, see Materials for Procedure A.)

50 mL commercial soap bubble solution

250 mL wide-mouth gas collecting bottle

ring stand, with clamp

3-holed rubber stopper to fit gas collecting bottle

2 platinum foil electrodes, 1 cm × 1 cm, with leads sealed in glass (e.g., Sargent-Welch no. S-29125-30)

45-degree glass bend, with outside diameter of 7 mm and length of each arm ca. 5 cm

dc power supply, capable of delivering 1 ampere at 20 volts, with clip leads

ca. 30-cm length of plastic or rubber tubing to fit 7-mm glass bend

scissors

100-mL beaker

candle

matches

piece of aluminum foil, 10 cm square

ear plugs

metal tablespoon

PROCEDURE A

Preparation

Pour 1.5 liters of 1M Na_2SO_4 solution into a 2-liter beaker. Add 60 mL of indicator solution (bromothymol blue or litmus) and stir the mixture. If necessary, adjust the color of the solution to the indicator's neutral color (green for bromothymol blue, purple for litmus). If the bromothymol solution or litmus solution is blue, add drops of 0.1M H_2SO_4. If the bromothymol solution is yellow or the litmus solution is red, add drops of 0.1M NaOH.

Close the stopcocks on the arms of the Hoffman apparatus (see Figure 1). Fill the bulb on the apparatus with the colored Na_2SO_4 solution. Open the stopcock on one arm of the apparatus a small amount until the arm is filled with the solution. Add more solution to the bulb if necessary. Close the stopcock. Fill the other arm in the same fashion.

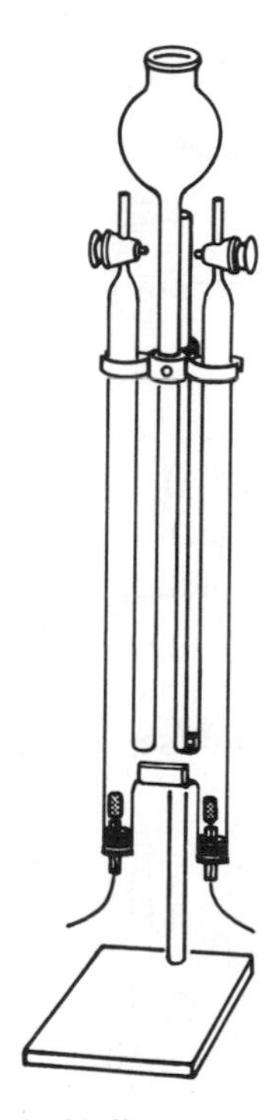

Figure 1. Hoffman apparatus.

Connect the wire leads from the dc power supply to the platinum electrodes of the Hoffman apparatus. Label the electrode connected to the negative terminal of the power supply "cathode." Label the other electrode "anode." Tape the piece of white poster board to the back of the Hoffman apparatus.

Presentation

Turn on the power supply, and adjust the potential to about 25 volts. Bubbles of gas will form rapidly in both arms of the apparatus. The gas will collect at the top of each arm after it has displaced the liquid, which flows from the arms into the central tube and bulb. The volume of gas above the cathode is twice that above the anode. The solution in each arm changes to a different color. (If bromothymol blue is used, the solution in the anode arm will turn yellow, and that in the cathode arm will turn blue. If litmus is used, the solution in the anode arm will turn red, and that in the cathode arm will turn blue.)

When the cathode arm is about two-thirds full of gas, turn off the power supply and disconnect the wire leads. Light the candle with a match. Hold the test tube inverted over the stopcock on the tube holding the smaller volume of gas (the anode tube). Open the stopcock and allow the trapped gas to escape into the test tube. Seal the test tube with the thumb. Light the wooden splint in the candle. Blow out any flame on the splint, leaving glowing embers. Insert the glowing splint into the test tube; the splint will burst into flame. Similarly, fill the other test tube with the gas from the other arm of the apparatus. Insert the glowing splint into this test tube. The gas will explode with a loud "pop."

Drain the solution from the Hoffman apparatus into an empty 2-liter beaker. After the solution has mixed, its color will return to the neutral color (green or purple). (If the solution does not have its neutral color, rinse the inside of the apparatus with some of the solution, and return this to the beaker.)

PROCEDURE B

Preparation

Pour 500 mL of 1M Na_2SO_4 solution into a 1-liter beaker. Add 20 mL of indicator solution (bromothymol blue or litmus) and stir the mixture. If necessary, adjust the color of the solution to the indicator's neutral color (green for bromothymol blue, purple for litmus). If the bromothymol solution or litmus solution is blue, add drops of 0.1M H_2SO_4. If the bromothymol solution is yellow or the litmus solution is red, add drops of 0.1M NaOH.

Clamp the U-tube to the ring stand (see Figure 2). Tape the piece of white poster

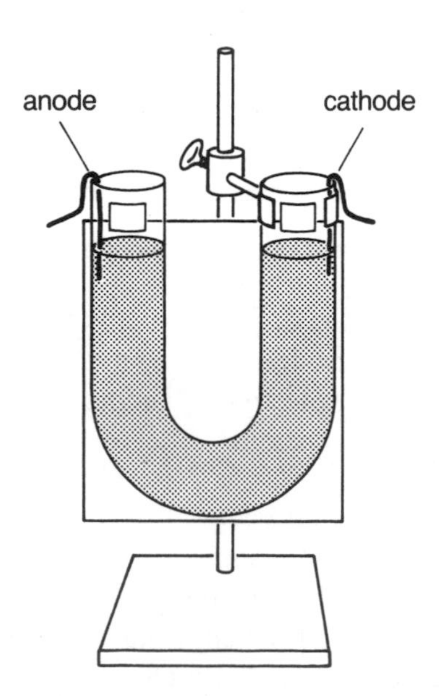

Figure 2. U-tube filled with sodium sulfate solution colored with pH indicator.

board to the back of the U-tube. Bend the platinum wires into a narrow U shape and hook one wire over the rim of each arm of the U-tube. Fill the U-tube with the colored Na_2SO_4 solution. Adjust the platinum wire electrodes so about 3 cm of the wires are immersed in the solution. Clip a wire lead from the dc power supply to each platinum wire. Label the electrode connected to the negative terminal of the power supply "cathode." Label the other electrode "anode."

Presentation

Turn on the power supply and adjust its potential to 20 volts. Bubbles of gas will form around each electrode. The solution in each arm changes to a different color. (If bromothymol blue is used, the anode arm turns yellow and the cathode turns blue. If litmus is used, the anode arm turns red and the cathode blue.)

After 10–15 minutes, turn off the power supply and disconnect the wire leads. Pour the solution from the U-tube into an empty 1-liter beaker. Pour the solution back into the U-tube to rinse it, and pour it back into the beaker. After the solution has been mixed, its color will return to the neutral color (green or purple).

PROCEDURE C

Preparation

Pour 50 mL of 1M Na_2SO_4 solution into a 150-mL beaker. Add 2 mL of indicator solution (bromothymol blue or litmus) and stir the mixture. If necessary, adjust the color of the solution to the indicator's neutral color (green for bromothymol blue, purple for litmus). If the bromothymol solution or litmus solution is blue, add drops of 0.1M H_2SO_4. If the bromothymol solution is yellow or the litmus solution is red, add drops of 0.1M NaOH.

Bend the two platinum wires over the rim of the petri dish, so one end of the wire reaches to the bottom of the dish and the other end is flat against the table top (see Figure 3). Use tape to secure the wires to the outside of the petri dish.

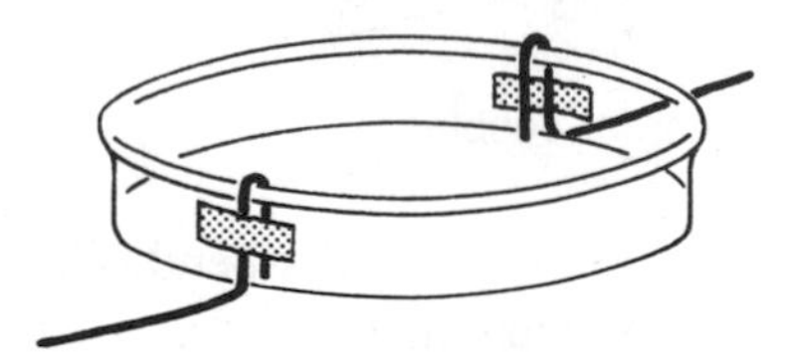

Figure 3. Petri dish with platinum wire electrodes.

Presentation

Set the petri dish with platinum wires on the stage of the overhead projector. Clip the leads from the power supply to the platinum wires extending outside the dish. Fill the dish half full with the colored Na_2SO_4 solution.

Turn on the power supply and adjust its potential to 20 volts. Bubbles form around each electrode. The solution around the electrodes will change color. Stir the mixture, and it will return momentarily to its original color. Turn off the power supply. Carefully swirl the petri dish, rinsing its sides with the solution to recover the solution spattered on the sides. If little or none of the solution is lost, the color will return to its original state.

PROCEDURE D

Preparation

Assemble the apparatus shown in Figure 4. Pour 1M Na_2SO_4 solution into a gas collecting bottle to within 1.5 cm of the rim. Clamp the bottle to a ring stand. Starting at the narrow end of the stopper, insert the glass tubes of the two platinum electrodes through two of the holes. Adjust the electrodes so they do not touch. From the wide end of the stopper, insert the glass bend in the remaining hole. Seat the stopper assembly firmly in the mouth of the bottle. Connect the leads from the power supply to the platinum electrodes.

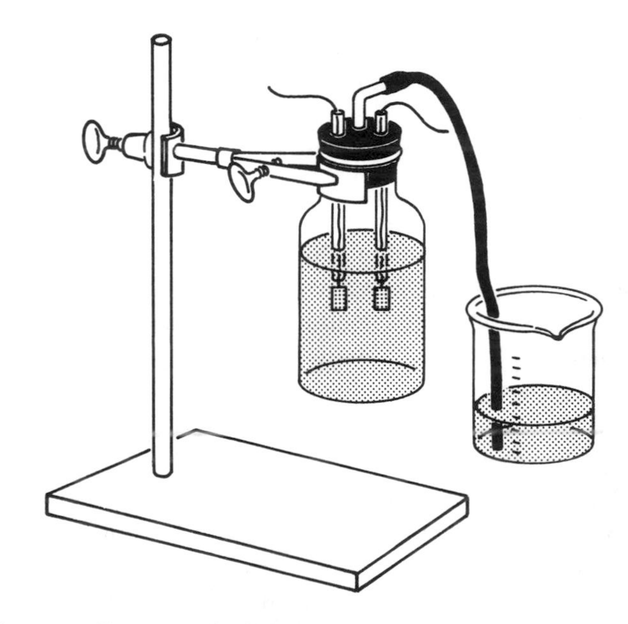

Figure 4. Generator for bubbles filled with hydrogen and oxygen.

Attach a piece of rubber tubing to the glass bend. Cut the tubing to the length at which the end of the tube will just reach the bottom of a 100-mL beaker placed next to the stand. Pour 50 mL of soap bubble solution into the beaker.

Presentation

Place the apparatus at least 2 m from any open flames or spark sources. Turn on the power supply. Within a few minutes, soap bubbles will fill the beaker. While the beaker is filling with bubbles, at a location at least 2 m from the gas generator, light a candle, drip some hot wax onto the center of the aluminum foil, and press the base of the candle into the hot wax. (The foil will protect the table top from wax drips.)

Caution the audience to cover their ears, and insert ear plugs in your own ears. Scoop up a spoonful of bubbles in a metal tablespoon. Carry the bubbles from the beaker to the candle flame. The first spoonful of bubbles may merely break in the flame. Repeat with successive spoonfuls of bubbles. After several spoonfuls, the bubbles will ignite with a loud explosion.

Remove the rubber tubing from the soap solution. Then, turn off the power supply.

HAZARDS

A serious electric shock may result if both leads of the power supply are touched simultaneously.

Hydrogen gas is very flammable and yields explosive mixtures with oxygen and air. Do not allow the leads of the power supply to touch each other; resulting sparks may ignite the hydrogen.

DISPOSAL

The colored sodium sulfate solution may be stored in a sealed container for repeated presentations of the demonstration. Alternatively, it may be flushed down the drain with water.

DISCUSSION

This demonstration shows the effect of passing an electric current through an aqueous solution of sodium sulfate. Bubbles form at both electrodes. Color changes of a pH indicator show that the solution at one electrode becomes basic, while that at the other electrode becomes acidic. When the solution is mixed, its pH returns to neutral. In Procedure A, the electrolysis is performed in a Hoffman apparatus. The Hoffman apparatus is designed to collect the gases formed at the electrodes separately. This shows that the volume of gas generated at one electrode is twice that at the other electrode. The gas produced in one arm of the apparatus (oxygen) causes a glowing splint to burst into flame. The gas produced in the other arm (hydrogen) explodes when exposed to a flame. The solution around the electrode that produces the larger volume of gas becomes basic, and the solution around the other electrode becomes acidic. In Procedure B, the electrolysis is executed in a U-tube. The bubbles and color changes can be observed, but the gas is not collected, so the volume relationships are not apparent. An overhead projector is used to display the bubbles and color changes in Procedure C. The gases produced in the electrolysis are trapped in soap bubbles in Procedure D. These bubbles are exposed to a flame, and they explode.

When a dc current is passed through an aqueous sodium sulfate solution, water is oxidized at the anode, and it is reduced at the cathode.

$$\text{anode:} \quad 2\ H_2O(l) \longrightarrow O_2(g) + 4\ H^+(aq) + 4\ e^-$$

$$\text{cathode:} \quad 4\ H_2O(l) + 4\ e^- \longrightarrow 2\ H_2(g) + 4\ OH^-(aq)$$

These equations show that the number of moles (amount) of hydrogen gas generated is twice that of oxygen. This accounts for the volume of gas generated at the cathode being twice that of the gas formed at the anode. The equations also show that hydrogen ions are formed at the anode and hydroxide ions at the cathode. This accounts for the decrease in pH around the anode and the increase around the cathode. Furthermore, the number of hydrogen ions formed is the same as that of hydroxide ions. Therefore, when the solution is mixed, they completely neutralize each other, and the pH of the solution returns to neutral.

The best choice for an indicator in this demonstration is one that changes from one color to another at a pH of around 7. Bromothymol blue is one such indicator, changing from blue in solutions with pH below 6 to yellow above 7.4 [*3*]. Litmus changes from red at pH values below 5.6 to blue at values above 7. The advantage of an indicator that changes color in this range is that it has a third color in neutral solution. At a pH of 7, bromothymol blue is green, because it contains a mixture of the blue and yellow forms. At a pH of 7, a solution of litmus is purple. Mixed indicators can be used to produce even more than three different colors during the electrolysis of water. However, a mixture of indicators usually does not return to its initial color when the solution is mixed. This may be the result of some oxidation or reduction of the indicators themselves, changing the ratio of indicators in the solution. In addition, loss of even a small amount of the solution by spattering around the electrodes will cause the pH of the mixed solution to be somewhat off neutral. A small deviation from neutrality can have a noticeable effect on the color of mixed indicators. This effect is more pronounced when the demonstration is done on a small scale, such as in Procedure C. Small losses of acid or base due to spattering are about 20 times more significant in the 25 mL of solution used in the petri dish of Procedure C than in the 500 mL of solution used in the U-tube of Procedure B.

The potentials used in the electrolysis of water are much larger than the minimum needed to cause the reaction. The standard potential for the reduction of water is -0.83 volt and for the oxidation is -1.23 volts. Thus, a potential of 2.06 volts would cause the electrolysis of water at standard conditions. However, the cell is not at standard conditions. The cell contains a solution of sodium sulfate, the concentrations of hydrogen ions and hydroxide ions are much less than 1.0M, and the pressure of the gases is less than 1.0 atmosphere. However, even compensating for these factors does not account for the high potentials used. Higher potentials are used to achieve a high current, so the reaction occurs quickly. The current is proportional to the voltage.

$$V = IR$$

where V is the potential applied (volts),
I is the current (amperes),
and R is the resistance of the cell (ohms).

In Procedures A and B, a potential of between 20 and 25 volts is required to drive a current of 0.5 ampere. If a lower potential is used, the rate of electrolysis is too slow for the demonstration. Standard electrode potentials apply to equilibrium conditions. For this demonstration, the cell needs to be quite far from equilibrium to produce gases at a convenient rate.

Pure water does not contain a very high concentration of ions. Autoionization produces a total concentration of hydrogen ions and hydroxide ions of about 2×10^{-7}M. At such a low ion concentration, water is a poor conductor of electricity (its resistance is high). Therefore, to electrolyze water, it is necessary to lower its resistance by adding ions from an electrolyte. Furthermore, the ions added must not themselves react at the electrodes. Sulfuric acid is often used as an electrolyte in the electrolysis of water. However, with sulfuric acid it is not possible to observe the pH changes at the electrodes with indicators. Therefore, the electrolyte used in this demonstration is sodium sulfate, whose solution is neutral.

In Procedure D a stoichiometric mixture of hydrogen and oxygen is produced by electrolysis and trapped in soap bubbles. These soap bubbles are exploded in a flame. The first few spoonfuls of bubbles may not explode because they are filled with the air initially filling the apparatus. Once the air has been swept out by the generated hydrogen-oxygen mixture, the bubbles will be filled with the explosive gas mixture. The gases produced in a Hoffman apparatus, such as is used in Procedure A, can also be combined and exploded. This has been done for an exhibit at the Chicago Museum of Science and Industry [*4*]. Other explosions of hydrogen and oxygen mixtures are described in Demonstration 1.42 in Volume 1 of this series.

The electrolysis of water can also be displayed with an overhead projector by using a flat, enclosed cell designed especially for this purpose [*2*]. The enclosed cell minimizes agitation of the solution during electrolysis. This allows the advancing front of indicator color change to be observed more clearly. However, it is difficult to extract all the solution from the cell to show that the mixed solution is neutral. Furthermore, the cell is difficult to clean.

REFERENCES

1. G. Bodner and T. Greenbowe, personal communication.
2. C. Mortimer, *Educ. in Chem.* 1987 (March): 46.
3. R. C. Weast, Ed., *CRC Handbook of Chemistry and Physics,* 66th ed., p. D-148, CRC Press: Boca Raton, Florida (1985).
4. D. A. Ucko, R. Schreiner, and B. Z. Shakhashiri, *J. Chem. Educ.* 63:1081 (1986).

11.15

Electrochemical Production of an Explosive Gas Mixture

An electric current is passed through a colorless solution in a sealed jar. A greenish gas is produced and collected in a test tube. When a bright light is shined on the test tube, the gas explodes, sending a cork flying across the room [*1*].

MATERIALS

150 mL 0.1M sodium hydroxide, NaOH (To prepare 1 liter of solution, dissolve 4 g of NaOH in 600 mL of distilled water and dilute the resulting solution to 1.0 liter.)

250 mL 6M hydrochloric acid, HCl (To prepare 1 liter of solution, pour 500 mL of concentrated [12M] HCl into 250 mL of distilled water and dilute the resulting mixture to 1.0 liter.)

2 10-cm carbon rods, ca. 0.8 cm in diameter (Suitable carbon rods can be obtained from welding supply stores.)

3-holed rubber stopper to fit gas-generator bottle

250-mL gas-generator bottle (e.g., Central Scientific catalog no. 40014)

2 5-cm lengths of glass tubing, with outside diameter of ca. 7 mm

2 0.5-m lengths of rubber tubing to fit 7-mm glass tubing

2-holed rubber stopper to fit thick-walled test tube

15-cm length of glass tubing, with outside diameter of ca. 7 mm

thick-walled test tube, 2.5 cm × 20 cm, wrapped with one thickness of transparent tape

250-mL Erlenmeyer flask

dc power supply capable of delivering 2 amperes at 5 volts, with clip leads

waxed cork to fit test tube

stand, with clamp

ear plugs or other hearing protection

slide projector, with bulb of at least 300 watts

PROCEDURE

Preparation

Assemble the apparatus as shown in the figure. Insert the two carbon rods through two of the holes in the three-holed rubber stopper. Adjust the rods so they reach nearly to the bottom of the gas-generator bottle when the stopper is seated in the mouth of the bottle. Be sure the rods do not touch each other. Insert one of the 5-cm pieces of glass tubing through the other hole and attach a 0.5-m piece of rubber tubing to the glass tubing. Insert the remaining 5-cm piece of glass tubing through one hole in the two-holed stopper, and insert the 15-cm piece through the other hole. Seat this stopper in the mouth of the test tube, and adjust the 15-cm length of glass tubing so it reaches nearly to the bottom of the test tube. Connect the free end of the rubber tubing from the gas-generator bottle to the longer glass tube in the test tube. Attach a 0.5-m piece of rubber tubing to the shorter piece of glass tubing in the test tube, and insert the free end of the rubber tubing in the 250-mL Erlenmeyer flask.

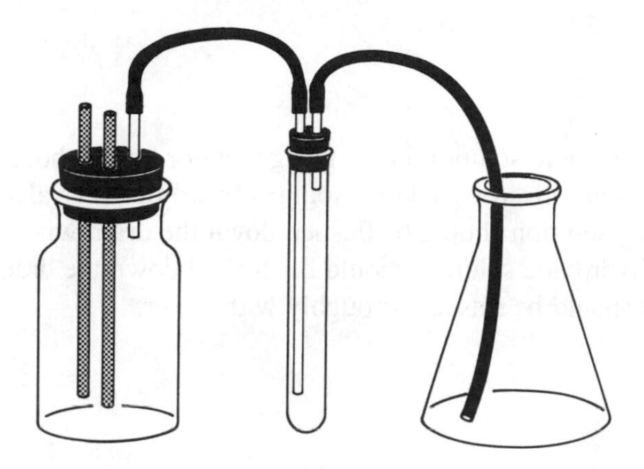

Pour 150 mL of 0.1M NaOH into the Erlenmeyer flask. This solution will trap excess chlorine.

Presentation

This demonstration should be presented only in a well-ventilated area.
Fill the gas-generator bottle with 6M HCl. Clip the leads from the dc power supply to the carbon rods in the gas-generator bottle. Turn on the power supply. Gas bubbles will appear around the two carbon electrodes. Bubbles will slowly emerge from the open end of the rubber tubing in the 0.1M NaOH solution in the Erlenmeyer flask.

Allow the reaction to proceed until the gas within the test tube is visibly green (about 20 minutes). Turn off the power supply. Remove the two-holed stopper assembly from the test tube, and quickly seal the tube with the solid cork. Clamp the test tube to the stand with its mouth angled up and away from anyone or any breakable objects. Put in ear plugs or other hearing protection and caution the students to cover their ears. From a distance of 1 m or less, aim the light beam of the slide projector at the bottom of the test tube. Within a second or two, a loud noise is produced, and the cork is ejected forcefully from the tube.

HAZARDS

Concentrated hydrochloric acid can irritate the skin. Its vapors are extremely irritating to the eyes and respiratory system. It should be handled only in a well-ventilated area.

The chlorine gas produced in the gas generator can irritate the eyes and mucous membranes. If inhaled, it can cause severe lung irritation.

The hydrogen gas produced in the gas generator and released from the chlorine trap is very flammable. It produces explosive mixtures with air and with chlorine. Because this demonstration releases hydrogen gas, it should be presented away from sparks and open flames.

The hydrogen chloride gas produced by the explosion of the hydrogen and chlorine in the test tube is a toxic and corrosive gas, highly irritating to the eyes and respiratory tract.

In small, enclosed areas the noise from the explosion can cause ringing in the ears. Even in large areas, the demonstrator should wear ear protection, and the observers should be cautioned to protect their ears.

DISPOSAL

The hydrochloric acid solution in the gas-generator bottle should be poured into a 1-liter beaker and neutralized by adding sodium bicarbonate ($NaHCO_3$) until fizzing stops. The resulting solution should be flushed down the drain with water.

The sodium hydroxide solution should be flushed down the drain with water.

The test tube should be rinsed thoroughly with water.

DISCUSSION

A current is passed through a 6M hydrochloric acid solution. At the anode, chloride ions are oxidized to chlorine gas.

$$2\ Cl^-(aq) \longrightarrow Cl_2(g) + 2\ e^-$$

At the cathode, hydrogen ions are reduced to hydrogen gas.

$$2\ H^+(aq) + 2\ e^- \longrightarrow H_2(g)$$

The product of the electrolysis of hydrochloric acid is a 1:1 mixture of chlorine and hydrogen gases. This is precisely the stoichiometric mixture for the formation of hydrogen chloride gas.

$$H_2(g) + Cl_2(g) \longrightarrow 2\ HCl(g)$$

This is a reaction which can be initiated photochemically, as shown in the demonstration and described in Demonstration 1.45 in Volume 1 of this series.

Chlorine is a greenish gas, and it can be seen as it collects in the test tube. Actually, the air initially in the test tube is swept out by the mixture of hydrogen and chlorine produced by the electrolysis. Excess hydrogen is allowed to escape by bubbling it through an aqueous solution of sodium hydroxide. This solution, however, traps the chlorine. It reacts with chlorine, producing a mixture of chloride ions and hypochlorite ions.

$$Cl_2(g) + 2\ OH^-(aq) \longrightarrow Cl^-(aq) + OCl^-(aq) + H_2O(l)$$

Because hydrogen escapes into the air, the demonstration must be performed away from flames or sparks, which could cause the escaping hydrogen to explode.

The standard potential for the oxidation of chloride ions is −1.36 volts [2]. The standard potential of the oxidation of hydrogen ions is defined as 0. Therefore, the minimum voltage which must be applied to the electrodes to generate hydrogen and chlorine from hydrochloric acid at standard conditions is 1.36 volts. However, in this demonstration the conditions are not standard. Because the concentration of ions is greater than 1M and the pressure of the gases is less than 1 atmosphere, the voltage required is less than 1.36 volts. To increase the rate at which the gases are generated, a potential greater than the minimum is applied.

REFERENCES

1. R. W. Ramette, *J. Chem. Educ.* 61:722 (1984).
2. R. C. Weast, Ed., *CRC Handbook of Chemistry and Physics,* 66th ed., p. D-152, CRC Press: Boca Raton, Florida (1985).

11.16

Electrolytic Cells in Series: A Red, White, and Blue Electrolysis

Three beakers containing different solutions (two colorless, one red) are connected to each other by salt bridges and to a power supply by electrodes. After a voltage is applied, the solutions turn red, white, and blue.

MATERIALS

500 mL 1M potassium nitrate, KNO_3 (To prepare 1 liter of solution, dissolve 101 g of KNO_3 in 800 mL of distilled water and dilute the resulting solution to 1.0 liter.)

1 mL thymolphthalein indicator solution (To prepare 100 mL of solution, dissolve 0.04 g of thymolphthalein in 50 mL of 95% ethanol and dilute the resulting solution to 100 mL with distilled water.)

2 mL phenolphthalein indicator solution (To prepare 100 mL of solution, dissolve 0.05 g of phenolphthalein (3,3-bis(*p*-hydroxyphenyl)phthalide) in 50 mL of 95% ethanol, and dilute the resulting solution to 100 mL with distilled water.)

ca. 2 mL 0.1M sodium hydroxide, NaOH, solution (To prepare 1 liter of solution, dissolve 4.0 g of NaOH in 500 mL of distilled water and dilute the resulting solution to 1.0 liter.)

ca. 1 mL 0.1M sulfuric acid, H_2SO_4 (To prepare 1 liter of stock solution, pour 3.7 mL of concentrated [18M] H_2SO_4 into 500 mL of distilled water and dilute the resulting solution to 1.0 liter.)

3 250-mL tall-form beakers

3 200-mm carbon rods, 8 mm in diameter

25-cm bare copper wire, 14 gauge

10-cm length of rubber tubing, with outside diameter of 5 mm

1-holed #6 rubber stopper

ring stand, with clamp

2 U-tubes, 100 mm tall, with outside diameter of 13 mm

4 balls of absorbent cotton to fit U-tube openings

3 20-cm glass stirring rods

adjustable dc power supply capable of producing 1 ampere at 10 volts, with clip leads (An automotive battery charger is suitable.)

white poster board, 10 cm × 40 cm

PROCEDURE

Preparation

Place three 250-mL tall-form beakers in a row. Pour 150 mL of 1.0M KNO_3 into each. Rest one of the carbon electrodes in the center beaker. Wrap the end of the copper wire tightly around the middle of one of the other carbon electrodes. Slide a 10-cm piece of rubber tubing over the bare copper wire. Wrap the other end of the copper wire around the middle of the remaining carbon electrode. Arrange this carbon-electrode assembly so one electrode is in the left beaker and the other is in the right beaker (see figure). Be sure this assembly does not touch the center electrode.

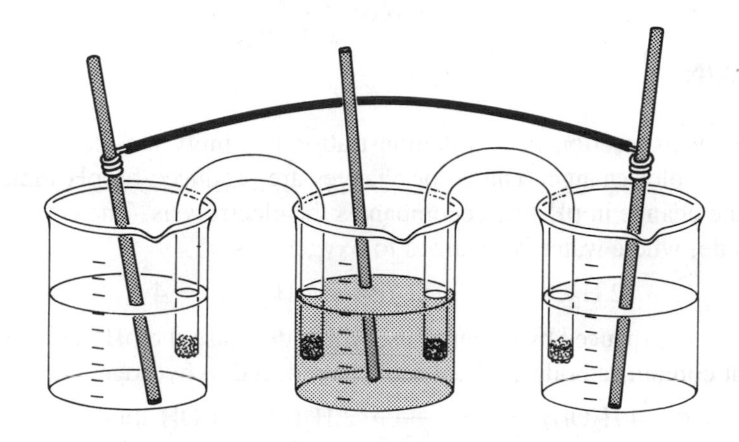

Fill the two U-tubes with 1M KNO_3 solution. Plug each arm of the U-tubes with a wad of cotton. Invert one of the U-tubes and hook it over the rims of the left and center beakers, so one arm is immersed in the solution in each beaker. (If a bubble appears in the U-tube, the U-tube was not sufficiently filled with solution; remove it and put more KNO_3 solution into it.) Similarly, hook the other U-tube over the rims of the center and right beakers. Place a glass stirring rod in each of the beakers.

With the power supply turned off, connect its positive (+) terminal to the central electrode and the negative (−) terminal to one of the outer electrodes.

Place the piece of white poster board behind the three beakers. Add 1 mL of thymolphthalein to the beaker on the right and 1 mL of phenolphthalein to each of the other two beakers. Add 1 mL of 0.1M NaOH solution to the center beaker. Stir the solutions in each beaker. The solutions from left to right will appear colorless, red, and colorless.

Presentation

Turn on the power supply and adjust it to 10 volts. Within 15 minutes, turn off the power supply. Stir each solution. The solutions will appear red, colorless, and blue.

The solutions can be returned to their original colors. Reverse the connections to the power supply. Turn on the power supply and adjust it to 10 volts. After 20 minutes turn off the power supply. Stir each solution. The solutions will return to colorless, red, and colorless. Alternatively, the solutions can be returned to their original colors by adding 0.1M H_2SO_4 to the outside beakers until their contents just turn colorless and by adding 0.1M NaOH to the middle beaker until its contents just turn red.

HAZARDS

An electric shock can result if the exposed metal parts of the apparatus are touched while the power supply is turned on.

DISPOSAL

The solutions may be stored in sealed containers for repeated presentation of the demonstration. If the colors become faint, add more indicator solution to each. The solutions may be disposed of by flushing them down the drain with water.

DISCUSSION

The chemical reaction in this demonstration is simply the electrolysis of water. The effect is a color change. The color changes are produced by pH indicators which respond to the change in pH that accompanies the electrolysis. The center beaker contains the anode, where water is oxidized to oxygen gas.

$$2\ H_2O(l) \longrightarrow O_2(g) + 4\ H^+(aq) + 4\ e^-$$

The reaction also produces hydrogen ions, which decrease the pH. The beakers on the left and right contain cathodes, where water is reduced to hydrogen gas.

$$4\ H_2O(l) + 4\ e^- \longrightarrow 2\ H_2(g) + 4\ OH^-(aq)$$

The reaction also produces hydroxide ions, which increase the pH. The indicators have been selected to produce solutions with the patriotic colors of red, "white," and blue.

Three beakers are used. Each contains a solution of potassium nitrate. The left and center beakers also contain phenolphthalein indicator. Phenolphthalein is colorless in solutions with a pH below 8 and magenta at a pH above 10 [*1*]. The potassium nitrate solution is neutral, so the phenolphthalein in it is colorless. The solution in the center beaker is made basic by the addition of sodium hydroxide, coloring the solution in this beaker magenta. The right beaker contains thymolphthalein indicator. Thymolphthalein is colorless in solutions with a pH below 8.5 and blue in solutions with a pH above 9.5 [*1*]. In the potassium nitrate solution, it is colorless.

The center beaker is joined to the left and right beakers with inverted U-tubes filled with potassium nitrate solution. These salt bridges essentially join all three solutions, allowing ion exchange between them. An inert carbon electrode is placed in each beaker. The electrodes in the left and right beakers are connected by a wire, so they have the same electric potential. Whatever reaction occurs at one also occurs at the other.

Initially, the electrode in the center beaker is connected to the positive terminal of the power supply, and the electrodes in the outer beakers are connected to the negative terminal. When the power supply is turned on, the electrode in the center beaker becomes the anode, and the electrodes in the outer beakers become cathodes. At the anode in the center beaker, water is oxidized to oxygen, producing hydrogen ions as well. These hydrogen ions decrease the pH of the solution in the center beaker. Eventually, the pH of the solution decreases enough so the phenolphthalein changes from magenta to colorless. The electrodes in the outer beakers are cathodes. Water is reduced to hydrogen, and hydroxide ions are formed. The hydroxide ions raise the pH of the solutions in the outer beakers. Eventually the phenolphthalein in the left beaker turns magenta, and the thymolphthalein in the right beaker turns blue.

After the color changes have occurred, the connections to the electrodes are reversed. Then the outer electrodes become anodes, and the center electrode becomes the cathode. The pH changes are also reversed. The pH of the solution in the outer beakers decreases, and the pH of the solution in the center beaker increases. Eventually, the solutions return to their original colors.

The rate of the electrode reactions depends on the current. The current depends on the applied voltage. When a potential of 10 volts is used, the color changes occur within 20 minutes.

REFERENCE

1. R. C. Weast, Ed., *CRC Handbook of Chemistry and Physics,* 66th ed., p. D-148, CRC Press: Boca Raton, Florida (1985).

11.17

Electrolysis of Potassium Iodide Solution

A carbon electrode is inserted in a colorless solution in each arm of a U-tube. The electrodes are connected to a dc power supply, and the solution around one electrode becomes magenta, while around the other it becomes brown (Procedure A). A similar effect is produced in a petri dish on the stage of an overhead projector (Procedure B). A carbon electrode is inserted in a colorless solution in each arm of a U-tube. The electrodes are connected to a dc power supply. The solution around one electrode becomes brown; the solution around the other remains colorless. When starch is added to the solutions in each arm, the brown solution turns dark blue. When phenolphthalein is added to the solutions, the colorless solution turns magenta (Procedure C).

MATERIALS FOR PROCEDURE A

1 liter 1.0M potassium iodide, KI (To prepare 1 liter of solution, dissolve 166 g of KI in 750 mL of distilled water and dilute the resulting solution to 1.0 liter.)

1 mL phenolphthalein indicator solution (To prepare 100 mL of solution, dissolve 0.05 g of phenolphthalein (3,3-bis(*p*-hydroxyphenyl)phthalide) in 50 mL of 95% ethanol, and dilute the resulting solution to 100 mL with distilled water.)

ca. 10 mL 1M sodium thiosulfate, $Na_2S_2O_3$ (To prepare 1 liter of solution, dissolve 250 g of $Na_2S_2O_3 \cdot 5H_2O$, in 600 mL of distilled water and dilute the resulting solution to 1.0 liter.)

ca. 10 mL 1M nitric acid, HNO_3 (To prepare 1 liter of solution, pour 60 mL of concentrated [16M] HNO_3 into 600 mL of distilled water and dilute the resulting solution to 1.0 liter.)

U-tube, 30 cm tall, with outside diameter of 4 cm

ring stand, with clamp

2-liter beaker

stirring rod

2 2-holed rubber stoppers to fit U-tube

sharp knife

2 10-cm carbon rods, ca. 0.8 cm in diameter (Suitable carbon rods can be obtained from welding supply stores.)

white poster board, ca. 15 cm × 30 cm

ca. 20 cm adhesive tape

12-volt dc power supply, or 12-volt automobile battery, with clip leads

illuminated magnetic stirrer, with stir bar

2 droppers

MATERIALS FOR PROCEDURE B

overhead projector

50 mL 1M potassium iodide, KI (For preparation, see Materials for Procedure A.)

1 mL phenolphthalein indicator solution (For preparation, see Materials for Procedure A.)

1 mL 1% starch solution (To prepare 100 mL of solution, bring 50 mL of distilled water to a boil in a 150-mL beaker. In a 50-mL beaker, make a slurry of 1 g of soluble starch in 2 mL of distilled water. Pour the slurry into the boiling water and boil the mixture for 5 minutes. Place 50 g of ice or 50 mL of very cold water in a 250-mL beaker and pour the hot starch mixture into the 250-mL beaker. After the ice has melted, dilute the mixture to 100 mL.)

2 terminal connectors for 9-volt batteries

2 9-volt batteries

ca. 30 cm adhesive tape

either

2 10-cm carbon rods, ca. 0.8 cm in diameter (Suitable carbon rods can be obtained from welding supply stores.)

or

2 wood-clad graphite pencils, with graphite exposed at each end

2 alligator clips

index card, 4 inches × 6 inches

petri dish

2 droppers

MATERIALS FOR PROCEDURE C

All the materials listed under Materials for Procedure A

and

1 mL 1% starch solution (For preparation, see Materials for Procedure B.)

2 dropping pipettes

4 test tubes, 18 mm × 150 mm, with solid rubber stoppers

PROCEDURE A

Preparation

Clamp the U-tube to the stand (see Figure 1). Pour 1 liter of 1.0M KI solution into the 2-liter beaker. Add 1 mL of phenolphthalein indicator solution to the beaker and stir the mixture. Fill the U-tube three-fourths full with this mixture.

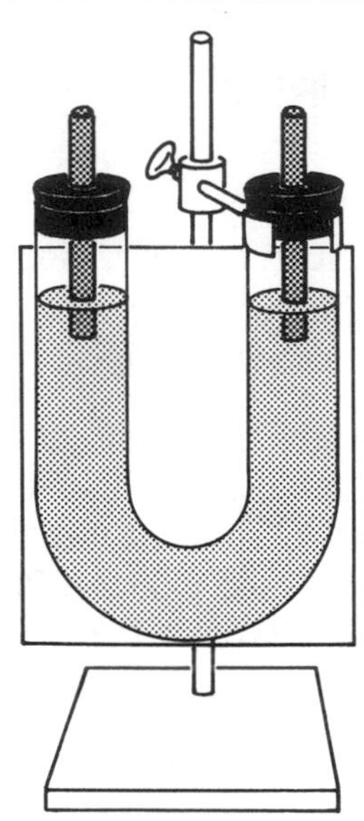

Figure 1. U-tube filled with potassium iodide solution.

Cut a slit in each of the two-holed rubber stoppers by inserting a sharp knife in one of the holes and cutting through to the outside. Insert a carbon electrode in the slit in each of the two stoppers. Seat one of these electrode assemblies in each arm of the U-tube. Adjust the carbon rods so that about 6 cm of each electrode are immersed in the solution. Tape the white poster board to the back of the U-tube.

Presentation

Clip one of the power supply terminals to each of the carbon rods. Turn on the power supply (if it is adjustable, set it to deliver 12 volts). The solution around the carbon rod connected to the positive (+) terminal of the power supply will turn brown. A stream of brown solution will drift down from this electrode to the bottom of the U-tube. Bubbles will form at the carbon rod connected to the negative (−) terminal, and the solution around this electrode will turn red. After 10–15 minutes, the solution in the U-tube will have three distinct regions, a red region around one electrode, a colorless region below the red region, and a brown region around the other electrode. Turn off the power supply after these regions are apparent.

Remove the electrode assemblies from the U-tube. Unclamp the U-tube and pour the colored solution into a 2-liter beaker containing a magnetic stir bar. Set the beaker on an illuminated magnetic stirrer. While stirring the mixture, add 1M $Na_2S_2O_3$ drop by drop until the solution just changes from brown to magenta. Then add 1M HNO_3 drop by drop until the solution just turns colorless. (If too much acid is added, the solution will become pale yellow. If this happens, add just enough drops of 1M $Na_2S_2O_3$ to return the solution to colorless.)

PROCEDURE B [*1–3*]

Preparation

Assemble the apparatus shown in Figure 2. Clip a terminal connector to each of two 9-volt batteries. Tape the 9-volt batteries together, side by side. Tape the two carbon rods to the batteries. Use the alligator clips to fasten the two black wires (from the negative terminals of the batteries) to one carbon rod and the two red wires (from the positive terminals) to the other carbon rod.

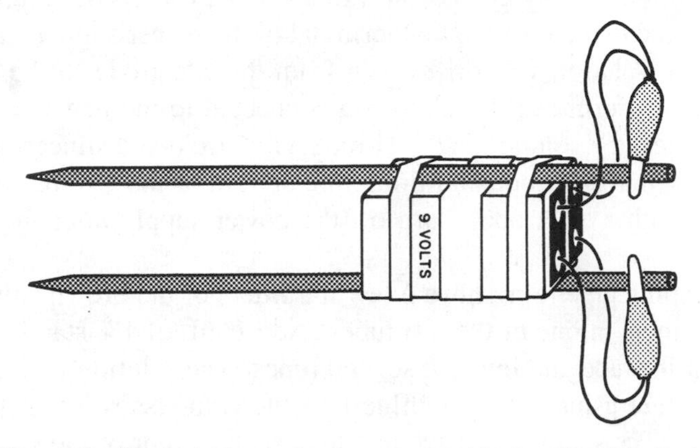

Figure 2. Electrolysis electrodes powered by two 9-volt batteries.

Prepare a dish divider as follows. From the index card, cut two strips 1 cm wide and 1 cm longer than the diameter of the petri dish. Place these strips side by side and wrap pieces of masking tape around them one-half centimeter from each end (see Figure 3). Spread the ends of the cards apart and place the assembly in the petri dish as a divider.

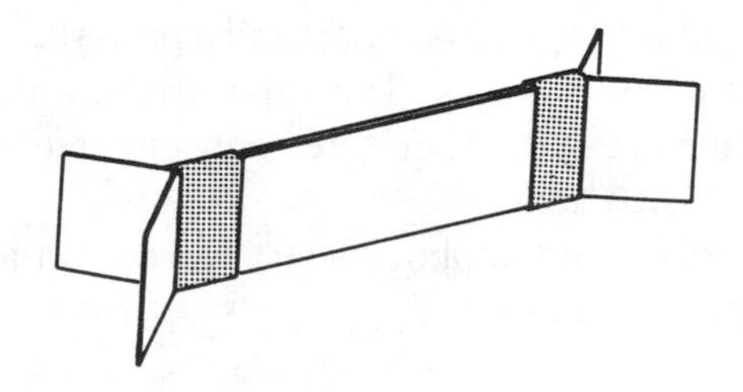

Figure 3. Petri-dish divider made from an index card.

Presentation

Place the petri dish with the divider on the stage of the overhead projector. Fill the dish three-fourths full with 1M KI solution. Immerse the electrodes of the battery assembly in the KI solution, one on each side of the divider. The solution near the carbon electrode connected to the positive terminal of the batteries will turn yellow-brown. Bubbles will appear at the other electrode. After about 45 seconds, add a drop of phenolphthalein solution to the solution around each electrode. The solution around the electrode connected to the negative battery terminals will turn red. Add a few drops of 1% starch solution to the solution around each electrode. The solution around the electrode connected to the positive battery terminals will turn deep blue.

PROCEDURE C

Preparation

Follow the instructions for the preparation for Procedure A, but do not add phenolphthalein solution.

Presentation

Turn on the power supply and adjust it to about 12 volts. The solution around the carbon rod connected to the positive (+) terminal of the power supply will turn brown. A stream of brown solution will drift down from this electrode to the bottom of the U-tube. Bubbles will form at the carbon rod connected to the negative (−) terminal. After 10–15 minutes, the solution in the U-tube will have two distinct regions, a brown region surrounding one electrode and filling the bottom of the U-tube, and a colorless region around the other electrode. Turn off the power supply after these regions are apparent.

Using a dropping pipette, remove a few milliliters of the brown solution from the U-tube and place them in one of the test tubes. Add 1 mL of 1% starch solution to the test tube. Stopper the tube and invert it several times. The solution will turn deep blue. Repeat this procedure using a few milliliters of the colorless solution from the other arm of the U-tube. The starch will have no effect on the color of the solution.

Repeat the procedure of the previous paragraph using phenolphthalein indicator solution in place of starch solution. The phenolphthalein will have no effect on the color of the brown solution, but will turn the colorless solution deep magenta.

HAZARDS

Concentrated nitric acid is both a strong acid and a powerful oxidizing agent. Contact with the skin can cause severe burns. The vapor irritates the respiratory system, eyes, and other mucous membranes. Therefore, concentrated nitric acid should be handled only in a well-ventilated area.

The 12-volt power supply or automobile battery can deliver a severe electric shock if both terminals are touched simultaneously.

DISPOSAL

The colorless potassium iodide solution from Procedure A may be stored in a sealed container for further presentations of the demonstration. Alternatively, it may be discarded by flushing it down the drain with water.

The waste solutions from Procedure B should be flushed down the drain with water.

The potassium iodide solution from Procedure C may be processed to make it suitable for repeated presentations of the demonstration. Pour the colored potassium iodide solution into a 2-liter beaker. While stirring the solution, add 1M sodium thiosulfate drop by drop until the solution just turns from brown to magenta. Then add 1M nitric acid drop by drop until the solution just becomes colorless. (Adding too much acid will cause the solution to turn yellow. If this happens, add drops of 1M sodium thiosulfate until the solution returns to colorless.) The solution is now ready to be used again. If the solution is not to be used again, it should be flushed down the drain with water. The solutions in the test tubes should be flushed down the drain.

The carbon rods should be cleaned by soaking them in 0.1M sodium thiosulfate for 15 minutes, then in distilled water for an hour.

DISCUSSION

This demonstration shows the effect of passing a direct current through an aqueous solution of potassium iodide. At one electrode (the anode), iodide ions are oxidized to molecular iodine.

$$2\,I^-(aq) \longrightarrow I_2(aq) + 2\,e^-$$

The iodine turns the colorless potassium iodide solution brown. At the other electrode (the cathode), water is reduced to hydrogen gas.

$$2\,H_2O(l) + 2\,e^- \longrightarrow H_2(g) + 2\,OH^-(aq)$$

Also produced at this electrode are hydroxide ions, which raise the pH of the solution. The increase in pH changes the color of a pH indicator. In Procedure A, the reaction occurs in a large U-tube, and the colorless potassium iodide solution contains phenolphthalein indicator. At one electrode the solution turns brown, and at the other electrode the solution turns magenta. In Procedure B, the potassium iodide solution is contained in a petri dish on an overhead projector. Here, phenolphthalein is added after the current has been flowing for a while, turning the solution around one electrode magenta. Starch solution is added too, turning the solution around the other electrode dark blue. Procedure C is similar to Procedure A, but the potassium iodide solution does not initially contain phenolphthalein. The indicator solution and starch solution are added after the electrodes have been removed from the U-tube.

During the electrolysis, the solution around the anode turns brown. It is actually aqueous triiodide ions rather than iodine that produce the brown color. Triiodide ions are formed by the reaction of molecular iodine with iodide ions.

$$I_2(aq) + I^-(aq) \longrightarrow I_3^-(aq)$$

As current flows through the solution, iodide ions move toward the electrode where they are reduced. This causes the concentration of triiodide ions around the electrode to increase, which increases the density of the solution. Then the dense, brown triiodide solution sinks in the surrounding, less dense potassium iodide solution. Adding a starch solution to a solution containing molecular iodine and iodide ions causes the solution to turn dark blue. This color is produced by a complex between starch and pentaiodide ions [*4*].

$$\text{starch} + 2\,I_2(aq) + I^-(aq) \longrightarrow \text{starch-}I_5^-\text{ complex}$$

As the electrolysis proceeds, a sharp interface develops between the brown solution and the colorless solution. This sharp interface is a result of a reaction between products of the reactions at the two electrodes. When hydroxide ions are formed at one electrode (the cathode), they move away from the electrode into the bulk of the solution. Similarly, triiodide ions move away from the other electrode (the anode) where they are formed. Where these ions meet, a reaction takes place. In basic solution, triiodide ions disproportionate into iodate ions and iodide ions.

$$3\ I_3^-(aq) + 6\ OH^-(aq) \longrightarrow IO_3^-(aq) + 8\ I^-(aq) + 3\ H_2O(l)$$

This reaction destroys the triiodide ions, which color the solution brown, converting them to colorless iodate ions. The hydroxide ions formed at the cathode prevent the solution around the cathode from turning brown.

The products of the electrolysis can be destroyed and the solution returned to near its initial concentration. This is accomplished by reducing the triiodide ions back to iodide with thiosulfate ions.

$$I_3^-(aq) + 2\ S_2O_3^{2-}(aq) \longrightarrow 3\ I^-(aq) + S_4O_6^{2-}(aq)$$

The brown color of the triiodide makes it easy to detect when all of it has been consumed. The sharp color change that occurs when the last of triiodide ions has been reduced makes it possible to titrate the triiodide ions quantitatively with thiosulfate ions. This titration is the basis for the iodine coulometer (Demonstration 11.19), which is used to determine the amount of current passed through the solution.

REFERENCES

1. G. S. Newth, *Chemical Lecture Experiments*, p. 314, Longmans, Green and Co.: New York (1928).
2. K. E. Kolb and D. K. Kolb, *J. Chem. Educ.* 63:517 (1986).
3. K. E. Kolb and D. K. Kolb, *J. Chem. Educ.* 64:891 (1987).
4. R. C. Teitelbaum, S. L. Ruby, and T. J. Marks, *J. Am. Chem. Soc.* 102:3322 (1980).

11.18

Electrolysis of Sodium Chloride Solution: The Disappearing Indicator

Each of two petri dishes on an overhead projector is divided into two compartments by a paper barrier, and both dishes are filled with a red solution. A compartment of one dish is connected to a compartment of the other dish with a wire. Wires in the two remaining compartments are connected to the terminals of a power supply. After the power supply is turned on, the solution in one compartment of each dish turns blue, and the other becomes colorless. When the solutions in the two compartments are mixed, the entire solution becomes colorless (Procedure A). A U-tube contains a red solution with an electrode immersed in the solution in each arm. Bubbles appear at both electrodes when they are connected to a power supply. Around one electrode the solution turns blue, and around the other the solution becomes colorless. When the solution in the U-tube is mixed, the entire solution becomes colorless (Procedure B).

MATERIALS FOR PROCEDURE A

overhead projector

100 mL 1.0M sodium chloride, NaCl (To prepare 1 liter of solution, dissolve 58 g of NaCl in 800 mL of distilled water and dilute the resulting solution to 1.0 liter.)

1 g potassium iodide, KI (optional)

10 mL distilled water (optional)

2 mL 1% starch solution (optional) (To prepare 100 mL of solution, bring 50 mL of distilled water to a boil in a 150-mL beaker. In a 50-mL beaker, make a slurry of 1 g of soluble starch in 2 mL of distilled water. Pour the slurry into the boiling water and boil the mixture for 5 minutes. Place 50 g of ice or 50 mL of very cold water in a 250-mL beaker and pour the hot starch mixture into the 250-mL beaker. After the ice has melted, dilute the mixture to 100 mL.)

2 mL 1% litmus indicator (To prepare 100 mL of solution, grind 1 g of litmus to a powder, boil the powder in 100 mL of distilled water for 5 minutes, cool the solution, and reconstitute it to 100 mL.)

ca. 5 mL 1.0M hydrochloric acid, HCl (To prepare 1 liter of solution, slowly pour 85 mL of concentrated [12M] HCl into 600 mL of distilled water and dilute the resulting solution to 1.0 liter.)

ca. 5 mL 1.0M sodium hydroxide, NaOH (To prepare 1 liter of stock solution, dissolve 40 g of NaOH in 600 mL of distilled water and dilute the resulting, cooled solution to 1.0 liter.)

3 ca. 8-cm platinum wires, 20 gauge

2 10-cm petri dishes

ca. 20 cm masking tape

index card, 3 inches × 5 inches

adjustable dc power supply capable of delivering 20 volts, with clip leads

250-mL beaker

50-mL beaker (optional)

stirring rod

2 droppers

MATERIALS FOR PROCEDURE B

glycerine, in a dropper bottle

400 mL 1.0M sodium thiosulfate, $Na_2S_2O_3$ (To prepare 1 liter of solution, dissolve 248 g of $Na_2S_2O_3 \cdot 5H_2O$ in 600 mL of distilled water and dilute the resulting solution to 1.0 liter.)

500 mL 1.0M sodium chloride, NaCl (For preparation, see Materials for Procedure A.)

30 mL distilled water (optional)

3 g potassium iodide, KI (optional)

1 mL 1% starch solution (optional) (For preparation, see Materials for Procedure A.)

10 mL 1% litmus solution (For preparation, see Materials for Procedure A.)

11 mL 1.0M hydrochloric acid, HCl (For preparation, see Materials for Procedure A.)

10 mL 1.0M sodium hydroxide, NaOH (For preparation, see Materials for Procedure A.)

U-tube, ca. 20 cm tall, with outside diameter of 4 cm (e.g., U-form drying tube)

stand, with clamp

white poster board, 15 cm × 25 cm

ca. 15 cm transparent adhesive tape

2 2-holed rubber stoppers to fit U-tube

2 platinum electrodes, with leads sealed in glass (e.g., Sargent-Welch catalog no. S-29125-30)

2 90-degree glass bends, with outside diameter of 7 mm and length of each arm ca. 5 cm

2 ca. 50-cm lengths of rubber tubing to fit glass bends

2 250-mL Erlenmeyer flasks

3 1-liter beakers

test tube, 25 mm × 200 mm, with stopper (optional)

stirring rod

adjustable dc power supply capable of delivering 20 volts, with clip leads at least 50 cm long

PROCEDURE A [*1, 2*]

Preparation

Assemble the apparatus shown in Figure 1. Hook one of the platinum wires over the edge of a petri dish so one end just touches the bottom of the dish and the other extends several centimeters outside the dish. Attach the second platinum wire to the other dish in a similar fashion. Use tape to secure the platinum wires to the outside of each petri dish. Set the dishes on the overhead projector so they touch at the points opposite the wire electrodes. Clip the two dishes together with the third platinum wire, bending it over their rims so the ends of the wire just touch the bottom of each dish.

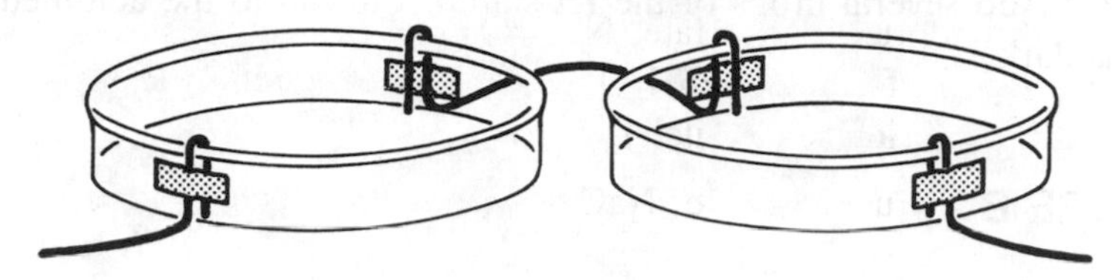

Figure 1. Two petri-dish cells in series.

Prepare two dish dividers as follows. From the index card, cut four strips 1 cm wide and 1 cm longer than the diameter of the dish. Place two of these strips side by side and wrap pieces of masking tape around them one-half centimeter from each end (see Figure 2). Spread the ends of the cards apart and place the assembly in one of the petri dishes as a divider between the platinum wires [*1*]. Repeat this with the remaining two strips, placing this second divider in the other petri dish.

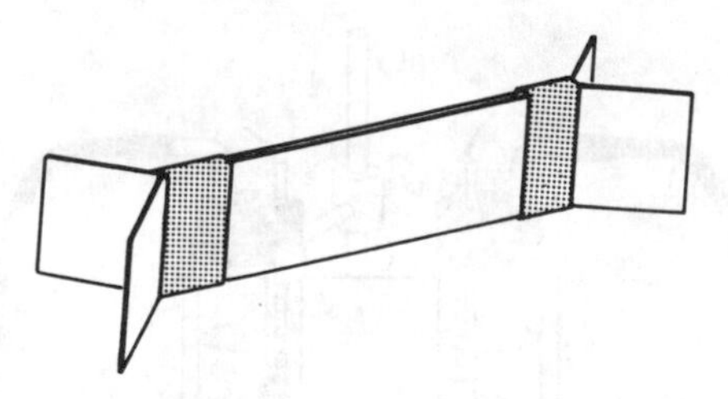

Figure 2. Petri-dish divider made from an index card.

Connect the leads from the power supply to each of the platinum wires extending from the dishes.

Pour 100 mL of 1.0M NaCl into the 250-mL beaker.

Optional: Dissolve 1 g KI in 10 mL of distilled water in a 50-mL beaker. Add 2 mL of 1% starch solution to the beaker and stir the mixture.

Presentation

Set the beaker containing 1.0M NaCl on the illuminated stage of the overhead projector. Add 2 mL of litmus indicator solution and stir the mixture. While stirring add a few drops of 1M HCl to the beaker to produce the distinct red color of acidified litmus. Fill both petri dishes with this red solution to a depth of about 1 cm.

Start the electrolysis by turning on the power supply and adjusting it to about 20 volts. Bubbles of gas appear at each electrode. The odor of chlorine gas (the odor of household bleach or a swimming pool) will develop around the apparatus. At the electrode connected to the negative terminal of the power supply, the solution turns blue. At the other electrode in this dish, the solution turns pale yellow. In the other dish, the solution around the electrode connected to the positive terminal of the power supply turns pale yellow. The solution at the other electrode turns blue.

After about a minute, turn off the power supply. Use a stirring rod to mix the blue solution in one compartment to produce a uniform color distribution. Similarly, mix the other three solutions. Remove the index card divider from one dish and stir the blue and pale yellow solutions together. The blue will disappear, and the entire mixture will become pale yellow. Repeat this with the other dish. Add a few drops of 1.0M HCl to one petri dish and a few drops of 1.0M NaOH to the other. The familiar red and blue color of litmus do not appear, the solutions remain pale yellow.

Optional: Add several drops of the KI-starch solution to the acidified dish. The mixture turns dark.

PROCEDURE B

Preparation

Assemble the apparatus shown in Figure 3. Clamp the U-tube to a ring stand. Tape a piece of white poster board to the back of the U-tube. Put a drop of glycerine in each hole in the two-holed stoppers. From the narrow end of the stopper, carefully insert the glass tube of a platinum-electrode assembly through one of the holes. From the wide

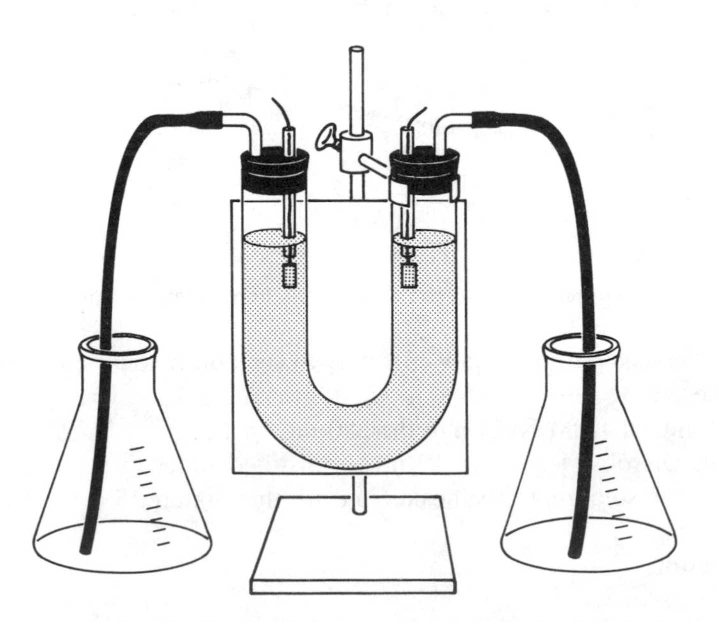

Figure 3. U-tube filled with sodium chloride solution.

end of the stopper, insert one arm of a glass bend through the other hole. In the same fashion, fit the other stopper with an electrode and a glass bend. Loosely seat the stopper assemblies in the arms of the U-tube. Attach a piece of rubber tubing to each glass bend. Cut the tubing so the free ends just reach the bench top. Place the free ends of the tubes in two 250-mL Erlenmeyer flasks. Pour 200 mL of 1M $Na_2S_2O_3$ solution into each flask. (These flasks serve as chlorine traps during the presentation of the demonstration.)

Pour 500 mL of 1.0M NaCl solution into one of the 1-liter beakers.

Optional: Half fill the 25-mm test tube with distilled water. Add 3 g of KI to the test tube, stopper it, and shake it until the KI has dissolved. Add 1 mL of 1% starch solution to the test tube and shake it again to mix its contents.

Presentation

While stirring the NaCl solution, add 10 mL of 1% litmus solution. Add 1 mL of 1.0M HCl solution to turn the litmus bright red. Remove one of the stoppers from the U-tube and fill the tube with the red solution. Reseat the stopper securely in the arm of the U-tube. Connect the insulated wire leads from the power supply to the electrodes.

Turn on the power supply and adjust its voltage to 20 volts. Bubbles of gas appear at the electrode connected to the positive terminal of the power supply, and the solution turns pale yellow around this electrode. Bubbles of gas form at the other electrode, and the solution turns blue.

After 10 minutes, turn off the power supply and remove the electrode assemblies from the U-tube. Remove the U-tube from the stand, and pour its two-colored contents into another of the 1-liter beakers. Upon mixing, the entire solution becomes yellow. Pour half of the solution into the remaining 1-liter beaker. Add 10 mL of 1.0M HCl to one beaker and 10 mL of 1.0M NaOH to the other beaker. The solution remains yellow; the familiar red and blue color of litmus do not appear.

Optional: Pour about 10 mL of the solution containing HCl into the test tube of KI-starch solution. The solution in the test tube immediately turns deep blue.

HAZARDS

Chlorine gas is toxic when inhaled. The amounts of chlorine gas generated in this demonstration are very small.

Concentrated hydrochloric acid can irritate the skin. Its vapors are extremely irritating to the eyes and respiratory system. It should be handled only in a well-ventilated area.

Solid sodium hydroxide and its concentrated solutions can cause severe burns to the eyes, skin, and mucous membranes. Dust from solid sodium hydroxide is very irritating to the eyes and respiratory system.

DISPOSAL

The waste solutions should be flushed down the drain with water.

DISCUSSION

In this demonstration an electric current is passed through an aqueous solution of sodium chloride that also contains litmus. The solution has been made slightly acidic to turn the litmus red. At one electrode, the solution turns blue. At the other electrode, the solution becomes pale yellow. When the solution is stirred after the current has flowed for several minutes, the entire solution turns pale yellow. Furthermore, adding acid or base no longer affects the color of the solution. The litmus has been bleached. When a solution containing potassium iodide and starch is added to the acidified electrolyte, the mixture turns dark blue. In Procedure A, this is done on an overhead projector, and in Procedure B, in a large U-tube.

When an electric current is passed through an aqueous solution of sodium chloride, bubbles of gas form at both electrodes. At the cathode, water is reduced to hydrogen gas.

cathode: $$2\ H_2O(l) + 2\ e^- \longrightarrow H_2(g) + 2\ OH^-(aq)$$

At the anode, chloride ions are oxidized to elemental chlorine.

anode: $$2\ Cl^-(aq) \longrightarrow Cl_2(g) + 2\ e^-$$

The electrolysis produces equal amounts of hydrogen and chlorine. It should therefore produce equal volumes of gases. However, chlorine is more soluble in water than hydrogen, so more hydrogen gas escapes during the electrolysis.

For a gas, chlorine is quite soluble in water, 0.092 mole/liter at 25°C [*3*]. The solubility of chlorine can be attributed to its reaction with water. In aqueous solutions, elemental chlorine disproportionates to chloride ions and hypochlorous acid.

$$Cl_2(aq) + H_2O(l) \rightleftarrows Cl^-(aq) + HOCl(aq) + H^+(aq)$$

In a saturated solution at 25°C, the concentrations of chloride ions and hypochlorous acid are 0.030M, and that of chlorine is 0.062M. Thus, a substantial fraction of the dissolved chlorine is not in the elemental state.

Litmus is a vegetable dye that behaves as an acid-base indicator. In the presence of acids, it is red. In the presence of bases, it is blue. During the electrolysis, the solution around the cathode becomes basic, and the litmus in it turns blue. This is a result of the formation of hydroxide ions. At the anode, the solution becomes acidic. This is a result of the disproportionation of chlorine. However, the litmus does not remain red. Instead it turns pale yellow.

The litmus turns pale yellow near the anode because it is reacting with hypochlorous acid. Hypochlorous acid and its salts (hypochlorites) are bleaches. That is, they destroy the color of many organic dyes, such as litmus. Most bleaches work by oxidizing the organic dyes to another form that is either colorless or less intensely colored. In the case of litmus, the hypochlorite oxidizes it to a pale yellow substance.

Organic dye molecules and hypochlorite molecules have properties that make them likely to react with each other. Organic dye molecules contain chains and rings of carbon atoms bonded together with alternating double and single bonds. It is this alternating pattern of bonds that allows them to absorb visible light, and it is because of this absorption that they appear colored. A double bond between two carbon atoms comprises four electrons that are held between the carbon atoms. This corresponds to a rather high concentration of negative charge between the atoms.

In the hypochlorite ion, the chlorine atom has a partial positive charge, because it is bonded to the more electronegative oxygen. This partial positive charge on the chlorine increases its affinity for electrons. When the chlorine atom of hypochlorite gets near a source of electrons, it is likely to react. It does just that when it nears one of the double bonds in the dye molecule. It removes electrons from (oxidizes) the dye. This destroys the alternating pattern of double bonds in the dye, and the dye can no longer absorb visible light. Then, the dye no longer appears colored. The dye can be oxidized to many different forms, depending on which of its many double bonds reacted with the hypochlorite. Because the organic molecules are no longer colored, they no longer function as pH indicators.

As an optional step in both Procedure A and Procedure B, the solution from the electrolysis is added to a mixture of potassium iodide and starch. The mixture turns dark blue immediately. The mixture of potassium iodide and starch is used as a test for elemental chlorine in the electrolyte. The appearance of the dark blue color indicates that there is elemental chlorine in the solution. Elemental chlorine reacts with iodide ions, oxidizing them to elemental iodine.

$$Cl_2(aq) + 2\,I^-(aq) \longrightarrow 2\,Cl^-(aq) + I_2(aq)$$

The elemental iodine reacts with iodide ions in the solution to form triiodide ions.

$$I_2(aq) + I^-(aq) \longrightarrow I_3^-(aq)$$

The iodine and triiodide ions in solution react with starch, forming a dark blue complex [*4*].

$$I_3^-(aq) + I_2(aq) + \text{starch} \longrightarrow \text{starch-}I_5^- \text{ complex}$$

It is necessary to use the acidified electrolyte in this test for chlorine, because in basic solution, iodine disproportionates to iodate and iodide ions. This disproportionation removes iodine from the mixture, and the colored starch complex does not form.

Both procedures call for platinum electrodes. Inert electrodes are needed to avoid complicating the mixture through the addition of material from the electrode itself. Carbon electrodes are not as effective as platinum. With carbon electrodes, the electrolysis must be allowed to run for a much longer time to produce enough chlorine to bleach the litmus.

In Procedure A, two separate electrolysis cells are used. They are connected by a platinum wire. The wire conducts electrons from one cell to another. Electrons enter the wire at one end, and this end of the wire is the anode in its cell. Electrons leave the wire at the other end, and this end is the cathode in its cell. If the two cells were joined with an ion conductor (salt bridge) rather than an electron conductor, the two solutions would form only one cell. Only oxidation would occur in the dish connected to the positive terminal of the power supply. Only reduction would occur in the dish connected to the negative terminal. If a salt bridge were used, the entire solution in one dish would turn blue, and the entire solution in the other dish would bleach. With the wire connecting the dishes, two complete cells are created. The index card dividers are used to prevent the mixing of the products of the two electrode reactions in each dish.

REFERENCES

1. G. S. Newth, *Chemical Lecture Experiments,* p. 314, Longmans, Green and Co.: New York (1928).

2. A. Joseph, P. F. Brandwein, E. Morholt, H. Pollack, and J. F. Castka, *Sourcebook for the Physical Sciences,* pp. 225–26, Harcourt, Brace, and World: New York (1961).
3. M. Windholz, Ed., *The Merck Index,* 10th ed., p. 294, Merck and Co.: Rahway, New Jersey (1983).
4. R. C. Teitelbaum, S. L. Ruby, and T. J. Marks, *J. Am. Chem. Soc.* 102:3322 (1980).

11.19

Coulometers: Measuring Charge by Electrolysis

Two lead plates are weighed and dipped into a colorless solution, and a power supply is connected to the lead plates. A current is passed through the solution for several minutes. The lead plates are dried and weighed again. One plate has increased in weight and the other has decreased by the same amount (Procedure A). Two platinum electrodes in a colorless solution are connected to a dc power supply, and a current is passed through the solution. After a short while the solution becomes brown. The number of triiodide ions in the solution is determined by a titration (Procedure B). A current is passed through a solution in a Hoffman electrolysis apparatus, and gas is produced and collected. The volume of each gas is recorded, along with its pressure and temperature (Procedure C).

MATERIALS FOR PROCEDURE A

200 mL 1.0M nitric acid, HNO_3 (To prepare 1 liter of stock solution, pour 62 mL of concentrated [16M] HNO_3 into 600 mL of distilled water and dilute the resulting solution to 1.0 liter.)

ca. 200 mL distilled water

200 mL acetone, CH_3COCH_3

200 mL 0.10M lead nitrate, $Pb(NO_3)_2$ (To prepare 1 liter of stock solution, dissolve 33 g of $Pb(NO_3)_2$ in 600 mL of distilled water, and dilute the resulting solution to 1.0 liter.)

25 g sodium sulfide nonahydrate, $Na_2S \cdot 9H_2O$ (See Disposal section for use.)

2 lead plates, 5 cm × 10 cm × ca. 2 mm thick

3 250-mL beakers

crucible tongs

balance accurate to 1 mg

constant-current dc power supply, with clip leads (An adjustable power supply with ammeter may be used in place of a constant-current model.)

MATERIALS FOR PROCEDURE B

7.4 g sodium thiosulfate pentahydrate, $Na_2S_2O_3 \cdot 5H_2O$

925 mL distilled water

0.24 g potassium iodate, KIO_3

25 g potassium iodide, KI

38 mL 1M hydrochloric acid, HCl (To prepare 1 liter of solution, pour 83 mL of concentrated [12M] HCl into 600 mL of distilled water and dilute the resulting solution to 1.0 liter.)

15 mL 1% starch solution (To prepare 100 mL of solution, bring 50 mL of distilled water to a boil in a 150-mL beaker. In a 50-mL beaker, make a slurry of 1 g of soluble starch in 2 mL of distilled water. Pour the slurry into the boiling water and boil the mixture for 5 minutes. Place 50 g of ice or 50 mL of very cold water in a 250-mL beaker and pour the hot starch mixture into the 250-mL beaker. After the ice has melted, dilute the mixture to 100 mL.)

10 g potassium nitrate, KNO_3

2 g agar

250 mL 0.1M nitric acid, HNO_3 (To prepare 1 liter of solution, pour 6 mL of concentrated [16M] HNO_3 into 600 mL of distilled water and dilute the resulting solution to 1.0 liter.)

50-mL buret, with stand

250-mL Erlenmeyer flask

25-mL volumetric pipette

2 250-mL volumetric flasks

250-mL beaker

hot plate

U-tube, 10 cm tall

3 400-mL beakers

2 1–5 cm platinum foil electrodes

constant-current dc power supply, with clip leads (An adjustable power supply with ammeter may be used in place of a constant-current model.)

MATERIALS FOR PROCEDURE C

0.10M sulfuric acid, H_2SO_4, sufficient volume to fill electrolysis apparatus (To prepare 1 liter of solution, slowly pour 5.5 mL of concentrated [18M] H_2SO_4 into 600 mL of distilled water and dilute the resulting solution to 1.0 liter.)

Hoffman-type electrolysis apparatus, with calibrated arms (e.g., Central Scientific catalog no. 81200-01)

dc power supply capable of delivering 1.0 ampere at 12 volts, with clip leads

meter stick

thermometer, $-10°C$ to $+110°C$

barometer

PROCEDURE A

Preparation

Bend each of the two lead plates into a C shape (Figure 1). Adjust the shape so both curved plates can be placed inside one of the 250-mL beakers without touching when the C's are facing away from each other (Figure 2).

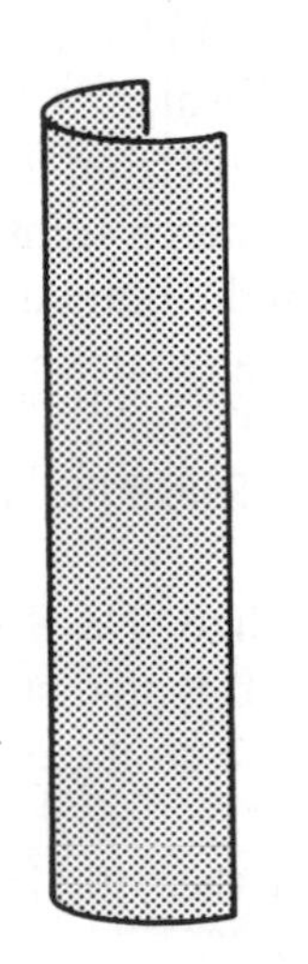

Figure 1. C-shaped lead electrode.

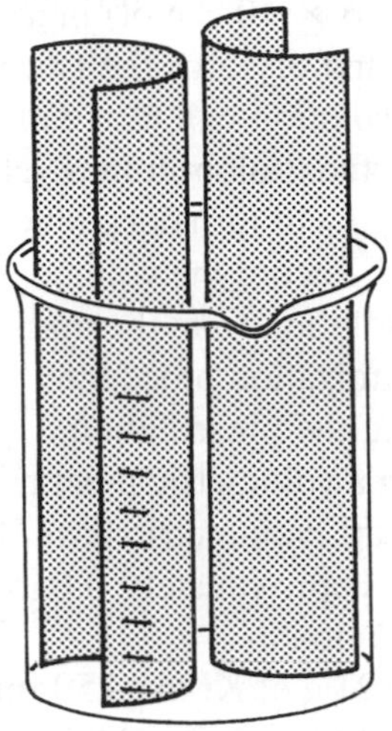

Figure 2. Lead coulometer cell.

Pour 200 mL of 1.0M HNO_3 into one of the 250-mL beakers and 200 mL of acetone into another. Clean the lead plates by immersing them in the 1.0M HNO_3 solution for 1 minute. Rinse the plates with distilled water. Holding the plates with tongs, dip them into the acetone. Allow the plates to dry in the air.

Pour 200 mL of 0.10M $Pb(NO_3)_2$ solution into the remaining 250-mL beaker.

Presentation

Handling the clean, dry plates with tongs, weigh each to the nearest milligram and record the masses. Place the two lead plates in the beaker containing the 0.10M $Pb(NO_3)_2$ solution, arranging them so the C's face away from each other and the two plates do not touch (Figure 2). Clip one of the wire leads from the dc power supply to each plate. Turn on the power supply and adjust it to produce a current of about 0.2 ampere. Allow it to run for 25 minutes.

After turning off the power supply, disconnect the wires from the lead plates. Using tongs, gently remove the plates from the solution and dip them in the acetone. Be careful not to break any of the newly formed lead crystals off the cathode. Allow the plates to dry in the air for several minutes.

Weigh the plates to the nearest milligram. Compare the present masses with their initial values. Calculate the mass gained by the cathode and the mass lost by the anode. Use these data to calculate the amount of charge passed through the cell (see the Discussion section).

PROCEDURE B

Preparation

Dissolve 7.4 g of $Na_2S_2O_3 \cdot 5H_2O$ in 300 mL of distilled water in a 400-mL beaker. This solution is about 0.10M $Na_2S_2O_3$. Rinse a 50-mL buret with this solution and fill the buret with it. Clamp the buret to the stand. Record the initial buret reading.

To standardize the solution, weigh to the nearest milligram approximately 0.12 g of KIO_3 (5.6×10^{-3} mol) in a 250-mL Erlenmeyer flask, and record the weight. Add 2 g of KI and dissolve the two solids in 25 mL of distilled water. Add 19 mL of 1M HCl. The solution will turn dark brown. Add $Na_2S_2O_3$ solution from the buret until the solution in the flask becomes yellow. Then add 5 mL of 1% starch solution to the flask. The solution will turn deep blue. Slowly add $Na_2S_2O_3$ from the buret until the blue just disappears. Record the volume of $Na_2S_2O_3$ solution used. Calculate the molarity of the $Na_2S_2O_3$ solution from the mass of KIO_3 as described in the Discussion section. Repeat this standardization procedure, and average the two results. Rinse the flask so it can be used during the presentation.

Using a volumetric pipette, place 25 mL of the standardized $Na_2S_2O_3$ solution in a clean 250-mL volumetric flask. Dilute the solution to 250 mL with distilled water. This solution is approximately 0.01M $Na_2S_2O_3$. Rinse the pipette so it can be used during the presentation.

Place 21 g of KI in a 250-mL volumetric flask, dissolve the solid in distilled water, and dilute the solution to 250 mL with more distilled water.

Prepare a salt bridge. Dissolve 10 g of KNO_3 in 100 mL of distilled water in a 250-mL beaker. Add 2 g of agar to the solution and heat the mixture to boiling. After the agar has dissolved, pour the hot mixture into the U-tube. Allow the filled tube to cool until its contents have solidified.

Presentation

Place the two 400-mL beakers side by side. Pour the 250 mL of the KI solution from its volumetric flask into the beaker on the left. Pour 250 mL of 0.1M HNO_3 into the other beaker. Join these solutions with the salt bridge (see Figure 3).

Suspend a platinum electrode in each beaker. Attach the negative terminal of the constant-current power supply to the electrode in the HNO_3 solution. Connect the positive terminal of the power supply to the electrode in the KI solution.

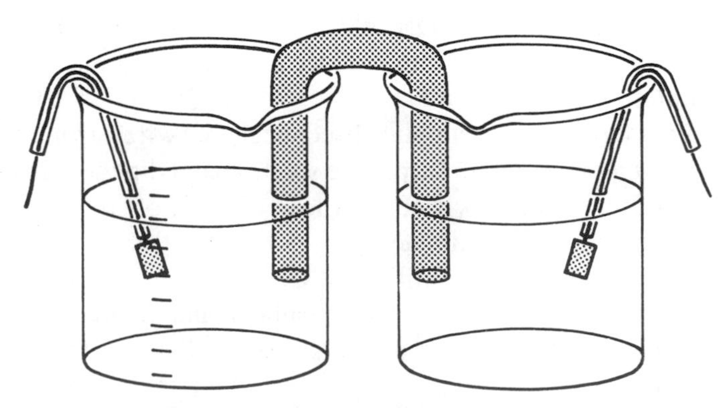

Figure 3. Iodine coulometer assembly.

Set the power supply for a current of about 50 milliamperes and turn it on. After the cell has been operating for a few minutes, the solution in the beaker containing the KI will become brown. Disconnect the power supply, and record the time. Remove a 25 mL aliquot with the volumetric pipette and place it in the clean Erlenmeyer flask. Titrate with the diluted $Na_2S_2O_3$ solution to a yellow endpoint. Add 5.0 mL of the 1% starch solution and continue titration until the blue color just disappears.

Calculate the number of moles of I_2 in the sample and the number of moles of I_2 produced by the electrolysis (see the Discussion section).

PROCEDURE C

Preparation

Fill the Hoffman apparatus with 0.10M H_2SO_4. Clip one of the wire leads from the dc power supply to one electrode and the other lead to the second electrode.

Presentation

Turn on the dc power supply. Bubbles of gas will form at each electrode and rise into the arms of the Hoffman apparatus. Once the apparatus has generated enough gas to nearly fill one arm of the Hoffman apparatus, turn off and disconnect the power supply.

Adjust the position of the reservoir so the level of liquid in the reservoir matches the level of liquid in one arm of the electrolysis apparatus. Record the volume of gas in that arm. Readjust the reservoir so its liquid level matches that in the other arm of the apparatus. Record the volume of gas in that reservoir. (If the position of the reservoir cannot be adjusted, measure the differences between the height of the liquid in the reservoir and its height in each arm.) Record the temperature of the liquid in the reservoir. Record the value of atmospheric pressure.

HAZARDS

Concentrated nitric acid is both a strong acid and a powerful oxidizing agent. Contact with combustible materials can cause fires. Contact with the skin can result in severe burns. The vapor irritates the respiratory system, eyes, and other mucous membranes, and therefore, concentrated nitric acid should be handled only in a well-ventilated area.

Lead compounds are harmful when taken internally. The effects of exposure to small concentrations can be cumulative, causing loss of appetite and anemia.

The power supply can deliver a serious shock if the electrodes are touched while it is turned on.

DISPOSAL

The lead nitrate solution may be saved for repeated presentations of the demonstration. To dispose of the solution, the lead nitrate should be converted to lead sulfide by the addition of 25 g of sodium sulfide nonahydrate ($Na_2S \cdot 9H_2O$) until no further precipitate forms. The lead sulfide precipitate should be separated by filtration and buried in a landfill designed for heavy metals. (Consult local regulations regarding the disposal of hazardous wastes.) The filtrate should be flushed down the drain with water.

The solutions from Procedures B and C should be flushed down the drain with water.

DISCUSSION

A coulometer is a device for measuring an amount of electric charge. The three types of coulometers featured in this demonstration are electrolytic cells used to determine the amount of electric charge passed through them. The amount of product in an electrolytic process is proportional to the quantity of electricity passed through the cell. In Procedure A, the coulometer relies on the change in the mass of an electrode when a metal is dissolved or deposited at the electrode. In Procedure B, the composition of the cell electrolyte changes when an electric current passes through it. In Procedure C, a gas is generated when an electric current passes through the electrolyte, and the gas is collected and its volume measured.

The amount of substance produced in an electrochemical process and the total charge passed through the electrochemical cell during the process are related. The relationship is revealed by the balanced electrode reactions. In Procedure A, an electric current is passed through an aqueous solution of lead nitrate. The electrodes are lead metal. At one electrode, lead ions from the solution are reduced to lead metal.

$$\text{cathode:} \qquad Pb^{2+}(aq) + 2\,e^- \longrightarrow Pb(s)$$

This metallic lead deposits on the electrode, increasing its mass. At the other electrode, lead metal of the electrode is oxidized to lead ions.

$$\text{anode:} \qquad Pb(s) \longrightarrow Pb^{2+}(aq) + 2\,e^-$$

This electrode decreases in mass as the current flows through the cell. These changes in mass are used to calculate the amount of electric charge that passed through the cell.

For the lead coulometer, the chemical equation for the cathode reaction shows that 2 moles of electrons produce 1 mole of lead. At the anode, 2 moles of electrons produce 1 mole of lead ions. To calculate the total charge passed through the solution, the amount (number of moles) of lead deposited at the cathode or dissolved at the anode must be determined. These quantities can easily be determined by comparing the mass of each electrode before electrolysis with its mass after electrolysis.

The cathode in the electrolysis gains mass and the anode loses mass. These mass changes, in theory, are the same. In 25 minutes at a current of 0.2 ampere, the change in mass is about 0.3 g. This corresponds to about 1.5×10^{-3} moles of lead. The number of electrons required to cause this reaction is twice the number of moles of lead, namely, 3×10^{-3}. The total charge passed through the cell is the charge of this number of electrons. The charge of a mole of electrons is given by the experimentally determined Faraday constant, 96,485 coulombs/mole e^-. Therefore, the total charge (Q) that passed through the solution is

$$\begin{aligned} Q &= (96{,}485 \text{ C/mol } e^-)(3 \times 10^{-3} \text{ mol}) \\ &= 300 \text{ coulombs} \end{aligned}$$

Lead is particularly suited for use in a coulometer. Its high molar mass provides a relatively large mass change for a given charge passed through the cell. The changes in mass of the two electrodes in this demonstration are seldom exactly the same. The difference is most often a result of loss of deposited lead from the cathode. At relatively high current, the deposited lead will form in needles. These needles are very easy to break off during cleaning and drying of the electrodes. This causes the mass change of the cathode to be less than that of the anode. If the newly deposited surface of the cathode is coarse, it may trap some of the electrolyte. This can cause the apparent mass change to be greater than that of the anode. The electrodes are bent into a C shape to promote smooth and even deposition of lead at the cathode.

In Procedure B, an electric current is passed through an aqueous solution of potassium iodide. The products of this are molecular iodine and hydrogen gas. An iodometric titration is performed on an aliquot of the electrolyte to determine the amount of iodine produced. This is related to the amount of current that passed through the cell during the electrolysis.

The reaction that occurs at the cathode when an electric current is passed through an aqueous potassium iodide solution is

anode: $$2\ I^-(aq) \longrightarrow I_2(aq) + 2\ e^-$$

The elemental iodine is produced in a solution of potassium iodide. Molecular iodine forms triiodide ions in this solution.

$$I_2(aq) + I^-(aq) \longrightarrow I_3^-(aq)$$

These triiodide ions color the solution yellow.

In the iodine coulometer, the amount of charge passed through the iodide solution is determined by titrating the triiodide ions it produces. As the equation for the anode reaction shows, 2 moles of electrons produce 1 mole of triiodide ions. To calculate the total charge passed through the solution, the number of moles of triiodide produced must be determined.

The triiodide ion concentration in the electrolyte can be determined by titrating it with a sodium thiosulfate solution [2]. Thiosulfate ions reduce triiodide ions to iodide ions.

$$I_3^-(aq) + 2\ S_2O_3^{2-}(aq) \longrightarrow S_4O_6^{2-}(aq) + 3\ I^-(aq)$$

The concentration of a solution of sodium thiosulfate cannot be accurately determined from the weight of the solid used to prepare it. Solid sodium thiosulfate is hygroscopic; that is, it absorbs water from the air. The amount of water absorbed will vary, depending on temperature and humidity, among other things. Therefore, weighing solid sodium thiosulfate does not accurately determine the amount of sodium thiosulfate. A solution of only approximate concentration is prepared. The accurate concentration of this solution is determined by using it to titrate the iodine produced when an excess of potassium iodide reacts with an accurately known quantity of potassium iodate. Analytical potassium iodate solutions can be prepared directly from a weighed amount of the solid, because solid potassium iodate can be obtained in a very pure and stable form; that is, potassium iodate is available as a "primary standard." Sodium thiosulfate is unsuitable for use as a primary standard. The procedure for standardizing sodium thiosulfate with potassium iodate involves several steps. First, an accurately weighed amount of potassium iodate is mixed with an excess of potassium iodide in an acidic solution. All the iodate ions are converted to triiodide ions in a ratio of one iodate ion to three triiodide ions.

$$IO_3^-(aq) + 8\ I^-(aq) + 6\ H^+(aq) \longrightarrow 3\ I_3^-(aq) + 3\ H_2O(l)$$

The triiodide ions color the solution deep brown. This mixture is titrated with the sodium thiosulfate solution. Thiosulfate ions reduce triiodide ions to iodide ions.

$$I_3^-(aq) + 2\ S_2O_3^{2-}(aq) \longrightarrow S_4O_6^{2-}(aq) + 3\ I^-(aq)$$

Three moles of triiodide ions (as produced in the earlier reaction) require 6 moles of thiosulfate ions to reduce them. Therefore, each mole of iodate ions requires 6 moles of thiosulfate ions. As the concentration of triiodide ions decreases, the color of the solution fades from brown to yellow. When all the triiodide has been consumed, the solution is colorless. The point at which the solution becomes colorless is difficult to judge. To increase the sensitivity of this endpoint determination, starch is added to the yellow triiodide ion solution. Starch forms an intensely colored blue complex with iodine and iodide ions. The color of this complex is so intense that even a tiny amount of iodine in the solution will produce a readily visible coloring. When the intense blue has just faded to invisibility, virtually all the triiodide ions have been consumed. The number of iodate ions originally in the solution is one-sixth the number of thiosulfate ions added at this point. This number of thiosulfate ions in the added solution allows the calculation of the accurate concentration of the sodium thiosulfate solution.

The standardized sodium thiosulfate solution is used to titrate an aliquot of the electrolyte from the coulometer. The thiosulfate solution is diluted 10-fold before it is used. It is diluted because the concentration of triiodide ions generated during the short electrolysis is quite low. The amount of iodine produced by the electrolysis is calculated using these relationships:

$$\text{moles of } Na_2S_2O_3 = (\text{molarity } Na_2S_2O_3)(\text{volume of titrant})$$

$$\text{moles of } I_3^- = (\text{moles } Na_2S_2O_3)/2$$

$$\text{moles of } I_2 = \text{moles of } I_3^-$$

The number of coulombs passed through the solution (Q) is related to the amount of iodine in the solution by the equation for the electrode reaction and by the Faraday constant.

$$Q = (\text{moles of } I_2)(2\ \text{mol e}^-/\text{mol } I_2)(96{,}485\ \text{C/mol e}^-)$$

The calculated result can be compared with the measured charge passed through the solution. The total charge passed through the solution is the product of the current (I) multiplied by the time the current flowed (t).

$$Q = It$$

The constant-current power supply is designed to maintain its set current. It does this by varying the voltage it applies during the electrolysis, responding to changes in the resistance of the solution. If a standard power supply is used, the potential will need to be adjusted manually to maintain a constant current. The results of this method are not as accurate as those with a constant-current power supply. If the current is not constant, the previous equation does not apply and will not yield meaningful results.

In Procedure C, a current is passed through an aqueous sulfuric acid solution. This causes the electrolysis of water. At the anode, water is oxidized to oxygen gas.

$$2\ H_2O(l) \longrightarrow O_2(g) + 4\ H^+(aq) + 4\ e^-$$

At the cathode, water is reduced to hydrogen gas.

$$2\ H_2O(l) + 2\ e^- \longrightarrow H_2(g) + 2\ OH^-(aq)$$

These two gases are collected in separate graduated tubes, so their volumes can be measured. The volumes are related to the amount (number of moles) of gas produced. The amount of each gas produced is related to the total charge passed through the solution by the balanced electrode reactions. These equations indicate that a given number of electrons will produce twice as many molecules of hydrogen gas as oxygen gas.

The ideal gas law equation can be used to calculate the amount of each gas from its volume.

$$PV = nRT$$

The temperature of the gas is taken to be the same as that of the liquid in the apparatus. The pressure of the gas is related to atmospheric pressure. If the level of liquid in the reservoir is the same as that in one arm of the apparatus, then the pressure of the gas in that arm is equal to atmospheric pressure. Because the gas is wet, its pressure should be corrected for the vapor pressure of water. Vapor pressure tables can be found in chemistry references, such as the *CRC Handbook of Chemistry and Physics*. If the level of the liquid in the reservoir cannot be adjusted, the pressure of the gas is equal to atmospheric pressure corrected for the water pressure resulting from the difference in the liquid levels. To convert the difference in liquid levels to one corresponding to the mercury pressure scale (mm of Hg, or torr), the difference should be multiplied by the ratio of the density of the liquid to that of mercury. Because the liquid is mostly water, its density is 1.0 g/mL. The density of mercury is 13 g/mL. Therefore, in effect, the difference in liquid levels should be divided by 13. If the level in the reservoir is above that in the arm, then the pressure of the gas in the arm is greater than atmospheric pressure, and the correction should be added to atmospheric pressure. If the reservoir level is lower than that in the arm, the reverse is the case.

It is quite simple to compare the results of the three coulometers described in this demonstration. Any two or all three can be connected in series. Then, the charge that passes through any one is identical to that which passes through the others.

REFERENCES

1. G. D. Muir, Ed. "Hazards in the Chemical Laboratory," 2d ed., p. 299, The Chemical Society: London (1977).
2. B. Z. Shakhashiri and G. E. Dirreen, *Manual for Laboratory Investigations in General Chemistry,* p. 268, Stipes Publishing Co.: Champaign, Illinois (1982).

11.20

A Chemical Rectifier: Converting AC to DC

A lamp and electrolysis cell are connected in series and plugged into a 110-volt ac outlet. The lamp glows brightly, and there is no apparent change in the electrolysis cell. A beaker containing a copper strip and an aluminum strip dipping into a colorless solution is connected in the series. When the plug is inserted in the ac outlet, the lamp glows less brightly, and the solution around the electrodes in the electrolysis cell changes colors.

MATERIALS

350 mL saturated sodium tetraborate, $Na_2B_4O_7$ (To prepare, combine 15 g of $Na_2B_4O_7 \cdot 10H_2O$ and 350 mL of water in a 600-mL beaker. Stir the mixture periodically for several hours. Decant the saturated solution from the undissolved solid.)

1 liter 1.0M potassium iodide, KI (To prepare 1 liter of solution, dissolve 166 g of KI in 750 mL of distilled water and dilute the resulting solution to 1.0 liter.)

1 mL phenolphthalein indicator solution (To prepare 100 mL of solution, dissolve 0.05 g of phenolphthalein (3,3-bis(*p*-hydroxyphenyl)phthalide) in 50 mL of 95% ethanol, and dilute the resulting solution to 100 mL with distilled water.)

ca. 10 mL 1M sodium thiosulfate, $Na_2S_2O_3$ (See Disposal section for use.) (To prepare 1 liter of solution, dissolve 250 g of $Na_2S_2O_3 \cdot 5H_2O$ in 600 mL of distilled water and dilute the resulting solution to 1.0 liter.)

ca. 10 mL 1M nitric acid, HNO_3 (See Disposal section for use.) (To prepare 1 liter of solution, pour 60 mL of concentrated [16M] HNO_3 into 600 mL of distilled water and dilute the resulting solution to 1.0 liter.)

ca. 100 mL 0.1M sodium thiosulfate, $Na_2S_2O_3$ (See Disposal section for use.) (To prepare 1.0 liter of solution, pour 100 mL of 1M $Na_2S_2O_3$ [see above] into 600 mL of distilled water and dilute the resulting solution to 1.0 liter.)

aluminum strip, 5 cm × 20 cm × ca. 2 mm thick

copper strip, 5 cm × 20 cm × ca. 2 mm thick

400-mL beaker

block of wood, 5 cm × 5 cm × 15 cm (6-inch section of 2-by-2)

2 rubber bands

U-tube, 30 cm tall, with outside diameter of ca. 4 cm

ring stand, with clamp

2-liter beaker

stirring rod

2 2-holed rubber stoppers to fit U-tube

sharp knife

2 15-cm carbon rods, ca. 0.8 cm in diameter (Suitable carbon rods can be obtained from welding supply stores.)

white poster board, ca. 15 cm × 30 cm

ca. 20 cm adhesive tape

alligator clip

2-m ac line cord, with polarized plug

wire strippers

porcelain lamp base, with 150-watt incandescent light bulb

60-cm insulated wire, with alligator clip on one end and bare wire on the other

60-cm insulated wire, with an alligator clip on each end

oscilloscope, with input probe (optional)

PROCEDURE [*1, 2*]

Preparation

Construct the rectifier cell as shown in Figure 1. Set the aluminum and copper strips in the 400-mL beaker. Rest the block of wood on top of the beaker between the metal strips. Loop a rubber band around the strips above the block. Holding the metal strips firmly against the wooden block, remove the assembly from the beaker and loop another rubber band around the metal strips below the block. Pour 350 mL of saturated $Na_2B_4O_7$ solution into the beaker. Return the assembly to the beaker.

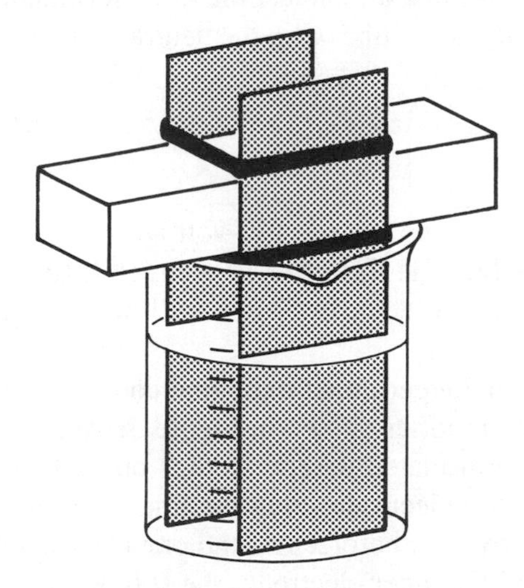

Figure 1. Rectifier cell.

Assemble a potassium iodide electrolysis cell (see Figure 2). Clamp the U-tube to the stand. Pour 1 liter of 1.0M KI solution into the 2-liter beaker. Add 1 mL of phenolphthalein indicator solution to the beaker and stir the mixture. Fill the U-tube three-

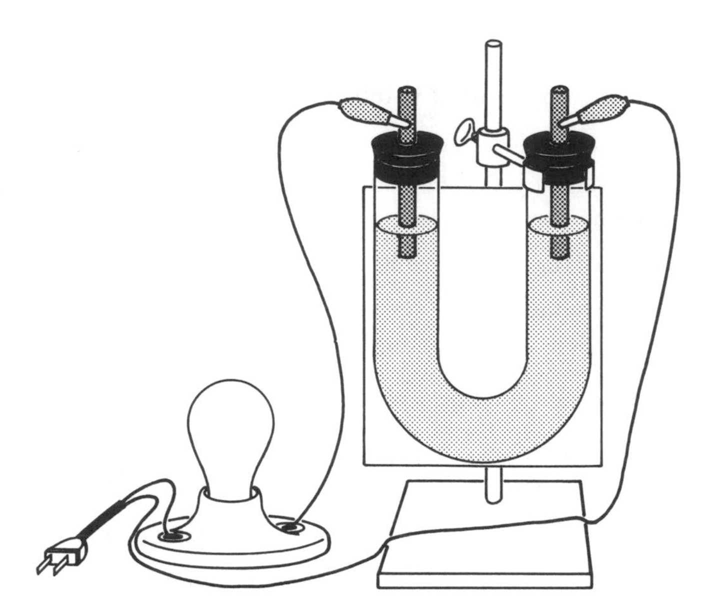

Figure 2. Connections between lamp and electrolysis cell.

fourths full with this mixture. Cut a slit in each of the two-holed rubber stoppers by inserting a sharp knife in one of the holes and cutting through to the outside. Insert a carbon electrode in the slit in each of the two stoppers. Seat one of these electrode assemblies in each arm of the U-tube. Adjust the carbon rods so that about 6 cm of each electrode is immersed in the solution. Tape the white poster board to the back of the U-tube.

Connect the electrolysis cell and lamp as shown in Figure 2. Strip about 1 cm of insulation from each wire of the line cord. Attach an alligator clip to the neutral wire of the polarized line cord (the wire attached to the wider prong of the polarized plug). Attach the other wire of the line cord to one terminal of the lamp base. Use the insulated wire with a clip on one end to connect the other terminal of the lamp base to one of the carbon electrodes in the U-tube. Clip the neutral wire of the line cord to the other carbon electrode.

Presentation

Plug the line cord into a polarized, 110-volt ac outlet. **Do not touch any metal part of the apparatus while the line cord is plugged into the outlet.** The light bulb will glow brightly, and the KI electrolysis cell will not change. After a few minutes, unplug the line cord.

With the line cord unplugged, unclip the line cord from the carbon electrode and insert the rectifier in the circuit as shown in Figure 3. Attach the line cord clip to the copper electrode. Use the insulated wire with clips on each end to connect the aluminum strip to the free carbon electrode. Plug the line cord into the ac outlet. The light bulb will glow less brightly than before. Around one electrode in the KI cell, the solution turns brown. Around the other electrode, the solution turns red. After 15 minutes the solution in the U-tube will be completely colored: red in one arm and brown in the other. While the plug is in the outlet, bubbles form around the copper and aluminum strips in the rectifier cell. The rectifier cell and the U-tube become slightly warm. Unplug the line cord.

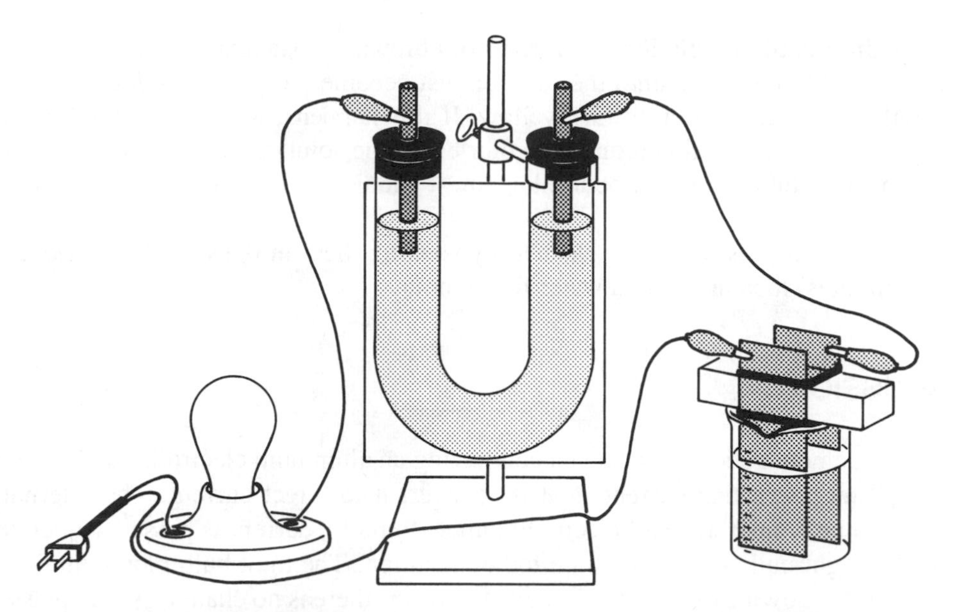

Figure 3. Connections between lamp, electrolysis cell, and rectifying cell.

With the line cord unplugged, reverse the connections to the rectifier cell; that is, connect the wire from the line cord to the aluminum strip, and connect the copper strip to the carbon electrode. Plug the line cord into the outlet. After about 15 minutes the red and brown colors in the arms of the U-tube will be interchanged. Unplug the line cord.

If an oscilloscope is available, with the line cord unplugged, connect the signal probe of the oscilloscope to the terminal of the lamp base connected to the line cord. (This terminal is connected to the line or "hot" side of the electrical outlet.) Adjust the oscilloscope to display dc voltage of 110-volts with a sweep time of about 50 milliseconds. Turn on the oscilloscope. Plug the line cord into the outlet. The oscilloscope will display a 110-volt, 60-Hertz sine-wave pattern. Unplug the line cord. Attach the oscilloscope probe to the metal strip connected to the KI electrolysis cell. Plug in the line cord. The oscilloscope will display a sine wave with half of each wave flattened.

HAZARDS

A severe, possibly fatal, electrical shock can result if any metal part of the apparatus is touched while the line cord is plugged into the 110-volt ac outlet. Therefore, while the line cord is plugged in, do not touch any part of the apparatus other than the plug.

Sodium tetraborate (borax) can be toxic if taken internally or inhaled.

DISPOSAL

The used potassium iodide solution may be processed to make it suitable for repeated presentations of the demonstration. Pour the colored potassium iodide solution into a 2-liter beaker. While stirring the solution, add 1M sodium thiosulfate ($Na_2S_2O_3$)

drop by drop until the solution just turns from brown to magenta. Then add 1M nitric acid (HNO_3) drop by drop until the solution just becomes colorless. (Adding too much acid will cause the solution to turn yellow. If this happens, add drops of 1M sodium thiosulfate until the solution returns to colorless.) The solution is now ready to be used again. If the solution is not to be used again, it should be flushed down the drain with water.

The carbon rods should be cleaned by soaking them in 0.1M sodium thiosulfate for 15 minutes, then in distilled water for an hour.

DISCUSSION

This demonstration shows that a cell having an aluminum electrode can be used to "rectify" an alternating current, that is, convert it to direct current. The alternating current is that from a standard electrical outlet. Initially, current is passed sequentially through a light bulb and a potassium iodide solution. The light bulb glows, indicating that current is flowing through the circuit. However, there is no change in the potassium iodide solution. Then, a cell containing a copper electrode and an aluminum electrode immersed in a sodium borate solution is connected in series in the circuit. Now the light bulb still glows, but not as brightly. Furthermore, the potassium iodide solution changes. The solution near one electrode turns brown, while around the other it turns magenta.

When a current passes through an aqueous solution of potassium iodide, a chemical reaction occurs at each electrode in the solution. At one electrode, iodide ions are oxidized to molecular iodine.

$$\text{anode:} \qquad 2\,I^-(aq) \longrightarrow I_2(aq) + 2\,e^-$$

The aqueous molecular iodine produced by this reaction turns the solution brown. At the other electrode, water molecules are reduced to hydrogen gas.

$$\text{cathode:} \qquad 2\,H_2O(l) + 2\,e^- \longrightarrow H_2(g) + 2\,OH^-(aq)$$

The hydroxide ions increase the pH of the solution around this electrode. If the current flowing through the solution is direct current (i.e., if it travels in only one direction), then the concentrations of aqueous molecular iodine ($I_2(aq)$) around one electrode and hydroxide ions ($OH^-(aq)$) around the other increase continuously. If the current is an alternating current that changes direction frequently, then the reactions can reverse when the current reverses. Therefore, an alternating current produces no discernible change in the potassium iodide solution if it reverses before much of the products can move away from the electrodes. The standard 60-Hertz alternating current reverses 120 times per second, which is often enough, as the demonstration shows, to cause no significant change in the potassium iodide solution when the current is passed through it for several minutes. Home steam (hot) vaporizers exploit this principle, passing alternating current between two electrodes immersed in water containing sodium chloride. The salt solution conducts electricity. However, because the current is alternating, no electrolysis products accumulate. The concentration of salt in the water is quite low, so the solution resistance is sufficiently high to cause the solution between the electrodes to become hot, hot enough to boil the water and produce steam.

In the late 19th century, it was discovered that certain metals (aluminum, tungsten, bismuth, magnesium, and tantalum) are nonreversible electrodes [*3*]. Current can pass in only one direction through these electrodes. The flow of current is restricted in the other direction by an oxide coating that forms on the electrode. This coating is somewhat porous, which allows tiny ions to reach the metal surface, but prevents large ions from approaching closely enough to react. Hydrogen ions can get to the metal to accept electrons and be reduced to hydrogen. Larger anions, however, cannot get near enough to release their electrons to the metal. Therefore, electrons flow only from the electrode into the solution, not in the other direction. Only if the potential applied to the electrode is very high (e.g., over 600 volts) can electrons jump across the oxide coating from the anions to the metal. Aluminum, used in this demonstration, was the first of the so-called valve metals to be discovered.

When displayed on the screen of an oscilloscope, the voltage from a 110-volt ac outlet has a sine-wave form (Figure 4). As this shows, the voltage has a total span of 150 volts, oscillating between +75 volts and −75 volts. The time average of this span is 110 volts. The light bulb glows at its full brightness at 110 volts. No net change occurs in the potassium iodide electrolysis cell, because the current flows equally in both directions, effectively reversing the reactions with each oscillation. When the aluminum rectifier is placed in the circuit, the oscilloscope trace changes (Figure 5). The bottom half of the sine wave is flattened. Now the voltage oscillates between +75 volts and about 0 volts. The time average of the span of the oscillations is just over 55 volts. The incandescent lamp in the circuit glows less brightly now, because the voltage has been nearly halved by the rectifier.

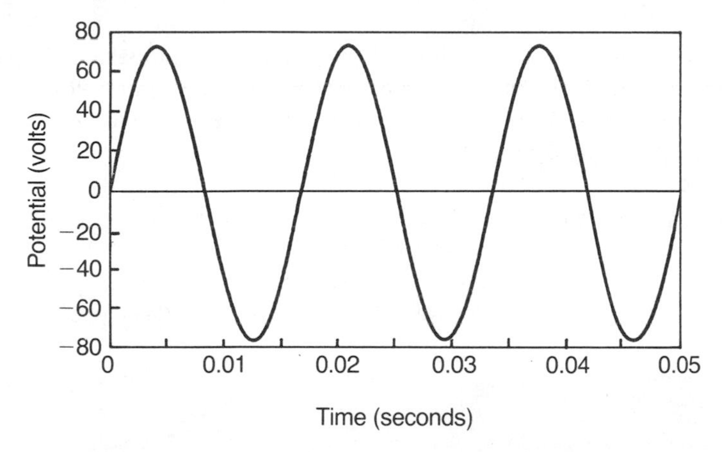

Figure 4. Potential versus time for standard 110-volt ac.

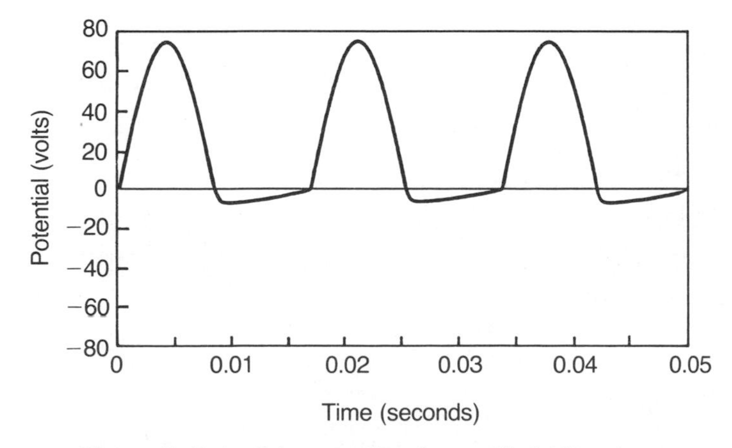

Figure 5. Potential versus time for rectified 110-volt ac.

Because most of the current flows in only one direction, the reactions in the potassium iodide electrolysis cell are not reversed, and a net change is produced. The voltage in the circuit dips slightly into the negative range in each cycle, because the electrodes are polarized during the positive swings of the potential. During the positive swings, extensive chemical reactions occur in the solution adjacent to the electrodes. These reactions change the composition of these regions of the solution. Therefore, the composition of the solution near the electrodes becomes quite different from the bulk of the solution. This difference is responsible for the slight negative voltage. When the circuit is disconnected, the potential gradually returns to 0 as the solution around the electrodes mixes with the bulk of the solution.

Early electric circuits used this nonreversible electrode phenomenon to rectify alternating current. Commercial rectifying cells contained a strip of aluminum for a cathode and graphite, iron, or lead for an anode. A solution of an alkali borate, tartrate, or citrate often served as the electrolyte. The aluminum rectifiers were of limited use, however, because the oxide coating gradually became thicker, causing an increase in the cell resistance [*4*]. These commercial electrolytic rectifying cells were replaced, first, with vacuum tube circuits and, more recently, with semiconductor devices.

REFERENCES

1. A. Joseph, P. F. Brandwein, E. Morholt, H. Pollack, and J. F. Castka, *A Sourcebook for the Physical Sciences,* p. 258, Harcourt, Brace, and World: New York (1961).
2. W. Koehler, *Principles and Applications of Electrochemistry,* Vol. 2, pp. 514–18, John Wiley and Sons: New York (1935).
3. G. Z. Schulze, *Trans. Faraday Soc.* 9:266 (1913).
4. G. Kortum, *Textbook of Electrochemistry,* Vol. 2, p. 464, Elsevier Publishing Co.: Amsterdam (1951).

11.21

Growing Metallic Crystals: Electrolysis of Metal Salts

A voltage is applied to two platinum electrodes immersed in a solution of a metal salt. Tree-like crystals form at one electrode (cathode), and either bubbles or different crystals form at the other electrode (anode). When the voltage is reversed, metallic crystals dissolve at one electrode and form at the other. The electrolysis is shown using an overhead projector [*1*].

MATERIALS

overhead projector

ca. 50 mL of *one* of the following:

1.0M tin(II) chloride (stannous chloride), $SnCl_2$ (To prepare 100 mL of solution, dissolve 22 g of $SnCl_2 \cdot 2H_2O$ in 80 mL of 1M HCl and dilute the resulting solution to 100 mL with 1M HCl. Filter the solution if it is cloudy. This solution should be tightly sealed and used within a few days of preparation.)

1.0M silver nitrate, $AgNO_3$ (To prepare 100 mL of solution, dissolve 17 g of $AgNO_3$ in 80 mL of distilled water and dilute the resulting solution to 100 mL.)

1.0M lead nitrate, $Pb(NO_3)_2$ (To prepare 100 mL of solution, dissolve 33 g of $Pb(NO_3)_2$ in 80 mL of distilled water and dilute to 100 mL with distilled water.)

10 mL concentrated (16M) nitric acid, HNO_3 (See Disposal section for use.)

10 g sodium chloride, NaCl (See Disposal section for use.)

ca. 2.5 g sodium sulfide nonahydrate, $Na_2S \cdot 9H_2O$, or ca. 5 mL 15% ammonium sulfide solution, $(NH_4)_2S$ (See Disposal section for use.)

4 D-cell batteries, with holder

double-pole, double-throw switch

board on which to mount switch

4 10-cm insulated wires, 18–24 gauge

2 30-cm insulated wire leads, each with an alligator clip on one end and bare wire on the other

2 ca. 8-cm platinum wires, 20 gauge (Nichrome wire may be substituted.)

or

2 ca. 10-cm single-strand insulated copper wires, 22 gauge

wire strippers

2 5-cm lengths of ⅛-inch heat-shrink tubing

2 5-cm graphite pieces removed from a wood-clad pencil

heat gun

petri dish, ca. 10 cm diameter × 1.5 cm deep

roll of masking tape

piece of cloth or paper towel, ca. 10 cm × 10 cm

PROCEDURE

Preparation

Assemble a reversible switch as follows. Attach the double-pole, double-throw switch to the board (Figure 1). Connect the terminal on one end of the switch to the terminal diagonally across from it on the other side of the switch. Connect the two remaining free switch terminals together. Attach the free end of one of the clip leads to one of the terminals on one end of the switch. Attach the free end of the other clip lead to the other terminal on the same end of the switch. Use insulated wire to connect one terminal of the battery holder to one of the center terminals of the switch. Connect the other battery-holder terminal to the other center terminal of the switch.

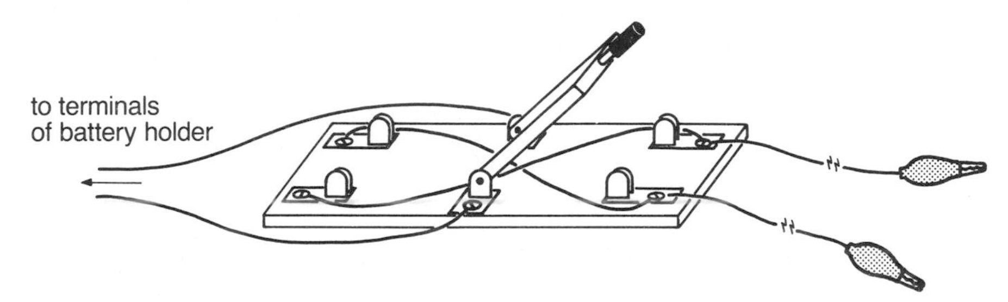

Figure 1. Wiring of double-pole double-throw switch to make reversing switch.

If platinum or Nichrome wires are not available, fabricate electrodes from graphite as follows. Strip about 1 cm of insulation from each end of a 10-cm piece of 22-gauge copper wire. Insert the wire about 4 cm in a 5-cm piece of ⅛-inch heat shrink tubing. Insert a 5-cm piece of graphite pencil lead about 2 cm in the other end of the heat-shrink tubing. (The lead and the copper should overlap and touch inside the tubing.) Use a heat gun to shrink the tubing tightly around the lead and the wire. Repeat this with the other pieces of pencil lead, copper wire, and heat-shrink tubing.

Hook a platinum wire over the edge of the petri dish so about 4 cm of the wire lies straight and flat against the bottom of the dish (Figure 2). (Or bend the copper wire over the edge of the petri dish so the pencil lead lies flat on the bottom of the dish.) Use tape to secure the wire to the outside wall of the petri dish. In a similar manner, secure another wire (or the lead) so it is parallel to and about 2.5 cm from the first piece across the bottom of the dish.

Place the petri dish with attached wires on the overhead projector. Clip the leads from the switch board to each of the wires at the ends extending outside the petri dish. Secure the wire leads to the overhead projector with tape. Make sure the switch is open.

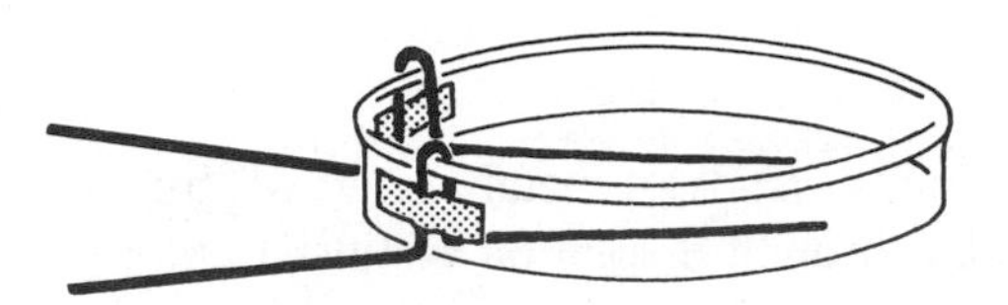

Figure 2. Electrodes positioned in petri dish.

Presentation

Fill the petri dish to a depth of about 1 cm with the solution of 1.0M $SnCl_2$, 1.0M $AgNO_3$, or 1.0M $Pb(NO_3)_2$.

Close the switch to the batteries. Crystals of metal will grow from the platinum wire connected to the negative terminal of the power supply. Bubbles or transparent crystals will grow at the other electrode. When the crystals have grown to a length of about 1 cm, flip the switch to the opposite closed position. The already-grown metal crystals shrink, and new metal crystals grow from the other wire.

HAZARDS

Tin(II) chloride is irritating to the skin and can have toxic effects if taken internally.

Lead compounds are harmful when taken internally. The effects of exposure to small concentrations can be cumulative, causing loss of appetite and anemia.

Silver nitrate is caustic and irritating to the skin and mucous membranes. When taken internally, it can cause irritation of the stomach lining severe enough to be fatal.

DISPOSAL

Any deposits on the platinum wires should be removed by carefully dipping them into concentrated nitric acid, rinsing them with water, then rubbing them with a cloth or paper towel to remove the coating.

Nichrome wires will gradually disintegrate. The fragments should be discarded in a solid waste receptacle.

The tin(II) chloride solution should be flushed down the drain with water.

The silver nitrate solution should be stored in the dark in a sealed glass bottle for repeated presentations of the demonstration. Alternatively, it may be disposed of as follows: Dissolve 10 g of sodium chloride in 250 mL of water and pour this solution into the silver solution. Filter the white precipitated silver chloride from the mixture. Dispose of the filtered solid in a landfill designated for such chemical wastes. (Consult local regulations regarding the disposal of hazardous wastes.) Flush the filtrate down the drain with water.

The lead nitrate solution should be stored in a sealed glass bottle for repeated presentations of the demonstration. To dispose of the lead nitrate solution, the lead nitrate should be converted to lead sulfide by the addition of sodium or ammonium sulfide until no further precipitate forms. The lead sulfide precipitate should be separated by filtration and buried in a landfill designed for heavy metals. (Consult local regulations regarding the disposal of hazardous wastes.) The filtrate should be flushed down the drain with water.

DISCUSSION

In this demonstration, crystals of metal are grown by an electrolytic process. An electric current is passed through a solution of tin(II) chloride, silver nitrate, or lead nitrate. Needle-like crystals of tin, silver, or lead grow from one of the electrodes in the solution. The electrolysis is carried out on the stage of an overhead projector, which magnifies the crystals to truly impressive proportions. When the direction of current flow is reversed, the crystals that grew on one of the electrodes will diminish, and new crystals will appear at the other electrode.

If tin(II) chloride solution is used, only relatively few crystals will grow, and these will be long and narrow, like blades of grass. As the tin crystals grow at the cathode, gas bubbles appear at the anode. These bubbles are chlorine gas. When the current is reversed, the crystals of tin rapidly dissolve, and new blades grow from the other electrode. As soon as all the original tin crystals have disappeared, gas evolution begins again.

If silver nitrate solution is used, the numerous short metal crystals will grow like fur on the cathode. The anode also will become covered with a coating of spiky crystals. The crystals at the anode are probably a complex compound containing silver in a higher oxidation state. A likely component of these crystals has the stoichiometry indicated by the formula AgO, but it is actually a mixed silver(I) and silver(III) oxide [*2*]. When the current is reversed, crystals of silver metal grow from the tips of the spikes, and the silver at the other electrode dissolves. After all the original silver has disappeared from the former cathode, crystals of higher silver oxides begin to grow there.

If lead nitrate is used, fine, branching threads of lead will grow at the cathode. At the anode, gas bubbles will form, along with dark flakes of lead dioxide. When the current is reversed, the threads quickly dissolve.

The tin(II) solution must be prepared in hydrochloric acid to suppress the hydrolysis of Sn^{2+} in water. This hydrolysis produces hydroxide ions, which cause the Sn^{2+} to precipitate from the solution [*3*].

$$Sn^{2+}(aq) + H_2O(l) \rightleftarrows Sn(OH)^+(aq) + H^+(aq)$$

$$Sn(OH)^+(aq) + Cl^-(aq) \rightleftarrows Sn(OH)Cl(s)$$

The hydrogen ions from the hydrochloric acid suppress the hydrolysis reaction. Furthermore, the chloride ions from hydrochloric acid form a complex with Sn^{2+}, which helps keep it in solution.

The results obtained in this demonstration depend on the concentration of the electrolyte solutions. If the tin(II) electrolyte is less concentrated, the tin crystals grow more slowly as a forest of fine fern-like branches, and the demonstration is less striking. If a dilute solution of silver nitrate is used, no crystals form at the anode, and the silver appears in dense tufts, which are not effectively displayed on the projector screen.

REFERENCES

1. E. J. Hartung, *The Screen Projection of Chemical Experiments,* pp. 212–14, Melbourne University Press: Carlton, Victoria (1953).
2. F. A. Cotton and G. Wilkinson, *Advanced Inorganic Chemistry,* 5th ed., p. 945, John Wiley and Sons: New York (1988).
3. F. A. Cotton and G. Wilkinson, *Advanced Inorganic Chemistry,* 5th ed., p. 296, John Wiley and Sons: New York (1988).

11.22

Electrolysis of Copper(II) Bromide Solution

Two graphite rods are dipped into a blue solution which is floating on a colorless liquid in a graduated cylinder. The electrodes are connected to a battery. The upper solution turns green within a few seconds, and the solution around one of the electrodes turns yellow-brown. After about 1 minute the two liquids are agitated by air bubbled up from the bottom through the two layers. When the agitation is stopped, the lower layer is brown. The carbon rods are removed from the solution, and one of them has a copper-colored coating.

MATERIALS

source of compressed air or nitrogen, N_2

100 mL dichloromethane, CH_2Cl_2

250 mL of 0.50M copper(II) bromide, $CuBr_2$ (To prepare 1 liter of solution, dissolve 112 g of $CuBr_2$ in 600 mL of distilled water and dilute the resulting solution to 1.0 liter.)

2 22-cm graphite rods, 7 mm in diameter

2 extension clamps and clamp holders

ring stand

2 50-cm wire leads, with alligator clips on both ends

several feet of plastic tubing, with outside diameter of ¼–⅜ inch

disposable capillary Pasteur pipette

500-mL graduated cylinder

dc power supply capable of delivering 3 volts, or 2 1.5-volt ignition cells

PROCEDURE

Preparation

Support the graphite electrodes by clamping each separately to a ring stand. Then attach a wire lead to each electrode with the alligator clips. Take the plastic tubing and attach one end to a compressed air (or N_2) outlet. Insert a capillary Pasteur pipette in the other end of the tube.

Presentation

When ready to begin the demonstration add 100 mL of dichloromethane to the graduated cylinder. Then add 250 mL of 0.50M $CuBr_2$ solution, noting the blue color.

Lower the carbon electrodes into the aqueous phase, and then insert the Pasteur pipette all the way to the bottom of the cylinder.

Electrolysis is initiated by attaching the leads from the graphite electrodes to the positive and negative terminals of the battery. Note that the electrode connected to the negative battery terminal takes on a shinier appearance, while the solution around the other electrode turns dark brown. After 1–2 minutes of electrolysis, disconnect the battery. Then agitate the two layers of liquids by passing air bubbles through the cylinder. Regulate the flow of air so that mixing occurs, but do not allow liquid to spill over the top. After a few seconds of mixing, stop the flow of air and allow the two layers to separate again. Note the colors of the two layers, and compare them with their original colors. The lower layer will be brown and the upper layer green. Now, pull out the electrode which was connected to the negative terminal of the battery, and point out the copper-colored coating.

HAZARDS

Dichloromethane is a volatile liquid whose vapors can have a narcotic effect at high concentrations.

Solid copper(II) bromide can irritate the skin.

DISPOSAL

The aqueous layer should be flushed down the drain with water. The dichloromethane layer should be poured into a flat dish and allowed to evaporate in a fume hood.

DISCUSSION

At the start of the demonstration, the cylinder contains two immiscible liquids. The lower layer is colorless dichloromethane, and the upper layer is blue 0.50M copper(II) bromide solution. Two electrodes are inserted in the copper(II) bromide solution, and an electric current is passed through the solution. During electrolysis, the blue solution around the anode turns yellowish brown. This color change results from the formation of bromine (Br_2) when bromide ions are oxidized in the electrolysis. The equation for the reaction occurring at the anode is

$$2\ Br^-(aq) \longrightarrow Br_2(aq) + 2\ e^-$$

When the interface between the two immiscible liquids is agitated with a stream of air bubbles, some of the bromine dissolves in the dichloromethane layer, turning it brown.

The copper-colored coating that forms on the cathode is copper metal. Copper metal is formed by the reduction of Cu^{2+} ions in the aqueous solution. The reaction occurring at the cathode is

$$Cu^{2+}(aq) + 2\ e^- \longrightarrow Cu(s)$$

Copper ion has a greater tendency to undergo reduction than water. This is indicated by its greater standard reduction potential [*1*].

$$Cu^{2+}(aq) + 2\,e^- \longrightarrow Cu(s) \qquad E° = 0.34V$$
$$2\,H_2O(l) + 2\,e^- \longrightarrow H_2(g) + 2\,OH^-(aq) \qquad E° = -0.83V$$

Therefore, in this electrolysis cell, copper metal is formed at the cathode and bromine at the anode. The overall cell reaction is

$$Cu^{2+}(aq) + 2\,Br^-(aq) \longrightarrow Cu(s) + Br_2(aq)$$

The minimum voltage that must be applied to produce the cell reaction by electrolysis can be estimated from the standard cell potential. The standard potentials of the electrode reactions (the half reactions) are

$$2\,Br^-(aq) \longrightarrow Br_2(aq) + 2\,e^- \qquad E°_{ox} = -1.07 \text{ volts}$$
$$2\,e^- + Cu^{2+}(aq) \longrightarrow Cu(s) \qquad E°_{red} = 0.34 \text{ volt}$$

The standard potential for the overall reaction is the sum of these half-reaction potentials: $E°_{cell} = -1.07$ volt $+ 0.34$ volt $= -0.73$ volt. For a spontaneous chemical reaction, the cell potential must be positive. Therefore, for this reaction to occur, a potential greater than 0.73 volt must be applied to the electrodes. The two batteries described supply 3.0 volts when connected in series.

The actual voltage required for electrolysis will be different from 0.73 volt for two reasons. First, the concentrations of the ions in a cell affect the potential of the cell. The standard reaction potentials apply to cells in which the concentrations of all solutes involved in the reaction are 1M. In this cell, the concentrations are not 1M. Second, the rate at which a cell reaction occurs is determined by the current that flows through the cell. All electrochemical cells possess some internal resistance to the flow of current. To overcome this internal resistance, a higher-than-minimum voltage must be applied to the cell. This higher-than-minimum voltage is called an overvoltage. The greater the overvoltage applied to an electrolysis cell, the faster the reaction in the cell, and the more quickly a noticeable change will occur.

The overvoltage required to produce a given current at an electrode depends on the electrode reaction. For the reduction of most metal ions to metal, the overvoltage is low. However, for the reduction of hydrogen ions to hydrogen gas, the required overvoltage is higher. Therefore, it is possible to deposit some metals from an aqueous solution of their ions, even though the reduction potentials of their ions are less than that of water.

REFERENCE

1. R. C. Weast, Ed., *CRC Handbook of Chemistry and Physics,* 66th ed., p. D-152, CRC Press: Boca Raton, Florida (1985).

11.23

Copper Leaves: Electroplating with Copper

A battery is connected with wires to a piece of silvery metal and a piece of copper immersed in a blue solution, and the silvery metal becomes copper-colored. When the wires are reversed, the copper color disappears and the metal becomes silvery again (Procedure A). A plant leaf is painted with silver paint, immersed in a blue solution, and connected to a dc power supply. After 20 minutes, the leaf has been coated with copper (Procedure B)†. A graphite-pencil sketch on a piece of acrylic sheet is connected to a power supply and dipped into a blue solution. After 30 minutes, a copper copy of the sketch can be lifted from the sheet (Procedure C)†.

MATERIALS FOR PROCEDURE A

105 g copper sulfate pentahydrate, $CuSO_4 \cdot 5H_2O$

ca. 450 mL distilled water

30 mL 6M sulfuric acid, H_2SO_4 (To prepare 1 liter of solution, set a 2-liter beaker, containing 400 mL of distilled water, in a pan of ice water. While stirring the water, slowly pour 330 mL of concentrated [18M] H_2SO_4 into the beaker. The mixture will become very hot. If all the ice melts, add more. Once the mixture has cooled to room temperature, dilute it to 1.0 liter with distilled water.)

13 mL 0.1M hydrochloric acid, HCl (To prepare 1 liter of solution, slowly pour 8.5 mL of concentrated [12M] HCl into 600 mL of distilled water and dilute the resulting solution to 1.0 liter.)

2 600-mL beakers

filter funnel, with filter paper

silver (or nickel) strip, ca. 5 cm × 18 cm × 1 mm thick

emery cloth or fine steel wool (if nickel strip is used)

copper strip, ca. 5 cm × 18 cm × 1 mm thick

#4 rubber stopper

1.5-volt D-cell battery, with holder, or 1.5-volt dc power supply

2 50-cm insulated wire leads, ca. 18 gauge, with alligator clips on both ends

† Procedures B and C were developed by Floyd Sturtevant and Kenneth Hartman of Ames Senior High School, Ames, Iowa.

MATERIALS FOR PROCEDURE B

840 g copper sulfate pentahydrate, $CuSO_4 \cdot 5H_2O$

ca. 3.75 liters distilled water

240 mL 6M sulfuric acid, H_2SO_4 (For preparation, see Materials for Procedure A.)

10.0 mL 1M hydrochloric acid, HCl (To prepare 1 liter of solution, slowly pour 85 mL of concentrated [12M] HCl into 600 mL of distilled water and dilute the resulting solution to 1.0 liter.)

20.0 mL commercial copper bath brightener†

or

0.4 g thiourea, $SC(NH_2)_2$

0.04 g dextrin, $(C_6H_{10}O_5)_n \cdot xH_2O$

ca. 50 mL acetone, CH_3COCH_3, in a wash bottle

1-gallon clear or translucent plastic container with air-tight lid, ca. 18 cm × 18 cm × 21 cm deep (e.g., Rubbermaid no. 71691 39221)

sheet of ¼-inch clear acrylic, ca. 4 cm wider than the mouth of the plastic container (e.g., 22 cm × 22 cm for the Rubbermaid container)

marking pen for acrylic sheet

straightedge

scrap wood, larger than acrylic sheet

drill, with 3⁄16-inch, ⅜-inch, and ¾-inch bits for acrylic sheet

saber or jig saw to cut acrylic sheet

2 copper electroplating electrodes, with phosphorus content of 0.02–0.08%, 15.2 cm × 7.6 cm × 1.3 cm‡

3⁄32-inch drill bit for copper

1 red banana binding post (e.g., Newark Electronics catalog no. 35N844, type 3-876-R§)

2 20-cm insulated solid copper wires, 14 gauge

wire strippers

sheet of ¼-inch acrylic, ca. 17 cm × 7.5 cm

copper alligator clip, with insulating sleeve that accepts a banana plug (e.g., Newark Electronics catalog no. 28F506)

† Small quantities of bright acid copper bath brightener may be obtained from a commercial electroplating shop. It may also be obtained from OMI International Headquarters, 21441 Hoover Road, Warren, MI 48089; catalog no. UBAC-1; and from the Hach Company, P.O. Box 907, 100 Dayton Avenue, Ames, Iowa 50010; catalog no. 23452-49.

‡ These are available from the Hach Company, catalog no. 23447-02. Quantities of at least 100 pounds are available from Talco Anodes, 5201 Unrich Avenue, Philadelphia, Pennsylvania 19135.

§ Newark Electronics has branch offices throughout the country. Local offices may be located by calling (312) 784–5100.

ca. 10 cm masking tape

magnetic stirrer, with 6-cm stir bar

or

45-cm length of stiff plastic tubing, with outside diameter of ¼ inch

pliers

1⁄16-inch drill bit

1-holed #2 rubber stopper

aquarium air pump

1-m flexible aquarium air tubing, with inside diameter of ca. ⅛ inch

4-liter graduated beaker

filter funnel, with filter paper

stirring rod

plant leaf with sturdy stem, having a total surface area of ca. 80 cm^2

15-cm bare solid copper wire, 16 gauge

50-cm bare solid copper wire, 22 gauge

10 mL silver conductive paint†

or copper conductive paint

or graphite conductive paint

small artist's paint brush

adjustable dc power supply capable of supplying a current of 1.5 amperes at 10 volts

ammeter, with a range of 0–2 ampere

3 1-m insulated wire leads, ca. 18 gauge

MATERIALS FOR PROCEDURE C

840 g copper sulfate pentahydrate, $CuSO_4 \cdot 5H_2O$

ca. 3.75 liters distilled water

240 mL 6M sulfuric acid, H_2SO_4 (For preparation, see Materials for Procedure A.)

10.0 mL 1M hydrochloric acid, HCl (For preparation, see Materials for Procedure B.)

20.0 mL commercial copper bath brightener‡

or

† Available from the Hach Company, catalog no. 23453-40. A substitute can be prepared by mixing silver powder with clear fingernail polish diluted 50% with nail-polish remover.

‡ Small quantities of bright acid copper bath brightener may be obtained from a commercial electroplating shop. It may also be obtained from OMI International Headquarters, catalog no. UBAC-1; and from the Hach Company, catalog no. 23452-49.

0.4 g thiourea, $SC(NH_2)_2$

0.04 g dextrin, $(C_6H_{10}O_5)_n \cdot xH_2O$

1-gallon clear or translucent plastic container with air-tight lid, ca. 18 cm × 18 cm × 21 cm deep (e.g., Rubbermaid no. 71691 39221)

sheet of ¼-inch clear acrylic, ca. 4 cm wider than the mouth of the plastic container (e.g., 22 cm × 22 cm for the Rubbermaid container)

marking pen for acrylic sheet

straightedge

scrap wood, larger than acrylic sheet

drill, with 3⁄16-inch and ¾-inch bits for acrylic sheet

saber saw to cut acrylic sheet

2 copper electroplating electrodes, with a phosphorus content of 0.02–0.08%, 15.2 cm × 7.6 cm × 1.3 cm†

3⁄32-inch drill bit for copper

1 red banana binding post (e.g., Newark Electronics catalog no. 35N844 type 3-876-R‡)

2 20-cm insulated solid copper wires, 14 gauge

wire strippers

sheet of ¼-inch acrylic, ca. 17 cm × 7.5 cm

copper alligator clip, with insulating sleeve that accepts a banana plug (e.g., Newark Electronics catalog no. 28F506)

ca. 10 cm masking tape

magnetic stirrer, with 6-cm stir bar

or

45-cm length of stiff plastic tubing, with outside diameter of ¼ inch

pliers

1⁄16-inch drill bit

1-holed #2 rubber stopper

aquarium air pump

1 m flexible aquarium air tubing, with inside diameter of ca. ⅛ inch

4-liter graduated beaker

filter funnel, with filter paper

stirring rod

sheet of 1⁄16- to ¼-inch acrylic, 10 cm × 15 cm

emery cloth or sandpaper

† Available from the Hach Company, catalog no. 23447-02.

‡ Newark Electronics has branch offices throughout the country. Local offices may be located by calling (612) 784–5100.

#1 graphite pencil

10 mL silver conductive paint†

or copper conductive paint

or graphite conductive paint

small artist's paint brush

adjustable dc power supply capable of supplying a current of 1.5 amperes at 10 volts

ammeter, with a range of 0–2 amperes

3 1-m insulated wire leads, ca. 18 gauge

single-edged razor blade

PROCEDURE A

Preparation

Dissolve 105 g of $CuSO_4 \cdot 5H_2O$ in 375 mL of distilled water in a 600 mL beaker. Add 30 mL of 6M H_2SO_4 and 13 mL of 0.1M HCl. Add enough distilled water to bring the volume of the solution to 500 mL. Filter the blue solution into another 600 mL beaker.

If a nickel strip is used, it should be cleaned by rubbing it with emery cloth or steel wool.

Presentation

Place both the copper strip and the silver (or nickel) strip in the beaker containing the blue $CuSO_4$ solution. Drop a rubber stopper between the strips to prevent the metal strips from touching. After about 30 seconds, remove each strip from the solution to show that no change has taken place. Return the strips to the solution. Use a wire lead with alligator clips to connect the copper strip to the positive terminal of the 1.5-volt battery. Similarly, connect the silver strip to the negative terminal. After 10 seconds disconnect the metal strips, remove them from the beaker, and note that the silver strip is now copper colored from a coating of copper metal.

Return the metal strips to the beaker and connect them to the battery in the opposite direction: copper to negative and silver to positive. After 15 seconds, disconnect the battery and examine the strips. The copper coating on the silver strip has been removed.

PROCEDURE B [*1, 2*]

Preparation

Fashion a cover for the electroplating vat as follows. Cover the 1-gallon container with the acrylic sheet. Use the marking pen to mark the locations of openings to be cut

† Available from the Hach Company, catalog no. 23453-40.

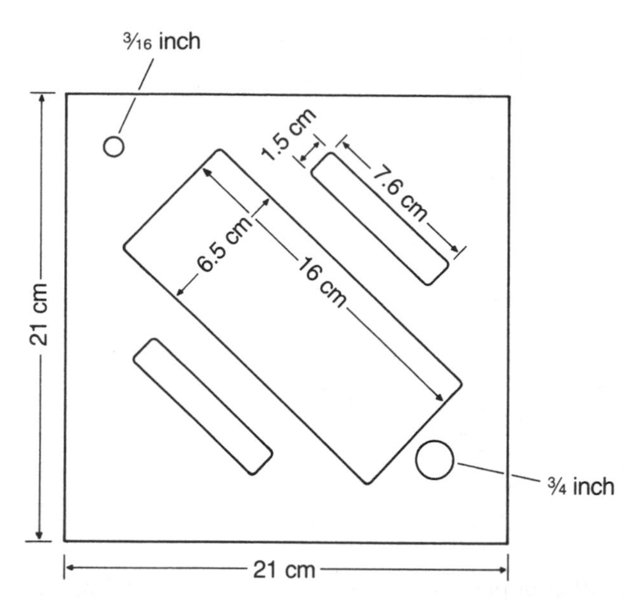

Figure 1. Cover for copper plating vat, cut from ¼-inch acrylic sheet.

through the sheet (Figure 1). These openings must be located over the mouth of the container when the sheet is centered over it. (The sizes of the rectangular holes indicated in Figure 1 are suggestions only. The smaller rectangular holes are openings in which the copper electrodes will be mounted and should be just large enough for the electrodes to pass through. The large rectangular opening at the center will accommodate the object to be plated and should be large enough for it to easily pass through.) After the locations of the openings have been marked, place the acrylic sheet on a piece of scrap wood and drill the 3⁄16-inch and ¾-inch holes. To make the rectangular holes, first drill a hole large enough to accept the blade of the saber saw within each of the three rectangular markings, and then use the saber saw to cut the final holes.

Drill a 3⁄32-inch hole through the largest face of each of the copper electrodes. The hole should be centered on a line about 4 mm from the edge of the electrode (Figure 2).

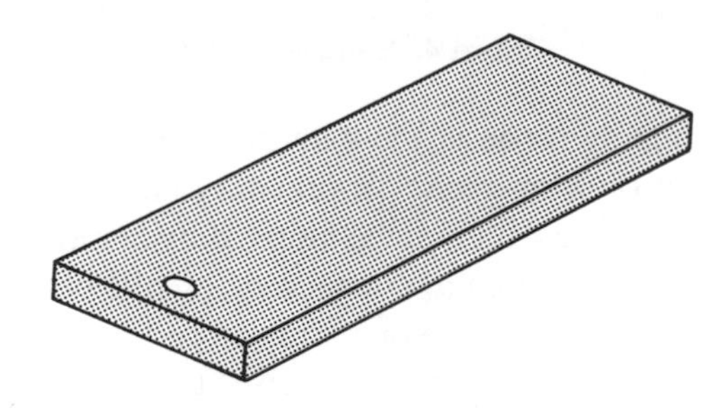

Figure 2. Copper anode with hole drilled near edge.

Place the prepared acrylic sheet on top of the plastic vat. Fasten the red banana binding post in the 3⁄16-inch hole of the acrylic sheet. Strip about 3 cm of the insulation from both ends of a 14-gauge solid copper wire. Insert one end through the hole in the copper electrode. Insert the copper electrode through one of the smaller rectangular holes until it is supported by the copper wire resting on the acrylic sheet. Bend the wire around so its free end can be attached to the binding post (Figure 3). In the same fash-

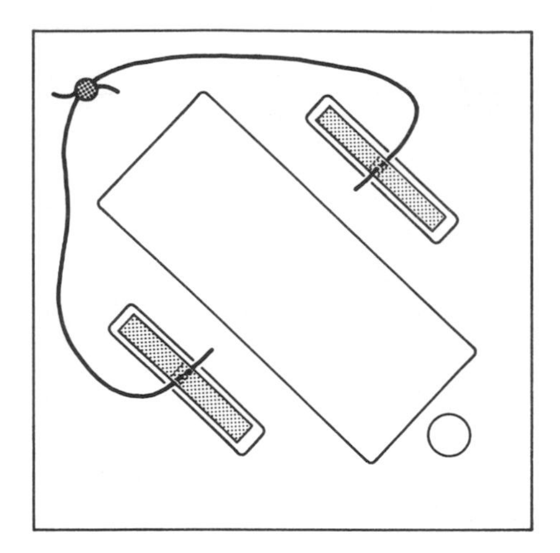

Figure 3. Copper anodes suspended with wires on vat cover.

ion, suspend the other copper electrode in the second small rectangular hole and connect the wire to the binding post.

Fashion a sample holder from a rectangle of acrylic sheet about 1 cm larger in each direction than the large rectangular opening in the acrylic vat cover (Figure 4). Drill a ⅜-inch hole in the center of this sheet. Insert the insulating sleeve of a copper alligator clip in the hole. The fit should be snug. If it is not, wrap masking tape around the sleeve before inserting it in the hole.

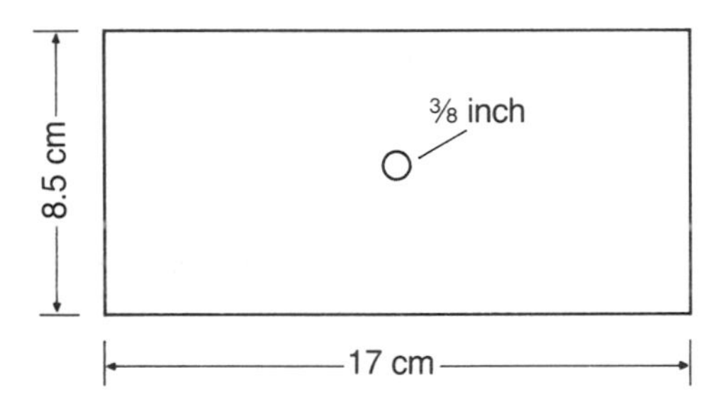

Figure 4. Sample holder.

During the electroplating process, the bath must be agitated. This can be accomplished by setting the vat on a magnetic stirrer and placing a large magnetic stir bar inside the vat. Alternatively, the bath may be agitated by aerating it. If aeration is used, prepare the aerator as follows. Soften a 45-cm piece of stiff plastic tubing in hot water. Make a 90-degree bend in the tubing, so one arm can rest diagonally (i.e., catercorner) across the bottom of the vat while the other arm rises up through the ¾-inch round hole in the acrylic cover. Seal the tube at the bottom end by softening the end in boiling water and squeezing it with pliers (Figure 5). Drill 1⁄16-inch holes every 2.5 cm along the portion of the tube that rests on the bottom of the vat. Insert the open end of the tube through the hole in a #2 rubber stopper and mount the aerator assembly in the ¾-inch round hole in the plastic top. Use a piece of flexible aquarium tubing to connect the air pump to the aerator tube.

Prepare the copper-plating solution. Dissolve 840 g of $CuSO_4 \cdot 5H_2O$ in 3.0 liters of distilled water in a graduated 4-liter beaker. Add 240 mL of 6M H_2SO_4 and 10.0 mL of 1M HCl. Add enough distilled water to bring the volume of the solution to 4.0 liters.

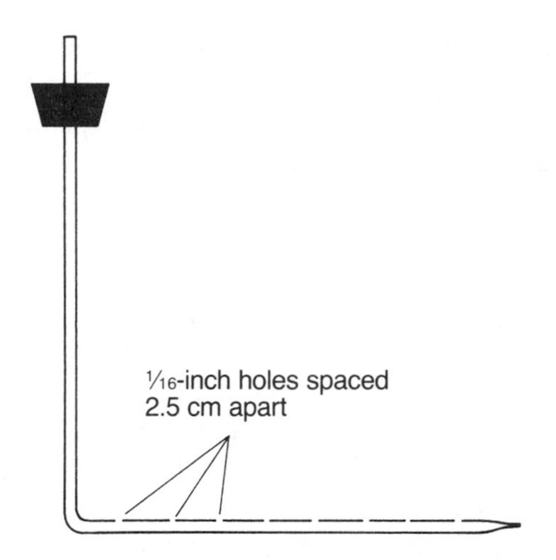

Figure 5. Aerating tube for copper-plating vat.

Filter the solution into the electroplating vat. Add 20 mL of commercial copper bath brightener (or 0.4 g of thiourea and 0.04 g of dextrin [*1*]) and stir the mixture. With the marking pen, indicate the level of the solution on the outside of the vat. Tightly seal the vat until ready for use.

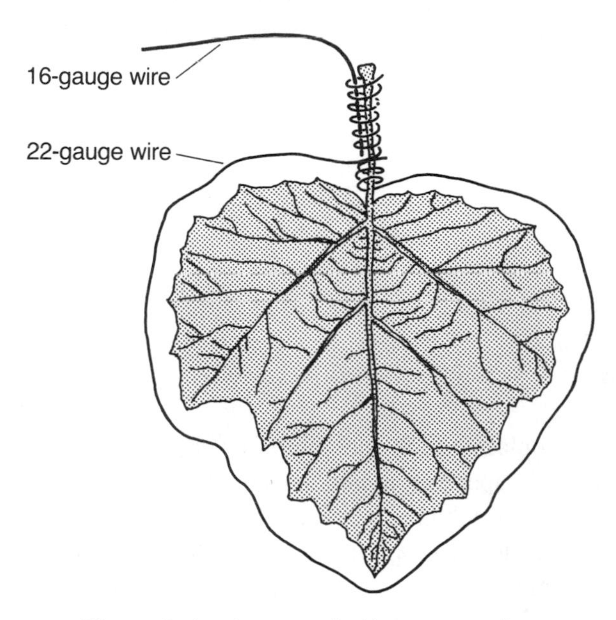

Figure 6. Leaf surrounded by copper wire.

Prepare the leaf to be plated with copper. From a living plant, select a leaf with a combined front and back area of about 80 cm^2. Attach a 15-cm length of bare 16-gauge solid copper wire to the leaf stem by placing one end of it alongside the stem and then wrapping one end of the 50-cm length of 22-gauge copper wire several times around both the 16-gauge wire and the stem (Figure 6). After the stem is attached, outline the leaf with the 22-gauge wire, staying 5–10 mm away from the edge of the leaf. (This wire outline should not touch the leaf.) Secure the free end of the 22-gauge wire by wrapping it, too, around the stem of the leaf. Using a small brush, paint the entire surface of the leaf, including the stem, with silver conductive paint. Thoroughly coat with paint the place where the leaf stem and copper wires join. After the paint dries, examine the silver coating for gaps. If any area of the leaf is exposed, give it additional coats to cover the leaf completely. After any additional paint is dry, the leaf with its copper wire can be stored between the pages of a discarded book until needed.

Presentation

Clip the 16-gauge copper wire, which extends from the stem of the painted leaf, to the sample holder so that when the sample holder is placed on the cover of the electrolysis cell, the leaf and its surrounding wire will be completely immersed in the electroplating solution, centered between and parallel to the two copper anodes.

With the power supply turned off, use an insulated wire lead to connect its positive terminal (red) to the positive (red) terminal of an ammeter. Use a second insulated wire lead to connect the negative (black) terminal of the ammeter to the red binding post on the cover of the electroplating tank. (This post is the one connected to the two copper anodes.) Connect a third insulated wire lead between the negative terminal of the power supply (black) and the terminal (clip) of the sample holder. If a magnetic stirrer is used, turn it on. If an aerator is used, turn on the aquarium pump so that air bubbles through the solution.

Turn on the power supply and adjust its voltage so the current is between 1.0 and 1.5 amperes. Although a higher current will deposit the copper coating more quickly, the most uniform coating can be obtained with a current of about 0.016 ampere per cm^2 of cathode (leaf) area. After about 30 minutes, turn off the power supply and remove the leaf from the vat by lifting the sample holder. Rinse the leaf first with tap water, then with distilled water, and finally with acetone. Allow the plated leaf to dry in the air for several minutes.

PROCEDURE C [*3*]

Preparation

Prepare the electroplating tank and solution as described in Procedure B.

Use a piece of sand paper to roughen the smooth surface of a piece of 10-cm × 15-cm acrylic sheet. On the roughened surface, with a #1 lead pencil, sketch a design in which all lines are connected. (Any isolated lines will not be coated with copper.) Draw a heavy pencil line from the design to the center of the shorter edge of the acrylic sheet. Use the small brush and silver conducting paint to cover this line. (After plating, this line will be removed.)

Presentation

Clip the acrylic sheet with the pencil sketch to the sample holder so the clip makes contact with the silver paint. Place the sample holder on the acrylic cover of the electroplating vat so that the pencil sketch is completely immersed in the solution. With the power supply turned off, use an insulated wire lead to connect its positive terminal (red) to the positive (red) side of an ammeter. Use a second insulated wire lead to connect the negative (black) side of the ammeter to the red binding post on the cover of the electroplating tank. (This post is the one connected to the two copper anodes.) Connect a third insulated wire lead between the negative terminal of the power supply (black) and the terminal (clip) of the sample holder. If a magnetic stirrer is used, turn it on. If an aerator is used, turn on the aquarium pump so that air bubbles through the solution. Turn on the power supply and adjust the voltage so that the current is about 0.3 ampere. After about 30 minutes, turn off the power supply and remove the pencil sketch by lifting the sample holder. Note that all the connected pencil lines have been covered with a copper coating.

Return the sketch to the vat, and plate the sketch for several hours. Remove the sketch, and rinse it first with tap water, and then with distilled water. Allow it to dry. Use a single-edged razor blade to lift the copper-coated sketch carefully from the acrylic sheet. The copper sketch may be strengthened by attaching it to the clip on the sample holder, immersing the sketch in the plating bath, and plating it for several more hours.

HAZARDS

Copper compounds can be toxic if taken internally, and dust from copper compounds can irritate mucous membranes.

Sulfuric acid is a strong acid and a powerful dehydrating agent, therefore it must be handled with care. Spills should be neutralized with sodium bicarbonate ($NaHCO_3$), and then wiped up.

The dc power supply is capable of delivering a serious shock. Do not touch any metal parts of the electrolysis apparatus while the power supply is turned on.

DISPOSAL

The copper and silver (or nickel) strips from Procedure A may be reused indefinitely. The copper solution from Procedure A may be stored in a sealed container for repeated presentations of the demonstration. Alternatively, it may be flushed with water down a drain connected to a sanitary sewer. (Consult local authorities for local regulations on disposal of copper sulfate solutions.)

The electrolysis solution used in Procedures B and C may be stored in the sealed vat and reused for repeated presentations of the demonstration. Replace any water lost through evaporation by adding enough distilled water to raise the level of the solution in the vat to its original level, as indicated by the mark drawn on the side of the vat. If the deposited copper is no longer bright, add another 20 mL of commercial brightener (or 0.4 g thiourea and 0.04 g dextrin). If this does not restore the bright plating, add 50 mL of 6% hydrogen peroxide and filter the solution through activated charcoal.

The brush used to apply the silver paint should be cleaned with methyl isobutyl ketone or methyl ethyl ketone.

DISCUSSION

This demonstration shows how a coating of copper can be formed by electrolysis. In Procedure A, copper is deposited on a strip of silver or nickel metal. The copper coating is easily distinguished from the silvery metal. In Procedure B, a leaf from a plant is coated with copper. The leaf is encased in copper, but the surface detail of the leaf is still visible. In Procedure C, a figure drawn in graphite on plastic is plated with copper. The figure can be removed from the plastic support, yielding an object made virtually completely of copper.

In the process of electroplating with copper, the object to be plated is made the cathode in an electrochemical cell. To be an electrode, the object must be electrically conductive. If the object itself is not electrically conductive, such as the leaf of a plant, it must be coated with something that is. In Procedure B, a leaf is made conductive by coating it with silver paint. Silver paint contains silver metal ground to a fine dust and suspended in a thin lacquer base. When the paint dries, it contains a network of tiny silver particles that conduct electricity. In Procedure C, a pattern is drawn with a graphite pencil on a sheet of plastic. The graphite is conductive, but the plastic is not. Therefore, copper deposits only on the graphite, producing a copper version of the pattern.

The electrochemical cell from which copper is deposited contains copper ions in the electrolyte. The copper ions conduct electricity through the solution. They are also reduced at the cathode to form copper metal. To prevent any other reduction at the cathode, copper ions must be the most easily reduced material in the electrolyte. When copper metal is deposited at the cathode, copper ions are removed from the solution. If these are not replaced, the concentration of copper ions will decrease, and the solution will become less conductive. When the solution becomes less conductive it requires a greater voltage to drive current through the cell. Furthermore, when the concentration of copper ions declines, the rate of deposition at the cathode also declines. To maintain the concentration of copper ions in the electrolyte, the anodes in the electroplating cell are made of copper metal. Copper from the anode is oxidized to copper ions, replacing those deposited at the cathode. The anodes used in Procedures B and C are a specially prepared alloy of copper containing phosphorus. This alloy dissolves more uniformly than pure copper. Pure copper tends to form tiny flakes that fall off the electrode and settle to the bottom of the cell, and the fallen copper is wasted.

The method used in Procedure B creates a bright, shiny copper plate on a leaf. This method can be used to produce jewelry. For jewelry, a fairly sturdy coating is required, which means the coating needs to be rather thick. The thickness of the coating depends on the current and the duration of the electrolysis. A current of 1 ampere deposits approximately 1.5 g of copper per hour. To copper plate a leaf with a surface area of 80 cm^2 so that it is suitable for jewelry requires between 6 and 15 hours. It is possible to start the electroplating in class one day and stop it in class the next. Gaps in the copper coating can sometimes be filled. Rinse the plated leaf with water, dry it, and cover the gap with silver paint. Then, electroplate the leaf again.

Electroplating is as much an art as a science. The effect on the plated product of the components of the electrolyte is frequently determined empirically. The manner in which the effect is produced is often not understood fully. The function of many of the components in the copper-plating bath can be described [*4*]. The copper sulfate is the immediate source of the copper which is deposited at the cathode. It also serves as an electrolyte, reducing the electric resistance of the solution. The copper anodes are the ultimate source of copper for the plating. While copper is deposited at the cathode, an equal amount dissolves from the copper anodes to replace it in the solution. Sulfuric acid in the solution is an electrolyte that reduces the electric resistance of the solution. In addition, it reduces anode polarization, prevents the precipitation of basic copper salts, and improves the "throwing power" of the plating solution. "Throwing power" is an expression of the uniformity of the plate formed by the solution. Solutions with a high throwing power produce a copper plate of uniform thickness. Hydrochloric acid in the solution provides chloride ions, which reduce both anode polarization and streaking in the deposits at high current.

REFERENCES

1. F. Sturtevant, *The Science Teacher* 41(7):42 (1974).
2. Floyd Sturtevant, Ames Senior High School, Ames, Iowa 50010, personal communication (1989).
3. R. R. Coleman, *Graphite Sketches Electroformed,* San Jose State University, P.O. Box 700, Angels Camp, California 95222.
4. *Metal Finishing and Guidebook Directory '86,* Metals and Plastics Publications, Inc.: Hackensack, New Jersey (1986).

11.24

Forming a Copper Mirror

A small amount of colorless liquid is poured into a beaker containing a deep blue solution, and this mixture is poured into a test tube immersed in a hot-water bath. After several minutes, the interior of the test tube becomes coated with a copper-colored mirror [*1*].

MATERIALS

ca. 2 liters distilled water

75 mL concentrated (16M) nitric acid, HNO_3

75 mL 6M sodium hydroxide, NaOH (To prepare 1 liter of solution, set a 2-liter beaker containing 600 mL of distilled water in a pan of ice water. While stirring the water, add 260 g of NaOH to the beaker. The mixture will become very hot. If all the ice melts, add more. Once the solution has cooled to room temperature, dilute it to 1.0 liter with distilled water.)

5 g copper acetate monohydrate, $Cu(C_2H_3O_2)_2 \cdot H_2O$

ca. 10 mL concentrated (15M) aqueous ammonia, NH_3

3 mL 35% aqueous hydrazine, N_2H_4

1-liter beaker

hot plate

thermometer, $-10°C$ to $+110°C$

gloves, rubber or plastic

test tube, 25 mm × 200 mm, with stopper

test-tube holder

500-mL beaker

250-mL beaker

stirring rod

dropper

PROCEDURE

Preparation

Fill the 1-liter beaker with water, place it on the hot plate, and heat the water to 90°C. Wear gloves whenever handling the test tube in the steps described in the remainder of this paragraph. Fill the 25-mm × 200-mm test tube with concentrated

HNO_3 and place it in the hot-water bath. Heat the acid-filled tube for 15 minutes. Remove the tube from the beaker and carefully pour the hot, concentrated HNO_3 into a 500-mL beaker. (**Be very careful not to spill any acid, because hot, concentrated HNO_3 is extremely caustic and a potent oxidizing agent.**) Rinse the tube several times with distilled water. Fill the tube with 6M NaOH solution and return the tube to the beaker of hot water. Heat the tube for 15 minutes. Remove the tube from the beaker and flush the hot 6M NaOH down the drain with flowing water. (**Be very careful handling the hot 6M NaOH; it is extremely caustic.**) Rinse the test tube several times with distilled water. Fill the tube with distilled water. Do not allow the interior of the tube to dry. If the tube is not to be used within several hours, it should be stoppered. The 90°C water bath will be needed for the presentation of the demonstration.

In the 250-mL beaker, prepare a copper(II) acetate solution by dissolving 5 g of $Cu(C_2H_3O_2)_2 \cdot H_2O$ in 50 mL of distilled water. The mixture may be warmed to hasten the dissolving. (This solution is 0.5M in $Cu(C_2H_3O_2)_2$.) While stirring the solution, add drops of 15M NH_3 until the pale blue precipitate that forms initially has just dissolved. (This will require about 8 mL of NH_3.) The solution will be deep blue. The solution should be no warmer than room temperature when it is used.

Presentation

Wearing gloves, slowly pour 3 mL of 35% N_2H_4 into the copper(II) solution in the 250-mL beaker. The mixture may froth when the N_2H_4 is added, and it will be dark brown. Empty the water from the test tube and pour the copper-hydrazine mixture into it. Place the test tube in the hot-water bath. The solution will bubble as it warms. After 3–4 minutes, the sides of the tube will darken. After 10 minutes, a copper mirror will have appeared in the tube. Remove the tube from the water bath, and pour its contents back into the 250-mL beaker. Gently rinse the inside of the tube with water and allow it to dry in the air. (The copper mirror will remain in the tube indefinitely.)

DISPOSAL

The concentrated nitric acid may be used to clean other test tubes. It may be disposed of by diluting it with 200 mL of water, neutralizing it by adding sodium bicarbonate ($NaHCO_3$) until fizzing stops, and flushing this mixture down the drain with water.

Flush the spent copper-hydrazine mixture down the drain with water.

HAZARDS

Concentrated nitric acid is both a strong acid and a powerful oxidizing agent. When it is hot, it is extremely corrosive. Contact with combustible materials can cause fires. Contact with the skin can result in severe burns. The vapor irritates the respiratory system, eyes, and other mucous membranes, and therefore concentrated nitric acid should be handled only in a well-ventilated area.

Concentrated solutions of sodium hydroxide can cause severe burns to the eyes, skin, and mucous membranes. Hot solutions are particularly caustic.

Hydrazine is highly toxic when inhaled or taken internally. It can also be absorbed through the skin. Hydrazine can cause liver and kidney damage. Hydrazine vapor can cause delayed eye irritation. Solutions of hydrazine should not be mixed with bases, because doing so will liberate hydrazine vapor. Hydrazine is a suspected carcinogen.

Copper compounds can be toxic if taken internally, and dust from copper compounds can irritate mucous membranes.

DISCUSSION

In this demonstration a test tube is mirrored with copper. The mirror forms when a thin film of metallic copper deposits on the inner surface of the test tube. The copper metal forms when hydrazine reduces an ammonia complex of copper(II) in basic solution. Hydrazine slowly reduces the complex to copper metal. When the surface of the test tube is very clean and wet, the copper metal will adhere to it, forming a copper-colored mirror.

The overall reaction that occurs in this demonstration can be represented by the equation

$$2\ [Cu(NH_3)_4]^{2+}(aq) + N_2H_4(aq) + 4\ OH^-(aq) \longrightarrow 2\ Cu(s) + N_2(g) + 4\ H_2O(l) + 8\ NH_3(aq)$$

The production of nitrogen gas accounts for the formation of bubbles during the reaction.

Because this process of plating with copper is performed without an external electrochemical circuit, it is called electroless plating. Quite a variety of metal coatings can be produced by electroless plating [*2*]. It is most often used to produce a metal coating on a surface that does not conduct electricity.

Smooth metal surfaces make good mirrors because they reflect most of the light that falls upon them. However, they are not equally reflective for all colors of light [*3*]. Gold, for example, is more reflective of yellow light than of blue light. Therefore, it appears yellow. Copper is more reflective of red than of other colors. Therefore, it has a reddish hue.

The copper coating on the interior of the test tube may be protected from oxidation and mechanical stress by giving it a protective coating. Either clear varnish or paint can be used to coat it. Paint will hide the back surface of the metal, but clear varnish will allow it to be seen. (Black paint, by blocking all light from the interior of the tube, will make the mirror appear more uniform.) In either case, the coating may be applied by pouring a small amount of water-based paint or varnish into the tube, turning the tube to coat its entire interior surface, pouring out the excess, and allowing the coating to dry.

The formation of a copper mirror by this procedure seems to require copper acetate as the copper source. Other salts of copper, such as the sulfate, nitrate, and chloride salts, do not work. If the acetate salt is not at hand, it can be easily prepared from these other salts by precipitating copper carbonate from a solution, filtering the solid, dissolving the solid in acetic acid, and evaporating the resulting solution to dryness. In addition, the hydrazine reducing agent must be used in its neutral form, not as the more stable and commonly available sulfate salt. The solution of the sulfate salt can be converted to a solution of the neutral form by precipitating the sulfate with a stoichiometric amount of barium hydroxide solution and separating the barium sulfate precipitate from the solution by filtration.

REFERENCES

1. J. R. Partington, *General and Inorganic Chemistry,* 4th ed., St. Martin's Press: New York (1966).
2. J. McDermott, *Electroless Plating,* Noyes Data Corporation: Park Ridge, New Jersey (1972).
3. L. Holland, *Vacuum Deposition of Thin Films,* Chapman and Hall: London (1966).

11.25

Nickel Plating: Shiny Nickel Leaves

A copper object and a silvery metal strip are connected to a dc power supply and immersed in a green solution. After several minutes, the copper object has become coated with a silvery metal.†

MATERIALS

either

20 mL commercial brightener for nickel plating solution

247 g nickel sulfate hexahydrate, $NiSO_4 \cdot 6H_2O$

33 g nickel chloride hexahydrate, $NiCl_2 \cdot 6H_2O$

or

225 g nickel sulfamate tetrahydrate, $Ni(H_2NSO_3)_2 \cdot 4H_2O$

11.3 g nickel chloride hexahydrate, $NiCl_2 \cdot 6H_2O$

28 g boric acid, H_3BO_3

ca. 750 mL distilled water

1-liter tall-form beaker

sheet of ¼-inch acrylic, 12 cm square

marking pen

30-cm straightedge

scrap of wood, ca. 12 cm × 12 cm × 1 cm

drill, with ⅜-inch bit

3 copper alligator clips (1 black, 2 red), with insulating sleeves that accept banana plugs (e.g., Newark Electronics catalog no. 28F506‡)

ca. 10 cm masking tape

15-cm insulated copper wire, 18 gauge

wire strippers

1-liter beaker

filter funnel, with paper

† This demonstration was developed by Floyd Sturtevant of Ames Senior High School, Ames, Iowa, and we thank him for providing information about the procedure.

‡ Newark Electronics has branch offices throughout the country. Local offices may be located by calling (312) 784–5100.

magnetic stir bar

piece of plastic wrap, ca. 30 cm square

adjustable dc power supply capable of supplying a current of 1.5 amperes at 10 volts

ammeter, with a range of 0–2 amperes

3 1-m insulated wire leads, ca. 18 gauge

stirring hot plate

thermometer, −10°C to +110°C

2 nickel strips, ca. 5 cm × 15 cm × ca. 0.2 mm thick

piece of copper, at most 10 cm × 10 cm × 1 cm (A suitable copper object is a copper-plated leaf prepared as described in Procedure B of Demonstration 11.23.)

PROCEDURE [*1*]

Preparation

Fashion a cover for the electroplating cell as follows. Cover the 1-liter tall-form beaker with the 12-cm-square sheet of ¼-inch acrylic. Use a marking pen and straightedge to lay out the pattern for the holes to be drilled in the acrylic sheet (see the figure). Be sure that the markings for the holes are located over the mouth of the beaker. Place a piece of scrap wood under the acrylic, and drill the three ⅜-inch holes along one diagonal. Insert the plastic sleeve of the black copper alligator clip in the center hole. If the fit is not tight, wrap the sleeve with a piece of masking tape to tighten it. In the same fashion, fit the two red alligator clips into the remaining ⅜-inch holes. Strip 2 cm of the insulation from both ends of the 15-cm piece of wire and attach one end to each of the red alligator clips.

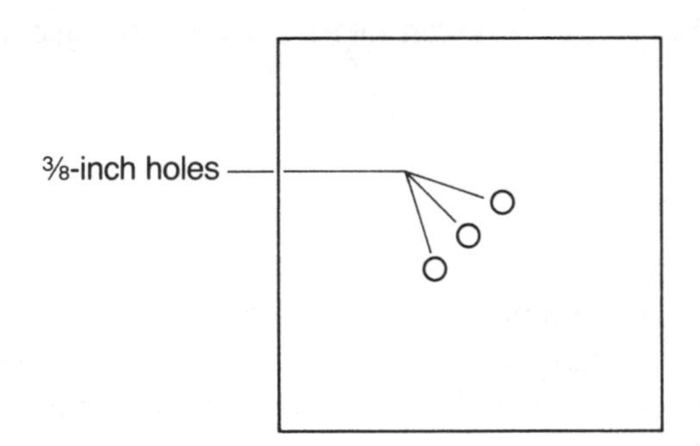

If commercial brightener is available, prepare the plating solution as follows. Dissolve 247 g of $NiSO_4{\cdot}6H_2O$, 33 g of $NiCl_2{\cdot}6H_2O$, and 28 g of H_3BO_3 in 500 mL of distilled water in a 1-liter beaker. Add enough distilled water to bring the volume of the solution to 750 mL. Filter the solution into the tall-form 1-liter beaker. Add 20 mL of commercial brightener. Place the magnetic stir bar in the beaker. Mark the level of the liquid on the outside of the beaker. Seal the beaker with plastic wrap until the demonstration is to be presented.

If the commercial brightener is not available, prepare the plating solution as follows. Dissolve 225 g of $Ni(H_2NSO_3)_2 \cdot 4H_2O$, 11.3 g of $NiCl_2 \cdot 6H_2O$, and 28 g of boric acid in 500 mL of distilled water in a 1-liter beaker. Add enough distilled water to bring the volume of the solution to 750 mL. Filter the solution into the tall-form 1-liter beaker. Place the magnetic stir bar in the beaker. Mark the level of the liquid on the outside of the beaker. Seal the beaker with plastic wrap until the demonstration is to be presented.

With the power supply turned off, use an insulated wire lead to connect its positive terminal (red) to the positive (red) terminal of an ammeter. Use a second insulated wire lead to connect the negative (black) terminal of the ammeter to one of the red alligator clips in the cell cover. Connect a third insulated wire lead between the negative terminal of the power supply (black) and the terminal (black clip) in the cell cover.

Place the tall-form beaker containing the plating solution on the stirring hot plate and turn on the stirrer. Heat the solution to a temperature between 50°C and 60°C.

Presentation

Attach a nickel strip to each of the outer, red alligator clips of the acrylic cover assembly. Attach the copper object to be plated to the center, black alligator clip. Adjust all three electrodes so they do not touch each other.

Rest the acrylic cover assembly on top of the beaker of nickel-plating solution, so that the electrodes are immersed in it. Turn on the power supply and adjust its voltage to produce a current between 0.5 and 1.0 ampere. After 1 minute, turn off the power supply, and examine the copper object. It will be coated with a silvery metal coating. Return the object to the solution and resume the electroplating. After 10 minutes, remove the plated object and rinse it with distilled water. The nickel coating will have made the object much stiffer.

HAZARDS

Nickel compounds are toxic if taken internally and are suspected carcinogens.

DISPOSAL

The nickel plating solution should be stored in a sealed glass container for repeated presentations of the demonstration. The concentration of nickel in the plating solution should remain constant, because the nickel anodes dissolve at the same rate that nickel is deposited at the cathode. The concentration may change as a result of evaporation of water from the solution. To compensate for evaporation, add enough distilled water to return the level of the solution to the initial mark on the tank. When the plating is no longer bright, add another 20 mL of brightener. If this does not return the bright plating, add 50 mL of 3% hydrogen peroxide to the plating solution, and filter it through activated charcoal.

To dispose of the nickel plating solution, add 800 g of $Na_2S \cdot 9H_2O$ to the plating solution and stir it periodically over a period of several hours. Filter the precipitated nickel sulfide from the mixture. Dispose of the filtered solid in a landfill designated for heavy metal wastes. (Consult local regulations regarding such disposals.) Flush the filtrate down the drain with water.

DISCUSSION

This demonstration shows how a copper object can be electrochemically plated with nickel to produce a bright, shiny, silvery coating. If the copper-plated object is somewhat flexible, it will become noticeably stiffer with the nickel plating.

An electrically conductive object is plated with nickel by making it the cathode in a cell containing nickel ions. The nickel ions are reduced to nickel metal on the cathode.

$$Ni^{2+}(aq) + 2\,e^- \longrightarrow Ni(s)$$

This process removes nickel ions from the electrolyte in the solution. To keep the concentration of nickel ions constant, a nickel anode is used in the cell. The anode itself provides nickel atoms, which are oxidized to nickel ions.

$$Ni(s) \longrightarrow Ni^{2+}(aq) + 2\,e^-$$

The fundamental components involved in nickel plating are a bath containing nickel ions as a source of nickel, a nickel anode to replenish the nickel ions, and an electrically conductive cathode on which the plate is formed.

The nickel electroplating bath contains several solutes. Each has an effect on the electroplating process [2]. Nickel sulfate is used as the primary source of nickel ions, because it is the least expensive salt of nickel with an anion that is not oxidized at the anode or reduced at the cathode. It is also very soluble, so a high concentration of the salt in the bath can be achieved. A high concentration of ions in the bath reduces its electric resistance. A low resistance allows a high current to flow through the solution, so the nickel plate can form quickly. A high concentration of nickel ions in the solution also improves the uniformity of the plating on the cathode. Nickel chloride is added to the bath to improve the dissolution of the anode. It also reduces the potential required to reduce nickel ions at the cathode. Boric acid promotes a smooth plate. Without it, the plate would be rough and pitted. It serves as a buffer to control the pH around the cathode. At the cathode, water may be reduced, forming hydroxide ions. These hydroxide ions can precipitate nickel ions from the bath in the form of nickel hydroxide. Boric acid keeps the pH of the bath low enough to prevent this precipitation.

Nickel is one of the most commonly used electroplates. Its hardness makes it desirable as a base for other plating, such as chromium. Nickel plating is used on steel, brass, zinc, copper, metallized plastics, and on aluminum and magnesium alloys. The nickel plate forms a hard, abrasion-resistant, and corrosion-protective finish on these materials. A thin coating of chromium may be applied to provide a permanently nontarnishing surface. Because nickel is hard, heavy nickel plating is also used on mechanical components to reduce the effects of wear.

REFERENCES

1. F. Sturtevant, *The Science Teacher* 41(7):42 (1974).
2. H. Brown and B. B. Knapp, "Nickel," in *Modern Electroplating*, 3d ed., ed. F. A. Lowenheim, pp. 287–341, John Wiley and Sons: New York (1974).

11.26

Chromium Plating

A lead strip and a nickel-plated strip are dipped into an orange solution, and both are connected to the terminals of an automobile battery. After about 1 minute, the nickel-plated strip is removed from the solution, and it has been coated with chromium. Buffing with a soft cloth makes the plate shiny [*1*].

MATERIALS

2.0 mL concentrated (18M) sulfuric acid, H_2SO_4

250 mL distilled water

63 g chromium(VI) oxide, CrO_3

100 g sodium bisulfite, $NaHSO_3$ (See Disposal section for use.)

250 g sodium sulfide nonahydrate, $Na_2S \cdot 9H_2O$ (See Disposal section for use.)

400-mL tall-form beaker

hot plate

thermometer, $-10°C$ to $+110°C$

lead strip, ca. 15 cm × 2 cm × 1 mm

light-duty automotive jumper cables

nickel strip, ca. 15 cm × 5 cm × 0.1 mm, or nickel-plated object of similar dimensions (See Demonstration 11.25 for preparation of nickel-plated copper.)

12-volt automotive battery, or dc power supply capable of delivering 2 amperes at 12 volts

soft cotton cloth, ca. 15 cm square

PROCEDURE

Preparation

Under a fume hood, pour 2.0 mL of concentrated H_2SO_4 into 250 mL of distilled water in a 400-mL tall-form beaker. Dissolve 63 g of CrO_3 in the acid solution.

Set the beaker on a hot plate, and heat the solution to 50°C.

Presentation

Place the lead strip in the beaker of warm solution, and clip it to the side of the beaker with the red clip of the jumper cable. Similarly fasten the nickel strip inside the beaker with the black clip. Be sure the nickel strip does not touch the lead strip. Attach the free red clip of the jumper cable to the positive terminal of the automobile battery. Attach the free black clip to the negative terminal of the battery. Bubbles will form at both electrodes, causing some spattering and mist formation. **Caution! Avoid inhaling the mist or coming in contact with spattered liquid.**

After 5 minutes detach the clips from the battery and from the strips. Remove the metal strips from the solution and rinse them with water. The area of the nickel strip that was in the solution now has a shinier surface than it did originally. Buff the surface of the chromium-plated nickel strip with a soft cloth to increase the shine of the new coating.

HAZARDS

Chromium(VI) oxide is highly corrosive and toxic; care should be taken to avoid coming in contact with spattered liquid or the mist produced near the electrodes. Chromium(VI) oxide is also a powerful oxidizing agent; avoid contact with reducing agents and organic materials. Compounds of chromium(VI) are also suspected carcinogens.

Sulfuric acid is a strong acid and a powerful dehydrating agent, therefore it must be handled with care. Spills should be neutralized with sodium bicarbonate ($NaHCO_3$), and then wiped up.

A 12-volt automotive battery is capable of delivering a severe electric shock. Do not touch both terminals of the battery simultaneously. Do not allow conductive objects connected to the terminals to come in contact with each other. This will produce sparks and could start a fire.

DISPOSAL

The chromium plating solution should be stored in a sealed glass container for repeated presentations of the demonstration. To dispose of the solution, combine 500 mL of crushed ice and 100 g of sodium bisulfite ($NaHSO_3$) in a 1-liter beaker, and in a fume hood, slowly pour the chromium plating solution over this ice-bisulfite mixture. This will reduce the orange chromium(VI) solution to a green chromium(III) solution. Add 250 g sodium sulfide nonahydrate ($Na_2S \cdot 9H_2O$) to the green solution and stir the mixture periodically for an hour. Filter the solid chromium(III) sulfide from the liquid. Dispose of the solid in a landfill specified for heavy metal wastes. Flush the liquid down the drain with water.

DISCUSSION

This demonstration shows the process of chromium plating and indicates several of its features. First, most chromium plating is done over a nickel base, because chromium adheres well to nickel. Second, chromium plating is done at an elevated temperature. Third, the very shiny surface we usually associate with chromium is not produced directly by electrolysis, but is achieved by polishing the electroplated chromium coating.

Chromium plating is performed for two purposes: decorative and industrial. Decorative chromium plating is usually thin, hard, and nontarnishing. Such plating is seen most frequently on plumbing fixtures. It was once quite common on automotive trim, but it has been largely replaced with aluminized plastics, because these are both less expensive and lighter. Industrial chromium plating is much thicker, and it is applied to reduce the friction between moving parts and to increase their resistance to heat and corrosion.

The bath used in this demonstration is a typical formula [1]. Chromium trioxide is a reddish brown material which forms chromic acid (H_2CrO_4) in aqueous solution. The reduction potential of chromium(VI) to metal is +0.4 volt, whereas that of chromium(III) to metal is −0.7 volt and chromium(II) to metal is −0.9 volt. The voltage needed to reduce chromium(VI) to metal is lower than that needed to reduce chromium(II) or chromium(III). For this reason, chromium(VI) is used in the plating bath. The chromium metal is deposited directly from the hexavalent state, rather than from an intermediate, lower-valent state.

Chromium does not plate from a bath containing only chromic acid. Sulfate ions are necessary as a catalyst in plating. The sulfate ion appears to aid in the formation of a film on the cathode, a film in which the chromium metal is deposited from an unknown intermediate.

A lead anode is used in this process, because the lead performs a useful function in the overall process occurring in the cell. The lead anode is oxidized to lead dioxide. This lead dioxide reacts with any chromium(III) that may be formed by incomplete reduction of chromium(VI) at the cathode, oxidizing it back to chromium(VI). Because only chromium(VI) forms the chromium metal plate, the lead anode maximizes the amount of chromium plate that can be obtained from a bath by returning any chromium(III) to chromium(VI). Furthermore, the presence of chromium(III) in the plating bath lowers the conductivity of the bath, and necessitates an increasing voltage to maintain the current [1].

The lead anodes, on standing in the chromium plating solution, may become passive, that is, nonfunctional. This is a result of the formation of lead chromate on the surface of the lead. Lead chromate insulates the electrode from the solution, hindering the passage of current. Therefore, the removal of anodes from the bath between platings is recommended.

The temperature of the plating bath has an effect on the brightness of the chro-

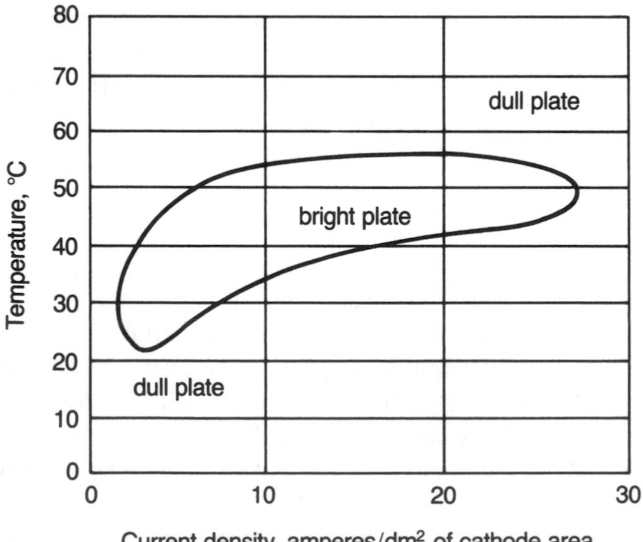

Dependence of chromium plate appearance on temperature and current density [1].

mium plate produced. The figure summarizes the variations in this effect. A temperature of 50°C is used in this demonstration, because this temperature yields the brightest chromium plate under the most variable current densities.

A potential much higher than that needed to cause the reduction of chromium is used to drive a high current, so the plating is more rapid. This higher potential, or overpotential, also generates hydrogen gas at the cathode and oxygen gas at the anode.

REFERENCE

1. F. Lowenheim, Ed., *Modern Electroplating,* 3d ed., pp. 90–107, John Wiley and Sons: New York (1974).

11.27

Silver Plating

A strip of copper and a carbon rod are immersed in a colorless liquid and connected to a battery. After about 15 minutes, the copper has a silver coating [*1, 2*].

MATERIALS

3.2 g silver nitrate, $AgNO_3$

64 g potassium iodide, KI

200 mL distilled water

6 drops concentrated (18M) sulfuric acid, H_2SO_4

ca. 250 mL 1.0M nitric acid, HNO_3 (To prepare 1 liter of solution, carefully pour 63 mL of concentrated [16M] HNO_3 into 600 mL of distilled water, and dilute the resulting solution to 1.0 liter.)

10 g sodium chloride, NaCl (See Disposal section for use.)

2 400-mL beakers

copper strip, ca. 15 cm × 1 cm × 0.8 mm

ca. 15-cm carbon rod, 5 mm in diameter

pad of fine steel wool

2 30-cm wire leads, with alligator clips on both ends

1.5-volt hobby battery

PROCEDURE

Preparation

Prepare the plating solution by dissolving 3.2 g of $AgNO_3$ and 64 g of KI in 200 mL of distilled water in a 400-mL beaker. Add 6 drops of concentrated H_2SO_4 to this solution.

Clean the copper strip by dipping it in 1.0M HNO_3 for 1 minute, and then scouring it with fine steel wool.

Presentation

Place the carbon rod in the beaker of plating solution. With one of the wire leads, clip the carbon rod to the positive terminal of the 1.5-volt hobby battery. With the other lead, connect the copper strip to the negative terminal of the battery. Place the copper strip in the plating solution, making sure it does not touch the carbon rod.

After 15 minutes, disconnect the battery from the copper strip. Remove the copper strip from the solution, rinse it with water, and examine its surface. Buff the white silver deposit on the copper with fine steel wool to polish the silver.

HAZARDS

Silver nitrate is irritating to the eyes and mucous membranes. If taken internally, silver nitrate can be toxic. Silver nitrate will stain the skin; these stains can be bleached by rinsing with an aqueous solution of sodium thiosulfate ($Na_2S_2O_3$) followed by water.

Sulfuric acid is a strong acid and a powerful dehydrating agent; therefore it must be handled with care. Spills should be neutralized with sodium bicarbonate ($NaHCO_3$), and then wiped up.

DISPOSAL

The silver plating solution should be stored is a sealed glass bottle for repeated presentations of the demonstration. Alternatively, it may be disposed of as follows: Dissolve 10 g of sodium chloride in 250 mL of water and pour this solution into the plating solution. Filter the white precipitated silver chloride from the mixture. Dispose of the filtered solid in a landfill designated for such chemical wastes (consult local regulations regarding the disposal of hazardous wastes), or retain it for recovery of the silver. Flush the filtrate down the drain with water.

DISCUSSION

This demonstration shows the process of electroplating a metal object with silver. The metal object is made the cathode in an electrolytic cell whose electrolyte contains silver(I) ions. These Ag^+ ions are reduced to silver metal, which adheres to the surface of the cathode.

When silver(I) ions are reduced electrolytically from a solution in which the concentration of Ag^+ is high, say, over 0.001M, the silver deposits in long needles rather than in a uniform coating on the cathode. If the concentration of Ag^+ is low, then a uniform plate forms. However, if all the silver(I) in the solution is in the form of Ag^+, the reduction of silver at the cathode will soon consume all of it, and the plating process will stop. To keep the concentration of Ag^+ low but the total concentration of silver(I) in the solution high, a complexing agent is added to the plating solution. In this demonstration, the complexing agent is iodide ions. These form $[AgI_2]^-$ and $[AgI_3]^{2-}$ with the silver(I) in solution. (For information concerning the stability of iodo-silver(I) complexes, see the Discussion section of Demonstration 4.4 in Volume 1 of this series.) Commercial silver-plating solutions use cyanide ions to complex the silver(I) in the form of $[Ag(CN)_2]^-$. These solutions produce a superior silver plate, but have not been used in this demonstration because cyanide compounds are extremely toxic.

Another reason to keep the concentration of free Ag^+ low is to prevent the precipitation of silver through the direct reaction of Ag^+ ions with the metal of the cathode. Silver is one of the noble metals; that is, it is less reactive than most metals. Therefore, when a more reactive metal (most other metals) is placed in a solution containing silver ions, this metal displaces the silver ions from the solution.

$$Ag^+(aq) + M(s) \rightleftarrows Ag(s) + M^+(aq)$$

Keeping the concentration of Ag^+ low assures that the production of silver metal by this reaction is small. Copper is used as the cathode in this demonstration to provide an easily visible contrast between the base metal and the silver coating.

Commercial plating solutions for decorative silver (e.g., flatware) contain brighteners to produce a shiny silver plate. Various brighteners are used. Carbon disulfide and sodium thiosulfate have been used, but the plating produced from these still requires buffing for a full shine. The commercial mirror-bright solutions are proprietary and contain a wide variety of substances. Some of the ingredients used include ketones, selenites, and polyalcohols [*3*]. The manner in which these ingredients produce a bright surface is not completely understood. The electroplaters' trade is as much an art as a science.

The amount of silver plated from solution is described by Faraday's law. The amount of silver plated is proportional to the quantity of electric charge passed through the solution. According to Faraday's law:

$$\text{mass of metal} = \frac{QZ_B A_r}{F}$$

where Q is the charge passed in coulombs,
Z_B is the charge on the metal ion,
A_r is the molar atomic mass of the ion,
and F is Faraday's constant (9.6485×10^4 C/mol).

The total charge passed through the solution can be estimated by:

$$Q = It$$

where I is the average current in amperes,
and t is the time in seconds.

The average current can be measured during the plating process by placing an ammeter in series with the battery, recording the current periodically, and averaging these values.

There are general standards applied to the silver plating of flatware and hollow ware. Flatware is specified by the number of troy ounces of silver that are on a gross (144) of teaspoons of the flatware. Some of the standards are:

Specification	Mass of silver per gross of teaspoons
federal specific plate	9 troy oz. (280 g)
quadruple plate	8 troy oz. (250 g)
triple plate	6 troy oz. (187 g)
double plate	4 troy oz. (124 g)
par plate	2 troy oz. (62 g)

The surface area of a typical teaspoon is about 50 cm^2. The density of pure silver is 10.5 g/cm^3 [*4*]. Therefore, the thickness of the silver plate on a quadruple-plate teaspoon is

$$\left[\frac{250 \text{ g Ag}}{144 \text{ teaspoons}}\right]\left[\frac{1 \text{ teaspoon}}{50 \text{ cm}^2}\right]\left[\frac{1 \text{ cm}^3}{10.5 \text{ g Ag}}\right] = 3 \times 10^{-3} \text{ cm}$$

A similar thickness is plated onto the other tableware items (forks, knives, tablespoons, etc.). Items that receive less use (e.g., serving pieces) are often given a thinner plate.

REFERENCES

1. H. Alyea, *Tested Demonstrations in Chemistry,* 6th ed., p. 193, Journal of Chemical Education: Easton, Pennsylvania (1965).
2. D. K. Alpern and S. Toprek, *Trans. Electrochem. Soc.* 74:321 (1938).
3. B. M. Luce and D. G. Foulke, "Silver," in *Modern Electroplating,* 3d ed., p. 367, John Wiley and Sons: New York (1974).
4. R. C. Weast, Ed., *CRC Handbook of Chemistry and Physics,* 66th ed., p. B-139, CRC Press: Boca Raton, Florida (1985).

11.28

Formation of a Silver Mirror

Two colorless liquids are poured into a large round-bottomed flask, and the flask is stoppered and swirled. Over a period of a few minutes, the flask darkens and gradually becomes reflective as a mirror.

MATERIALS

300 mL 0.10M silver nitrate, $AgNO_3$ (To prepare 1 liter of solution, dissolve 17 g of $AgNO_3$ in 600 mL of distilled water and dilute the resulting solution to 1.0 liter.)

ca. 25 mL concentrated (15M) aqueous ammonia, NH_3

150 mL 0.80M potassium hydroxide, KOH (To prepare 1 liter of solution, dissolve 45 g of KOH in 600 mL of distilled water and dilute the resulting solution to 1.0 liter.)

25 mL concentrated (16M) nitric acid, HNO_3

100 mL 0.5M dextrose, $C_6H_{12}O_6$ (To prepare 1 liter of solution, dissolve 90 g of $C_6H_{12}O_6$ in 600 mL of distilled water, and dilute the resulting solution to 1.0 liter.)

50 mL 1M nitric acid, HNO_3 (See Disposal section for use.) (To prepare 1 liter of solution, carefully pour 63 mL of concentrated [16M] HNO_3 into 600 mL of distilled water, and dilute the resulting solution to 1.0 liter.)

hot water to fill round-bottomed flask

600-mL beaker

stirring rod

dropper

12-liter round-bottomed flask, with rubber stopper (A smaller flask can be used, but the impact of the demonstration will be diminished.)

PROCEDURE

Preparation

This mixture should be prepared no more than 1 hour before use (**Caution! See the Hazards section**). Pour 300 mL of 0.10M $AgNO_3$ into the 600-mL beaker. While stirring the solution, add drops of 15M NH_3, until the brown precipitate which forms initially has just dissolved. Add 150 mL of 0.80M KOH to this mixture. If a precipitate forms again, add drops of 15M NH_3 ammonia until it dissolves.

Pour 25 mL of 16M HNO_3 into the round-bottomed flask and stopper it. Swirl the acid around to dampen the entire interior surface of the flask. Pour the acid from the flask, and flush it down the drain with water. Rinse the flask with water and stopper it.

Within 10 minutes of presenting the demonstration, fill the flask with hot water to warm it.

Presentation

Empty the water from the flask. Pour 100 mL of 0.5M $C_6H_{12}O_6$ into the flask. Add the contents of the beaker, and stopper the flask. Swirl the flask continuously to cover its entire interior surface with a thin coating of the liquid. Within about a minute, the flask will begin to darken as a film of metallic silver forms on its inside surface. Continue to swirl the flask until the entire interior of the flask is covered with a film of silver, and the flask looks like a mirror. Pour the liquid from the flask, and flush it down the drain with water.

HAZARDS

If mixtures of aqueous silver nitrate, ammonia, and potassium hydroxide are heated or allowed to stand for several hours, a highly explosive precipitate (silver nitride) may form [*1*, *2*]. Therefore, the mixture should be freshly prepared for each demonstration. Furthermore, the mixture used to silver the flask should be flushed down the drain with water immediately after it is used.

Concentrated nitric acid is both a strong acid and a powerful oxidizing agent. Contact with the skin can result in severe burns. The vapor irritates the respiratory system, eyes, and other mucous membranes; therefore concentrated nitric acid should be handled only in a well-ventilated area.

Solid potassium hydroxide and its concentrated solutions can cause severe burns to the eyes, skin, and mucous membranes. Dust from solid potassium hydroxide is very irritating to the eyes and respiratory system.

Silver nitrate is irritating to the eyes and mucous membranes. It will stain the skin; these stains can be bleached by rinsing with an aqueous solution of sodium thiosulfate ($Na_2S_2O_3$) followed by water. If taken internally, silver nitrate can be toxic.

Concentrated ammonia can irritate the skin, and its vapors are harmful to the eyes and mucous membranes.

DISPOSAL

The solution used to silver the flask and any unused solutions should be flushed down the drain with water.

After the demonstration, the silver can be removed from the inside of the flask by dissolving it in 50 mL of 1M nitric acid, and the flask may then be reused. Silver nitrate may be recovered from the acid solution by evaporating the acid.

DISCUSSION

The reaction used in this demonstration is the Tollen's test, the reaction used in qualitative organic analysis to identify aldehydes. Usually the Tollen's test is performed in a small test tube. To increase the effectiveness of this demonstration, a large flask is mirrored. The mirror forms when a thin film of metallic silver deposits on the inner surface of the flask. The silver metal is formed by the reduction of silver nitrate by an aldehyde. When an aldehyde is combined with an ammonia complex of silver in a basic aqueous solution, the aldehyde slowly reduces the complex to silver metal. When the surface of the glass is clean and wet, the silver metal adheres to the glass, forming a highly reflective surface.

The reaction which takes place in this demonstration is

$$CH_2OH(CHOH)_4\overset{\overset{\displaystyle O}{\|}}{C}{-}H + 2\,[Ag(NH_3)_2]^+ + 3\,OH^- \longrightarrow$$

$$2\,Ag + CH_2OH(CHOH)_4\overset{\overset{\displaystyle O}{\|}}{C}{-}O^- + 4\,NH_3 + 2\,H_2O$$

The aldehyde functional group (—CHO) of dextrose is oxidized to an acid ($—COO^-$). The silver(I) in the diamminesilver(I) ion is reduced to metallic silver. The metallic silver is deposited on the sides of the reaction vessel in the form of a "mirror." This reaction is characteristic of aldehydes. Other aldehydes, such as benzaldehyde, can also be used to produce a silver mirror. Dextrose is used in this demonstration, because it is soluble in water. A more uniform mirror is produced with a water-soluble aldehyde.

Dextrose is a sugar, that is, a soluble carbohydrate. Carbohydrates have a molecular formula of the form $C_x(H_2O)_n$. For dextrose, $x = 6$ and $n = 6$. Common table sugar is sucrose, a carbohydrate in which $x = 12$ and $n = 11$. Carbohydrates contain either an aldehyde group or a ketone group. Those that contain an aldehyde group produce a positive Tollen's test; that is, they reduce $[Ag(NH_3)_2]^+$ in basic solution to metallic silver. For this reason, they are called reducing sugars. Dextrose is a reducing sugar; sucrose is not.

Because this process of plating a flask with silver is performed without an external electrochemical circuit, it is called electroless plating. Quite a variety of metal coatings can be produced by electroless plating [*3*]. It is most often used to produce a metal coating on a surface that does not conduct electricity.

Smooth metal surfaces make good mirrors because they reflect most of the light that falls upon them. However, they are not equally reflective for all colors of light [*4*]. Gold, for example, is more reflective of yellow light than of blue light. Therefore, it appears yellow. Copper is more reflective of red than of other colors. Therefore, it has a reddish hue. Aluminum and silver reflect almost all the light that strikes them. Therefore, they do not appear colored; they look "silvery." However, silver is slightly more reflective of red than it is of blue. Aluminum reflects blue slightly more than red [*4*]. Therefore, light reflected by silver has a slightly pink tinge compared with that reflected by aluminum. The slight pink tinge of a silver mirror gives a pleasing effect to human skin. For this reason, cosmetic mirrors are usually made from silver, although aluminum mirrors would be less expensive. Optical mirrors, such as those in telescopes and cameras, are usually made from aluminum, because it is less expensive, and it does not tarnish as does silver.

The silver coating on the interior of the flask may be protected from oxidation and mechanical stress by giving it a protective coating. Either clear varnish or paint can be used to coat it. Paint will hide the back surface of the metal, but clear varnish will allow it to be seen. (Black paint, by blocking all light from the interior of the flask, will make the mirror appear more uniform.) In either case, the coating may be applied by pouring about 50 mL of water-based paint or varnish into the flask, turning the flask to coat its entire interior surface, pouring out the excess, and allowing the coating to dry.

REFERENCES

1. L. Bretherick, *Handbook of Reactive Chemical Hazards,* CRC Press: Cleveland, Ohio (1975).
2. *Manual of Hazardous Chemical Reactions,* National Fire Protection Association: New York (1971).
3. J. McDermott, *Electroless Plating,* Noyes Data Corporation: Park Ridge, New Jersey (1972).
4. L. Holland, *Vacuum Deposition of Thin Films,* Chapman and Hall: London (1966).

11.29

Galvanizing: Zinc Plating

A zinc strip and an iron nail are dipped into a colorless solution, and both are connected to the terminals of a battery. After about 2 minutes, the iron nail will be coated with zinc [*1*].

MATERIALS

ca. 350 mL distilled water

12 g ammonium citrate, $(NH_4)_2HC_6H_5O_7$

7.5 g ammonium chloride, NH_4Cl

30.0 g zinc sulfate heptahydrate, $ZnSO_4 \cdot 7H_2O$

large iron nail, ca. 10 cm long

fine emery cloth, or sandpaper

400-mL beaker

stirring rod

2 30-cm wire leads, with alligator clips on both ends

6.0-volt dry cell

zinc strip, ca. 15 cm × 1 cm × 0.2 mm thick

PROCEDURE

Preparation

Clean the iron nail by scouring it with the emery cloth until it is shiny. Rinse the nail with distilled water.

Prepare the zinc-plating solution by combining 12 g $(NH_4)_2HC_6H_5O_7$, 7.5 g NH_4Cl, 30.0 g $ZnSO_4 \cdot 7H_2O$, and 300 mL of distilled water in a 400-mL beaker, and stir the mixture until the solids dissolve.

Presentation

Use one of the wire leads to connect the zinc strip to the positive terminal of the 6.0-volt dry cell. Place the zinc strip in the beaker of solution. With the other lead, connect the iron nail to the negative terminal of the battery. Immerse the iron nail in the solution, taking care to assure that it does not touch the zinc strip. Maintain a separation of about 4 cm between the zinc and the iron.

After 2 minutes remove the nail from the solution, and disconnect it from the battery. Rinse the nail with water and dry it. The portion of the nail that was immersed in the solution will be coated with zinc.

HAZARDS

Zinc sulfate can irritate the eyes and mucous membranes.
Ammonium chloride is toxic when taken internally.

DISPOSAL

The zinc-plating solution should be stored in a sealed glass bottle for repeated presentations of the demonstration. Alternatively, it may be flushed down the drain with water.

DISCUSSION

This demonstration shows the process of galvanizing, that is, coating iron with zinc. Iron is galvanized to inhibit its rusting. Approximately 35% of the world consumption of zinc is used for the commercial galvanization of steel (an alloy of iron).

Iron is galvanized commercially by either of two processes: electrolysis or dipping. In the dipping process, the iron object is dipped into molten zinc. The zinc forms a coating on the iron, and the coating crystallizes when the zinc cools. This crystallization produces the "spangle" commonly seen on the surface of galvanized steel. When iron is galvanized by dipping, the product actually has three layers: the iron layer, the zinc layer, and a layer of zinc-iron alloy between them. When iron is galvanized by electrolysis (the less common practice), the zinc coating is uniform. The coating may be either dull or bright, but it does not show the crystal patterns of dipped galvanized iron.

During galvanization in this demonstration, zinc ions are deposited as zinc metal on the iron cathode. At the zinc anode, zinc metal dissolves to form zinc ions.

anode: $Zn(s) \longrightarrow Zn^{2+}(aq) + 2\ e^-$

cathode: $Zn^{2+}(aq) + 2\ e^- \longrightarrow Zn(s)$

Because the zinc ions removed at the cathode are replaced at the anode, the composition of the solution does not change. Galvanization can proceed as long as the zinc anode remains.

In the plating solution used for this demonstration, the ammonium chloride is added to the bath to improve solution conductivity. Without this added electrolyte, the resistance of the solution would require a significantly greater voltage to drive a current high enough to produce a noticeable zinc deposit within several minutes. The ammonium citrate serves as a buffer to maintain the pH of the solution at about 4.

A variety of zinc-plating baths are used commercially [2]. Cyanide baths and hydroxide baths are used for slow electrolysis, which produces a uniformly thick coating. Such uniformity is required on objects that must fit together with close tolerances, such as nuts and bolts. Acid baths are used in faster processes, where the uniformity of the coating is not critical, such as, screens and fencing. Other baths, such as neutral chloride, acid chloride, and borate baths, are used for specialized applications.

Iron is galvanized to inhibit rusting. When iron rusts, it reacts with oxygen of the air to form a hydrated iron(III) oxide with a composition represented approximately by the formula $Fe_2O_3{\cdot}H_2O$.

$$4\ Fe(s) + 3\ O_2(g) + 2\ H_2O(l) \longrightarrow 2\ Fe_2O_3{\cdot}H_2O(s) \quad (1)$$

Rusting is an electrochemical process. Where water, air, and iron come in contact, a voltaic cell develops. Oxygen from the air dissolves in the water. At the surface of the iron, this oxygen is reduced to hydroxide ions.

$$O_2(aq) + 2\ H_2O(l) + 4\ e^- \longrightarrow 4\ OH^-(aq) \quad (2)$$

The electrons for this reduction are provided by the iron, which is oxidized to iron(II) ions that dissolve in the water.

$$Fe(s) \longrightarrow Fe^{2+}(aq) + 2\ e^- \quad (3)$$

The Fe(II) ions combine with the hydroxide ions to form a precipitate of iron(II) hydroxide.

$$Fe^{2+}(aq) + 2\ OH^-(aq) \longrightarrow Fe(OH)_2(s) \quad (4)$$

Dissolved oxygen quickly oxidizes this precipitate to rust.

$$4\ Fe(OH)_2(aq) + O_2(aq) \longrightarrow 2\ Fe_2O_3{\cdot}H_2O(s) + 2\ H_2O(l) \quad (5)$$

This rusting process is inhibited if the iron is in electrical contact with a more active metal, such as zinc. The more active metal oxidizes in place of the iron to provide electrons to the oxygen; that is, the reaction of equation 3 is replaced by

$$Zn(s) \longrightarrow Zn^{2+}(aq) + 2\ e^- \quad (6)$$

The more active metal corrodes before the iron, protecting the iron from rusting. This kind of rust protection is called sacrificial protection, because another metal is "sacrificed" to corrosion in place of the iron.

REFERENCES

1. A. Joseph, P. F. Brandwein, E. Morholt, H. Pollack, and J. F. Castka, *A Sourcebook for the Physical Sciences,* p. 261, Harcourt, Brace, and World: New York (1961).
2. L. D. McGraw, "Acid Zinc," in *Modern Electroplating,* 3d ed., p. 453, John Wiley and Sons: New York (1974).

11.30

Electrodeposition of Metallic Sodium Through Glass

The tip of a light bulb is immersed in molten sodium nitrate in an iron crucible. The lamp and crucible are connected to a 120-volt ac outlet. Within several minutes, a mirror has formed on the inside of the bulb.

MATERIALS

400 mL distilled water (optional)

1 mL phenolphthalein indicator solution (optional) (To prepare 100 mL of solution, dissolve 0.05 g of phenolphthalein (3,3-bis(p-hydroxyphenyl)phthalide) in 50 mL of 95% ethanol, and dilute the resulting solution to 100 mL with distilled water.)

1 drop 0.10M sulfuric acid, H_2SO_4 (optional) (To prepare 1 liter of solution, slowly pour 5.5 mL of concentrated [18M] H_2SO_4 into 600 mL of distilled water and dilute the resulting solution to 1.0 liter.)

ca. 65 g sodium nitrate, $NaNO_3$

compressed air tap

drill, with ⅜-inch bit

block of wood, 8 cm × 20 cm × ca 1.5 cm thick

⅜-inch dowel, ca. 20 cm long

wood glue

ceramic incandescent-lamp base

screwdriver

2 screws to attach lamp base to wood block

3-conductor ac power cord, with 3-prong plug

wire strippers

30 cm copper wire, 18 gauge, with red insulation

2 large alligator clips

Bunsen burner

ring stand, with ring

wire gauze, 12 cm square

50-mL iron crucible

clamp and clamp holder for crucible

clamp holder for lamp holder
110-volt tubular unfrosted light bulb, 25- or 40-watt (e.g., aquarium lamp)
ca. 30 cm adhesive tape
variable autotransformer (e.g., Variac), 3-wire
2 400-mL beakers (optional)
2 droppers (optional)
60 cm rubber tubing to fit compressed air tap
hot plate (optional)
2 sheets of typing paper (optional)
hammer (optional)

PROCEDURE

Preparation

Assemble the lamp holder as shown in Figure 1. Drill two ⅜-inch holes through the large face of the wooden block, about 2.5 cm from the edges. Drill a ⅜-inch hole about 4 cm deep into the small face of the block nearest the edge containing the first pair of holes, but so as not to penetrate either existing hole. Glue the dowel in the hole with wood glue. Screw the ceramic lamp base to the wooden block near its center. Separate the three wires in the free end of the 3-conductor ac cord, freeing about 15 cm each of the green, white, and black insulated wires. Strip about 2 cm of the insulation from the end of each of the freed wires. From the side opposite the ceramic lamp base, thread the black wire through one of the holes in the block. Attach the end of the black wire to the nearest terminal on the lamp base. Thread the white wire through the other hole and attach it to the other terminal. Strip about 2 cm of the insulation from one end of the 30-cm piece of red wire. Thread it through the hole along with the white wire and attach it to the same terminal as the white wire. Strip about 2 cm of the insulation from the free end of the red wire. Attach one of the alligator clips to this wire and another clip to the free end of the green wire.

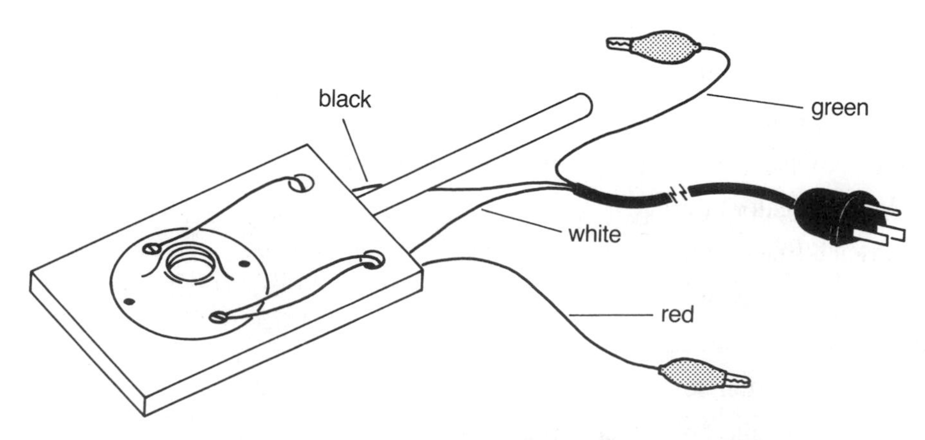

Figure 1. Lamp holder for electrolysis of sodium through glass.

Set the Bunsen burner on the base of the metal stand and clamp a ring to the stand about 7 cm above the top of the burner (Figure 2). Place the wire gauze on the ring. Clamp the iron crucible to the stand so it is centered over the gauze. Screw the light bulb into the base on the lamp holder. Use a clamp holder to attach the lamp holder to the top of the stand with the bulb pointing down. Clip both alligator clips (the one on the green wire and the one on the red wire) to the stand. Tape the slack portion of the wires to the top of the wooden block to protect them from the heat of the Bunsen burner.

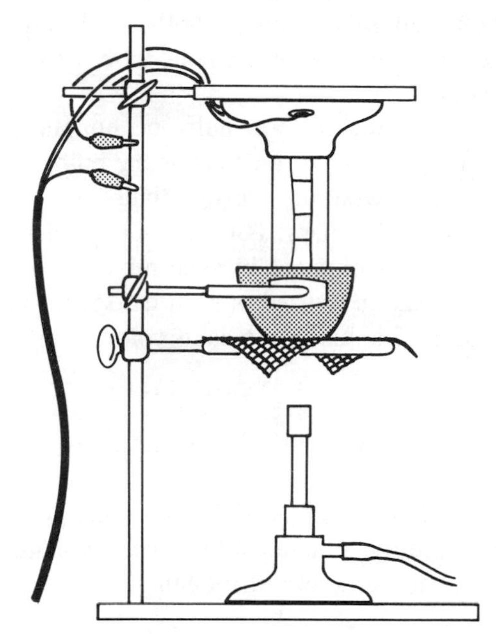

Figure 2. Lamp inverted in crucible for electrolysis of sodium through glass.

Make sure the autotransformer is turned off and its dial set to 0. Plug the line cord from the lamp holder into the autotransformer. Test the circuit by turning on the autotransformer and slowly increasing the voltage. The lamp will light if the apparatus has been assembled properly. (Note: This circuit will not work with an electrical outlet protected by a ground-fault detector.) Turn the autotransformer dial back to 0, and turn off the autotransformer.

For the optional portion of the demonstration, pour 200 mL of distilled water into each of the two 400-mL beakers. Add several drops of phenolphthalein to the water in each beaker. (If the solutions should appear pink, add a drop of 0.1M H_2SO_4 to make them colorless.)

Presentation

Pour about 65 g of $NaNO_3$ into the iron crucible. Light the burner and heat the crucible strongly until the salt is completely molten. (**Caution!** If the crucible contains a solid mass of $NaNO_3$ from a previous trial, molten $NaNO_3$ can erupt from cracks in the solid surface of the salt. Molten $NaNO_3$ is very hot [over 300°C] and extremely caustic. To avoid eruptions of molten $NaNO_3$, melt the surface of the $NaNO_3$ first by heating the side of the crucible before placing the burner beneath it.) After the $NaNO_3$ has melted, lower the flame so that it is just hot enough to keep the salt liquefied. Lower the lamp holder until the tip of the bulb dips into the molten salt and is slightly above the bottom of the iron crucible. Turn on the autotransformer and gradually increase the voltage to about the 85% level. The bulb will glow. (Bubbles will also appear at the sides and bottom of the crucible and rise through the molten salt.) Over the course of several minutes, the interior of the light bulb near its metal base will begin to darken and eventually appear mirrored. Once the mirroring is readily apparent, direct a cool stream of air from the rubber tubing connected to the compressed air tap onto a portion of the bulb which is not mirrored. That area will soon become mirrored. After about 10 minutes, turn off the Bunsen burner and the autotransformer. Raise the lamp holder to lift the bulb out of the molten salt. Allow the bulb and the crucible of $NaNO_3$ to cool. After the bulb has cooled, unscrew it from the lamp base, rinse it with water, dry it, and allow the audience to examine it.

As an optional step, warm the mirrored bulb on a hot plate. The mirrored area will move from the warmed area to a cooler area inside the bulb.

In an optional procedure, wrap the mirrored bulb in a sheet of typing paper and gently tap the paper to break the glass. Pour the shards of the broken bulb into the beaker of water containing phenolphthalein. The mixture will immediately turn red-purple. For comparison, break an unused bulb in the same manner and pour its fragments into the other beaker. The mixture remains colorless.

HAZARDS

Molten sodium nitrate is very hot (over 300°C) and a powerful oxidizing agent. It can cause severe burns. Care must be taken to avoid any contact with molten sodium nitrate. Keep water out of the crucible of molten sodium nitrate. Adding water can cause spattering.

To avoid the possibility of electric shock, do not touch bare metal parts of the assembly while the autotransformer is turned on.

DISPOSAL

The solidified sodium nitrate in the crucible can be left in the crucible for use in repeated presentations of the demonstration. However, see the caution in the Procedure section about reheating the solidified sodium nitrate. Alternatively, the crucible may be immersed in a beaker of water to dissolve the sodium nitrate, and the solution flushed down the drain with water.

The remains of the light bulb should be discarded in a solid-waste receptacle. The solutions in the beakers should be flushed down the drain with water.

DISCUSSION

This demonstration shows the result of sodium ion migration from a molten salt through the glass of a heated, energized light bulb. Sodium ions are reduced to sodium atoms, which deposit as a film of sodium metal inside the bulb. This film produces a mirror on the bulb.

When a potential is applied between the filament and the crucible, electrons ejected from the hot filament of the bulb are attracted to the positive iron crucible and collide with sodium ions on the inside wall of the glass. When the sodium ions on the inside glass surface combine with these electrons, they are reduced to metallic sodium. Because the glass is hot, the metallic sodium evaporates from the surface of the glass. It condenses on a cooler area of the inside of the bulb, near the inverted base of the bulb.

As atoms of metallic sodium evaporate from the inside surface of the glass, they leave holes in this surface of the bulb. Their departure leaves an excess of negative ions on the inside surface of the glass. This excess produces a potential difference between the inside and outside surfaces of the bulb. The potential difference causes sodium ions in the hot glass to drift inward, filling the holes at the inside surface, and eventually creating holes in the outer surface of the glass. These holes in the outer surface are filled by sodium ions from the molten sodium nitrate. The net effect is the migration of sodium ions from the molten salt to the inside of the bulb.

During the electrolysis, sodium ions are reduced to sodium atoms inside the bulb.

$$Na^+ + e^- \longrightarrow Na^0$$

A corresponding oxidation is occurring at the other electrode, the iron crucible. The evidence of this oxidation is the formation of bubbles in the molten sodium nitrate at the surface of the crucible. The identity of the gas in these bubbles and of the oxidation processes producing the gas is not known. However, there are some likely candidates. As positive sodium ions migrate toward the cathode inside the bulb, the negative nitrate ions move to the anode, the crucible. At the anode, the nitrate ions are oxidized. The oxidation of nitrate ions involves removing electrons from them. The nitrogen in nitrate ions is already fully oxidized (it is in the +5 oxidation state). Therefore, it is likely to be the oxygen atoms in nitrate ions that are oxidized. Oxidizing oxygen in nitrate ions would produce peroxides or molecular oxygen. The bubbles may contain oxygen gas. A reaction that accounts for the production of oxygen is

$$2\ NO_3^- \longrightarrow 2\ NO_2^+ + O_2(g) + 4\ e^-$$

The nitronium ion (NO_2^+) is well-known. Nitronium ions exist, for example, in pure dinitrogen pentoxide (N_2O_5), which is both a solid at room temperature and ionic, containing NO_2^+ and NO_3^-. Thus, the electrochemical production of NO_2^+ in the presence of NO_3^- could produce dinitrogen pentoxide, as well. However, dinitrogen pentoxide is not stable at the high temperature of molten sodium nitrate, and it would decompose to a mixture of oxygen and lower oxides of nitrogen. The bubbles of gas formed at the iron crucible may contain oxides of nitrogen as well as oxygen.

In this demonstration, an alternating-current source is used to cause an electrochemical reaction. When an alternating-current source is used, the potential applied to the electrodes alternates. At one time, one of the electrodes is positive and the other negative, and at another time their polarities are reversed. Generally, this produces oxidation and reduction at both electrodes, although not simultaneously. However, in this demonstration, reduction occurs only inside the bulb, and oxidation occurs at the surface of the crucible. This situation results because the glass between the electrodes acts as a rectifier; that is, it allows current to pass through in only one direction. Electric current in the form of sodium ions passes only from the outside to the inside of the bulb. With the apparatus used in this demonstration, a pulsating direct current of about 50 milliamperes flows from the crucible to the filament.

Some investigators have reported that cations of other molten salts (e.g., potassium ions and lithium ions) will enter the glass of a light bulb, but that after a short time, the bulb cracks. The glass in a light bulb is soda glass, which is 14–17% sodium oxide (Na_2O). Presumably, when ions of a size different from that of sodium ions enter the glass, they distort the arrangement of the other atoms in the glass, weakening the glass. Specially prepared potassium glass, however, will accept potassium cations. In fact, this ion-migration method has been used to prepare both extremely pure sodium and potassium metal for use in early photocells [*1*].

The method suggested in this demonstration reflects the experience of many investigators. Sodium nitrate is used because it has one of the lowest melting points (307° C) among anhydrous sodium salts. (The process will not occur if a hydrated salt is used.) The purpose of the autotransformer is to avoid the sudden surge of current that occurs when the full ac voltage is applied to the bulb. This surge frequently causes the filament to break. The use of a 110-volt alternating-current source as the sole source of power is much simpler than using an additional dc power supply in the circuit, as suggested by Joseph [*2*] and Sutton [*3*]. To minimize the possibility of electric shock, the entire ring stand is connected to the ground of the electrical outlet. This causes the neutral terminal of the outlet to be grounded as well. Such a connection creates a "ground fault," which will trigger the protective circuitry of an outlet equipped with a ground fault detector. Therefore, the apparatus in this demonstration cannot be used with a ground fault detecting outlet.

The first work on the migration of sodium ions through glass was published by Warburg in 1884 [*4*]. His method used a special bulb and dc current at over 1000 volts. In 1925, Burt offered a simplified method which uses an ordinary incandescent bulb, an iron crucible containing molten sodium nitrate salt, and 110-volt dc or ac current [*1*]. This is essentially the method used in this demonstration. A number of investigators have suggested that this demonstration can be used to verify Faraday's law [*1, 5, 6*]. Because of the difficulty in maintaining a constant current, the loss of water from the hydrated glass bulb at the high temperature of the molten salt, and the length of time necessary to accumulate a significant mass of sodium, the verification of Faraday's law by this method is far from simple. It is better done as a laboratory experiment than as a demonstration.

Several authors have reported the use of beakers or ceramic crucibles to contain the molten sodium nitrate [*1–3, 6*]. However, the iron crucible, recommended by Burt [*1*] and used in this demonstration, eliminates the possibility of breakage and the associated hazards.

REFERENCES

1. R. C. Burt, *J. Optical Soc. Amer.* 11:87 (1925).
2. A. Joseph, P. F. Brandwein, E. Morholt, H. Pollack, and J. F. Castka, *Sourcebook for the Physical Sciences,* pp. 225–26, Harcourt, Brace, and World: New York (1961).
3. R. M. Sutton, Ed., *Demonstration Experiments In Physics,* pp. 336–37, McGraw-Hill Book Co.: New York (1938).
4. G. W. Morey, *The Properties of Glass,* 2d ed., p. 465, Reinhold Publishing Corp.: New York (1954).
5. D. K. Alpern, *J. Chem. Educ.* 34:289 (1957).
6. H. N. Alyea and F. B. Dutton, Eds., *Tested Demonstrations in Chemistry,* 6th ed., p. 214, Journal of Chemical Education: Easton, Pennsylvania (1965).

11.31

Anodization of Aluminum

An oxide coating is formed on a piece of aluminum used as an anode. This oxide coating allows the aluminum to accept a dye (Procedure A). Three pieces of aluminum are shown: one anodized and sealed; one anodized and not sealed; and one only cleaned. Although all appear the same, simple tests are used to distinguish among the three pieces (Procedure B).

MATERIALS FOR PROCEDURE A

ca. 1250 mL distilled water

60 mL concentrated (18M) sulfuric acid, H_2SO_4

25 g oxalic acid, $H_2C_2O_4$

400 mL 0.2M nitric acid, HNO_3 (To prepare 1 liter of solution, pour 12.6 mL of concentrated [16M] HNO_3 into 600 mL of distilled water and dilute the resulting solution to 1.0 liter.)

400 mL 1M sodium hydroxide, NaOH (To prepare 1 liter of stock solution, dissolve 40 g of NaOH in 600 mL of distilled water and dilute the resulting, cooled solution to 1.0 liter.)

1.5 g fabric dye (e.g., Rit powder)

2 mL 1M acetic acid, $HC_2H_3O_2$ (To prepare 1 liter of solution, pour 57 mL of glacial [17.5M] $HC_2H_3O_2$ into 600 mL of distilled water, and dilute the resulting solution to 1.0 liter.)

ca. 10 mL acetone, CH_3COCH_3

sheet of aluminum, 13 cm × 25 cm × ca. 0.5 mm thick

600-mL beaker

piece of rubber-mesh sink matting, 12 cm × 25 cm

1-liter graduated beaker

plastic dish pan, half filled with ice water

crystallizing dish, 190 mm × 100 mm

ring stand, with clamp

4 400-mL beakers

2 hot plates

thermometer, $-10°C$ to $+110°C$

2 watch glasses to cover 400-mL beakers

2 aluminum strips, 5 cm × 20 cm × ca. 0.5 mm thick

cloth or soft paper towel

adjustable dc power supply capable of delivering 500 milliamperes at 15 volts, with voltmeter and clip leads

MATERIALS FOR PROCEDURE B

ca. 10 mL acetone, CH_3COCH_3

400 mL 1M sodium hydroxide, NaOH (For preparation, see Materials for Procedure A.)

400 mL 0.2M nitric acid, HNO_3 (For preparation, see Materials for Procedure A.)

ca. 1250 mL distilled water

60 mL concentrated (18M) sulfuric acid, H_2SO_4

25 g oxalic acid, $H_2C_2O_4$

1.5 g fabric dye (e.g., Rit powder)

2 mL 1M acetic acid, $HC_2H_3O_2$ (For preparation, see Materials for Procedure A.)

3 aluminum strips, 5 cm × 20 cm × ca. 0.5 mm thick

permanent marking pen (e.g., Sanford "Sharpie")

2 cloths or paper towels (one of them white)

sheet of aluminum, 13 cm × 25 cm × ca. 0.5 mm thick

3 600-mL beakers

piece of rubber-mesh sink matting, 12 cm × 25 cm

1-liter graduated beaker

plastic dish pan, half filled with ice water

crystallizing dish, 190 mm × 100 mm

ring stand, with clamp

2 400-mL beakers

2 hot plates

thermometer, $-10°C$ to $+110°C$

2 watch glasses to cover 400-mL beakers

110-volt light-bulb conductivity tester, with 2 graphite electrodes†

PROCEDURE A [*1–3*]

Preparation

Assemble the anodizing cell as follows. Roll the 13-cm × 25-cm sheet of aluminum into a cylinder 13 cm tall and just narrow enough to fit inside the 600-mL beaker (see figure). Roll the rubber mesh into a cylinder, and place it inside the aluminum cylinder in the beaker.

† A suitable conductivity tester is available from Sargent-Welch Scientific Company, 7300 North Linder Avenue, Skokie, Illinois 60077.

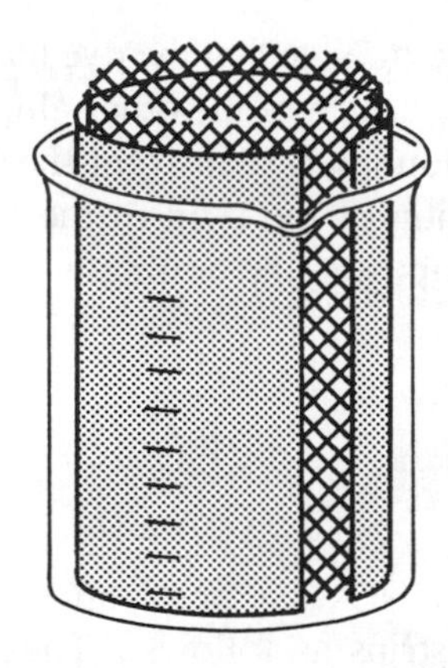

Anodization cell.

Set the graduated 1-liter beaker in the pan half filled with ice water. Pour 250 mL of distilled water into the beaker. Slowly pour 60 mL of concentrated (18M) H_2SO_4 into the beaker. Dissolve 25 g of $H_2C_2O_4$ in this solution. Dilute the resulting solution to 500 mL with distilled water. Pour this solution into the anodizing cell.

Set the anodizing cell in the center of the crystallizing dish. Fill the crystallizing dish with water, and set the dish and anodizing cell on the base of the ring stand.

Set a 400-mL beaker filled with 0.2M HNO_3 to the left of the stand. To the left of this beaker set a 400-mL beaker filled with 1M NaOH.

Fill a third 400-mL beaker with distilled water. Dissolve 1.5 g of fabric dye in the water. Add 2 mL of 1M acetic acid to the dye solution. Set the beaker of dye solution on a hot plate to the right of the stand. Adjust the hot plate to maintain the temperature of the dye solution between 60°C and 70°C. To the right of the hot plate, fill the fourth 400-mL beaker with distilled water, then set it on a second hot plate. Adjust the second hot plate to keep the distilled water at a rolling boil. Cover both the dye solution and the boiling water with watch glasses.

Presentation

Clean the bottom half of the two 5-cm × 20-cm aluminum strips. First, rub the aluminum with a soft cloth moistened with acetone. Then, immerse the ends of the strips in the 1M NaOH solution for 1 minute. After rinsing the strips with water, dip the cleaned end in the 0.2M HNO_3 for a few seconds. Rinse the strips again with water.

Place the cleaned end of one aluminum strip in the hot dye bath and leave it there.

The remaining aluminum strip will be anodized. Immerse the cleaned end of the remaining strip in the solution in the anodizing cell. Clamp the strip to the ring stand so the strip is centered in the beaker. With the power supply turned off, clip the negative lead wire to the cylinder of aluminum in the anodizing cell. Clip the positive lead to the aluminum strip. Turn on the power supply and slowly increase the voltage to about 15 volts. Keep the power supply connected for 10–30 minutes. (A longer time will produce a thicker anodized coating on the strip. The thicker coating will take more dye, producing a deeper color.) Check the voltage periodically and adjust it as necessary to keep the potential at 15 volts. After turning off the power supply, disconnect the wire lead from the aluminum strip. Avoid touching the anodized coating. Rinse the strip with water.

Remove the untreated aluminum strip from the hot dye solution, rinse it with water, and wipe it dry with a cloth. Note that the dye has had little effect on this piece of aluminum. Place the anodized surface of the other aluminum strip in the dye solution. After 5 minutes remove the aluminum strip from the beaker of dye and rinse it with water. Immerse the dyed portion of the strip in the beaker of rapidly boiling water. After 5 minutes remove the strip from the boiling water and dry it with a cloth.

PROCEDURE B

Preparation

Prepare three aluminum strips as follows. (The strips may be prepared several days in advance of the demonstration.) Using a permanent marking pen, number the three aluminum strips from 1 to 3, writing the numbers at one corner of each strip. Clean the unlabelled half of each of these three strips. First, rub the aluminum with a soft cloth moistened with acetone. Then, dip the unlabelled half of the strips in 1M NaOH solution for 1 minute. Rinse the strips with water, and dip the cleaned end in 0.2M HNO_3 for a few seconds. Rinse the strips again with water.

Assemble an anodizing cell as follows. Roll the 13-cm × 25-cm piece of aluminum into a cylinder 13 cm tall and just narrow enough to fit inside one of the 600-mL beakers (see figure shown in Procedure A). Roll the rubber mesh into a cylinder, and place it inside the aluminum cylinder in the beaker.

Set the graduated 1-liter beaker in the pan half filled with ice water. Pour 250 mL of distilled water into the beaker. Slowly pour 60 mL of concentrated (18M) H_2SO_4 into the beaker. Dissolve 25 g of $H_2C_2O_4$ in this solution. Dilute the resulting solution to 500 mL with distilled water. Pour this solution into the anodizing cell.

Set the anodizing cell in the center of the crystallizing dish. Fill the crystallizing dish with water, and set the dish and anodizing cell on the base of the ring stand.

Anodize strips 2 and 3 as follows. Immerse the cleaned end of strip 2 in the solution in the anodizing cell. Clamp the strip to the ring stand so the strip is centered in the beaker. With the power supply turned off, clip the negative lead wire to the cylinder of aluminum in the anodizing cell. Clip the positive lead to the aluminum strip. Turn on the power supply and slowly increase the voltage to about 15 volts. Keep the power supply connected for 20 minutes. Check the voltage periodically and adjust it as necessary to keep the potential at 15 volts. After turning off the power supply, disconnect the wire lead from the aluminum strip. Avoid touching the anodized coating. Rinse the strip with water. Repeat this anodizing process with strip 3.

Fill a 400-mL beaker with distilled water and set it on a hot plate. Adjust the hot plate to keep the distilled water at a rolling boil. Immerse the anodized portion of strip 3 in the boiling water for 5 minutes.

Pour 400 mL of distilled water into the other 400-mL beaker. Dissolve 1.5 g of fabric dye in the water. Add 2 mL of 1M $HC_2H_3O_2$ to the dye solution. Set the beaker of dye solution on the second hot plate, and adjust the hot plate to maintain the temperature of the dye solution between 60°C and 70°C. Cover the beaker of heated dye solution and the beaker of boiling water with watch glasses.

Presentation

Display the aluminum strips labelled "1," "2," and "3." All three appear identical. Place the three aluminum strips on a table top. Turn on the conductivity tester and touch its two carbon electrodes to the treated area on strip 1. (**Caution! Do not touch the electrodes or the aluminum strip while the tester is turned on.**) The lamp on the tester will glow. Touch the two electrodes to the treated area on strip 2. The lamp will not light. Touch the electrodes to the treated area of strip 3. Again the lamp will not glow. Touch the electrodes to an untreated area of strip 3 (near the number). The lamp will light.

Moisten a small area of a clean, white cloth with acetone and rub the treated area of strip 1. A black spot will develop on the cloth. Repeat this with the other two strips. Neither of these will turn the cloth black.

Place all three strips in the heated dye solution and leave them there for at least 5 minutes. Remove the strips from the dye, rinse them with water, and place them in the boiling water for at least 5 minutes. Remove the strips from the boiling water and dry them with a clean towel. Only strip 2 will be colored.

HAZARDS

Sulfuric acid is a strong acid and a powerful dehydrating agent; therefore it must be handled with care. Spills should be neutralized with sodium bicarbonate ($NaHCO_3$), and then wiped up.

Solid sodium hydroxide and concentrated solutions can cause severe burns to the eyes, skin, and mucous membranes. Dust from solid sodium hydroxide is very irritating to the eyes and respiratory system.

Glacial acetic acid can irritate the skin, and its vapors are irritating to the eyes and respiratory system. It should be handled only in a well-ventilated area.

Concentrated nitric acid is both a strong acid and a powerful oxidizing agent. Contact with combustible materials can cause fires. Contact with the skin can result in severe burns. The vapor irritates the respiratory system, eyes, and other mucous membranes, and therefore, concentrated nitric acid should be handled only in a well-ventilated area.

Acetone is extremely flammable and should be kept away from sparks and open flames.

Severe electric shock can result if the electrodes of the conductivity tester are touched while it is turned on.

DISPOSAL

The waste solutions should be flushed down the drain with water.

The aluminum strips should be discarded in a solid-waste receptacle.

DISCUSSION

This demonstration shows the process of anodizing aluminum. Procedure A shows the various steps in the process: the preparation of the aluminum object, the electrochemical treatment of the object, the dyeing of the treated surface, and the final heat treatment. Procedure B shows the effect of each step on the properties of the aluminum surface.

When aluminum was first used in the late 1920s for the fabrication of commercial products such as drinking glasses, railings, and knitting needles, consumers often complained about the grayish markings left on light-colored clothing and their hands. This problem was solved by anodizing the surface of the aluminum, a process for depositing an oxide coating, first discovered in 1923 [*4*].

Anodizing produces an oxide coating by making a cleaned piece of aluminum the anode in an electrolytic cell. When this is done at low temperature (−5°C to +5°C) a hard oxide coating is formed. If anodizing takes place at a higher temperature (20–30°C), the oxide coating is softer but more receptive to dyeing. As a final step, the oxide coating is sealed by surrounding the aluminum by steam or by immersing it in boiling water.

The first step in the anodization process is cleaning the aluminum surface with acetone or some other organic solvent to remove the greases and oils. An alkaline solution such as sodium hydroxide is then used to remove the thin oxide coating that forms on aluminum exposed to air. This coating is about 10 nm thick [*5*]. The alkali solution dissolves the aluminum oxide coating.

$$Al_2O_3(s) + 2\ OH^-(aq) + 3\ H_2O(l) \longrightarrow 2\ [Al(OH)_4]^-(aq)$$

The alkali cleaning also dissolves some of the aluminum metal.

$$2\ Al(s) + 2\ OH^-(aq) + 6\ H_2O(l) \longrightarrow 2\ [Al(OH)_4]^-(aq) + 3\ H_2(g)$$

If the aluminum object is made of an alloy containing copper, this alkali cleaning process will leave a black "smut" on the surface of the aluminum [*6*]. This discoloration can be removed by immersing the object in a 10–30% nitric acid bath. (The dilute nitric acid rinse suggested in this demonstration is used to neutralize any remaining sodium hydroxide solution.)

Aluminum is anodized in a sulfuric acid bath. The aluminum is the anode, and another piece of aluminum (or lead) is the cathode in an electrolysis cell. A potential of 12–18 volts is applied between the electrodes to produce a current of 1.0–1.5 amperes per dm^2 of anode surface area. For an even oxide coating on the anode, the cathode should completely surround the anode. Also, the surface area of the cathode should be at least as large as that of the anode. It is necessary to cool the anodizing solution, because the electrolysis produces heat. A good oxide coating cannot be produced when the temperature of the sulfuric acid solution goes above 25°C. However, the addition of oxalic acid to the anodizing solution permits anodizing at a temperature as high as 30°C. The anodizing process produces an oxide coating on the aluminum anode.

$$2\ Al(s) + 3\ H_2O(l) \longrightarrow 6\ e^- + Al_2O_3(s) + 6\ H^+(aq)$$

Hydrogen gas is produced at the cathode.

$$2\ H^+(aq) + 2\ e^- \longrightarrow H_2(g)$$

As the anode coating of aluminum oxide grows, the current will drop, because the coating is electrically insulating. The final coating is about 10,000 nm thick [*3*].

In practice, the anodization process produces more than 1 mole of aluminum oxide for every 6 moles of electrons passed through the solution. If the anode reaction occurs exactly as represented above, then the current efficiency is greater than 100%. Raijola and Davidson explain this by suggesting the formation of some Al^+ intermediate at the anode [*7*].

$$Al(s) \longrightarrow Al^+(aq) + e^-$$

This Al^+ is immediately oxidized by water to aluminum oxide.

$$2\ Al^+(aq) + 3\ H_2O(l) \longrightarrow Al_2O_3(s) + 2\ H_2(g) + 2\ H^+(aq)$$

The oxide coating produced by anodization is porous and has a very large surface area. When it is immersed in water, the oxide surface reacts with water molecules and develops a positive charge.

$$Al_2O_3(s) + H_2O(l) \longrightarrow Al_2O_2(OH)^+(s) + OH^-(aq)$$

This positive charge attracts anions. Because of this, the oxide can be colored with water-soluble organic dyes. These dyes contain large organic anions, which are responsible for their colors. When the dye is dissolved in water, these anions are attracted to the positively charged aluminum oxide surface and penetrate into the pores. The absorption of the dye is pH dependent, so a small amount of acetic acid is often added to the dye solution. However, sulfate ions interfere with dye absorption. Therefore, before it can be dyed, the anodized aluminum must be thoroughly rinsed to remove the sulfuric acid from the anodizing solution. The optimum temperature for dyeing is about 65°C. Temperatures above 70°C must be avoided to prevent sealing of the pores in the oxide surface. Once the pores are sealed, the dye cannot penetrate the coating.

After the dye is absorbed onto the surface of the anodized aluminum, the surface pores are sealed by heating the aluminum with steam or boiling water. Sealing the pores prevents the dye from leaching out or stains from entering the oxide coating. This sealing process forms an aluminum oxide monohydrate, which fills the openings of the pores. Sealing the surface also makes it more resistant to corrosion.

REFERENCES

1. R. C. Spooner and H. P. Godard, *J. Chem. Educ.* 25:340 (1948).
2. A. Doeltz, S. Tharaud, and W. Sheehan, *J. Chem. Educ.* 60:156 (1983).
3. R. G. Blatt, *J. Chem. Educ.* 56:268 (1979).
4. V. F. Henley, *Anodic Oxidation of Aluminium and Its Alloys,* 1st ed., Pergamon Press: New York (1982).
5. H. H. Uhlig, Ed., *Corrosion Handbook,* p. 858, John Wiley and Sons: New York (1948).
6. K. R. Van Horn, *Aluminum,* Vol. 3, *Fabrication and Finishing,* American Society for Metals: Metals Park, Ohio (1967).
7. E. Raijola and A. Davidson, *J. Amer. Chem. Soc.* 78:556 (1956).

11.32

The Mercury Beating Heart

A pool of mercury resting at the center of a watch glass and covered with a yellow solution is touched with a wire. The mercury pool begins to change shape, rhythmically pulsing between triangular heart forms [*1–3*].

MATERIALS

ca. 2 mL mercury

ca. 10 mL 6M sulfuric acid, H_2SO_4 (To prepare 100 mL of solution, slowly pour 33 mL of concentrated [18M] H_2SO_4 into 50 mL of distilled water and dilute the resulting cooled solution to 100 mL.)

1 mL 0.2M potassium chromate, K_2CrO_4 (To prepare 100 mL of solution, dissolve 3.8 g of K_2CrO_4 in 60 mL of distilled water and dilute the resulting solution to 100 mL.)

ca. 10 mL concentrated (18M) sulfuric acid, H_2SO_4, in dropper bottle

plastic tray (e.g., a cafeteria tray)

125-mm watch glass

ca. 20-cm iron wire, 16–24 gauge

balance weight, 50 or 100 g

PROCEDURE

Preparation and Presentation

Place the watch glass in the center of the plastic tray (to catch any mercury spills). Pour enough mercury into the 125-mm watch glass to make a pool about 2 cm in diameter. Add enough 6M H_2SO_4 to cover the mercury. Pour 1 mL of 0.2M K_2CrO_4 over the mercury. A film will appear on the surface of the mercury, and the pool will flatten slightly.

Wrap the iron wire around the neck of the weight, leaving about 10 cm of wire extending horizontally from the weight. Set the weight on the tray next to the watch glass, and adjust the wire so its tip just touches the edge of the mercury pool in the watch glass. Adjust the distance between the tip of the iron wire and the mercury pool, until the mercury begins to pulsate. Add drops of 18M H_2SO_4 on top of the mercury until the pulsing motion is maximized.

As needed, adjust the position of the wire to maintain the rhythmic motion of the mercury. The pool of mercury can continue to beat for over an hour if the iron wire is secured to the weight tightly enough so that its tip is held firmly in place at the edge of the mercury pool.

HAZARDS

Because sulfuric acid is a strong acid and a powerful dehydrating agent, it must be handled with great care. Spills should be neutralized with an appropriate agent, such as sodium bicarbonate ($NaHCO_3$), and then wiped up.

Potassium chromate is caustic to the skin and mucous membranes. Chromates are suspected carcinogens.

Mercury is extremely toxic by skin absorption and by inhalation of the vapor. Continued exposure to the vapor may result in severe nervous disturbances, insomnia, and depression. Continued skin contact can also cause kidney damage. Mercury should be handled only in well-ventilated areas. Mercury spills should be cleaned up immediately by using a capillary attached to a trap and an aspirator (see figure). Small amounts of mercury in inaccessible places should be treated with zinc dust to form a nonvolatile amalgam.

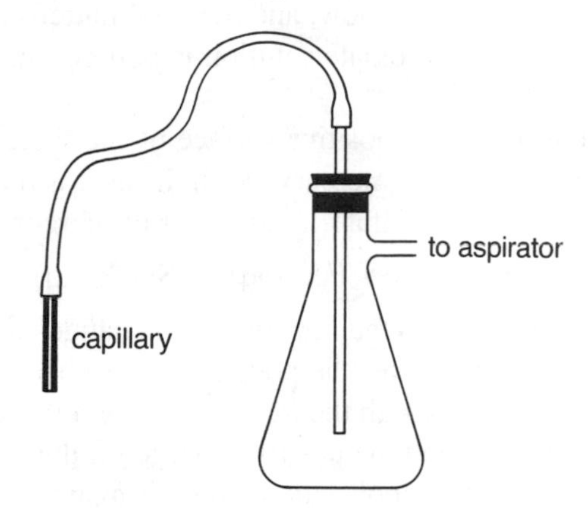

Device for recovering spilled mercury.

DISPOSAL

The mercury should be cleaned by pouring it into a 250-mL beaker filled with distilled water, decanting the water, and pouring the mercury through a pinhole in the center of a piece of filter paper suspended in a funnel. The mercury should be returned to a sealed container for storage.

To dispose of the potassium chromate solution, add sodium bisulfite ($NaHSO_3$) to it until the solution is green. Then add sodium sulfide ($Na_2S \cdot 9H_2O$) to the green solution until precipitation stops. Filter the solid chromium(III) sulfide from the liquid. Dispose of the solid in a landfill specified for heavy metal wastes. (Consult local regulations regarding the disposal of hazardous wastes.) Flush the liquid down the drain with water.

DISCUSSION

This demonstration shows several unique properties of mercury as a liquid metal. Because mercury is a liquid, it has a property called surface tension. Surface tension is the force that causes a liquid to form drops. When the surface tension is high, the liquid forms nearly spherical drops. When the surface tension is low, the drops are flat. The surface tension of mercury varies with the electric charge on its drops. When the charge is positive, the surface tension is low, and the drop is flattened. When the charge is negative, the surface tension is high, and the drops are spherical. This is called the electrocapillary effect [*4*].

Mercury can behave as a reducing agent when it is combined with an oxidizing agent. This is what happens when mercury is combined with chromate ions in an acidic solution. The chromate ions oxidize mercury to mercury(II) ions. In the sulfuric acid used in the demonstration, these mercury ions form an insoluble film of mercury(II) sulfate on the surface of the mercury [*5*].

$$2\,CrO_4^{2-}(aq) + 3\,Hg(l) + 16\,H^+(aq) + 3\,SO_4^{2-}(aq) \longrightarrow 2\,Cr^{3+}(aq) + 3\,HgSO_4(s) + 8\,H_2O(l)$$

The effect of this process is to carry electrons (negative charge) from the mercury pool into the solution. This leaves a positive charge on the mercury pool. The positive charge causes the surface tension to be low, and the pool flattens and spreads. This can be observed when the potassium chromate solution is poured onto the mercury pool in sulfuric acid.

When the iron wire touches the positive surface of the mercury drop, an electron transfer occurs from the iron to the mercury pool. Iron is a more active metal than mercury, so it reacts with mercury sulfate, reducing it to mercury [*6*].

$$Fe(s) + HgSO_4(s) \longrightarrow Fe^{2+}(aq) + SO_4^{2-}(aq) + Hg(l)$$

The transfer of electrons from iron to the mercury pool changes the charge on the pool. As the charge becomes more negative, the pool contracts, becoming more spherical. When it does this, it breaks contact with the iron wire. When this contact is broken, the mercury sulfate film regenerates and the positive charge on the pool gradually returns. As the positive charge builds, the pool flattens until it again makes contact with the iron wire.

The overall process in the mercury beating heart can be thought of as the mercury-catalyzed oxidation of iron to iron sulfate. In the first reaction, mercury metal is oxidized to mercury sulfate. In the second reaction, this mercury sulfate is reduced by iron back to mercury, producing iron sulfate. The overall reaction is

$$2\,CrO_4^{2-}(aq) + 3\,Fe(s) + 16\,H^+(aq) \longrightarrow 2\,Cr^{3+}(aq) + 3\,Fe^{2+}(s) + 8\,H_2O(l)$$

The color of the chromate solution fades more rapidly when the mercury heart is beating than it does when the iron and mercury are kept separated. Some of the energy released by this reaction drives the motion of the mercury beating heart.

REFERENCES

1. R. H. Adams, *J. Chem. Educ.* 10:512 (1933).
2. J. A. Campbell, *J. Chem. Educ.* 34:A105 (1957).
3. J. A. Campbell, *J. Chem. Educ.* 34:362 (1957).
4. J. O'M. Bockris and A. K. N. Reddy, *Modern Electrochemistry,* Vol. 2, pp. 688–92, Plenum Press: New York (1970).
5. J. Keizer, P. A. Rock, and S. W. Lin, *J. Amer. Chem. Soc.* 101:5637 (1979).
6. S. W. Lin, J. Keizer, P. A. Rock, and H. Stenschke, *Proc. Natl. Acad. Sci.* 71:4477 (1974).

11.33

Copper to Silver to Gold

Granular metal and a colorless solution are placed in a beaker and heated until the liquid boils. Two copper coins are placed on top of the metal granules. After a short time the copper coins become silvery. One of the silvery coins is heated in a burner flame, and the coin turns golden. (Procedure A uses zinc; Procedure B uses tin.)

MATERIALS FOR PROCEDURE A

25 g granular zinc, 40 mesh or coarser (Higher mesh number corresponds to finer grains. **Do not use powdered zinc; the reaction may be too vigorous to control.**)

25 mL 3M sodium hydroxide solution, NaOH (To prepare 100 mL of solution, dissolve 12 grams of NaOH in 100 mL of distilled water.)

250-mL beaker

Bunsen burner

ring stand

3 copper coins (U.S. cents prior to 1983 are suitable; those after 1983 are copper-coated zinc and may melt when heated in a Bunsen burner flame.)

steel wool, 000 grade

150-mL beaker

wire gauze to fit ring stand

tongs

paper towel

MATERIALS FOR PROCEDURE B

100 mL 1M hydrochloric acid, HCl (To prepare 100 mL of solution, pour 10 mL of concentrated [12M] HCl into 90 mL of distilled water.)

25 g granular tin, about 20 mesh

25 mL 3M sodium hydroxide solution, NaOH (For preparation, see Materials for Procedure A.)

250-mL beaker

Bunsen burner

ring stand

3 copper coins (U.S. cents prior to 1983 are suitable; those after 1983 are copper-coated zinc and may melt when heated in a Bunsen burner flame.)

steel wool, 000 grade
2 150-mL beakers
wire gauze to fit ring stand
tongs
paper towel

PROCEDURE A

Preparation

Put about 150 mL of tap water in the 250-mL beaker. Set the Bunsen burner under the ring stand and adjust the position of the ring so it is about 5 cm above the top of the burner. Clean the three cents by buffing them with fine steel wool until they are shiny.

Presentation

Pour 25 g of granular zinc into the 150-mL beaker. Pour 25 mL of 3M NaOH onto the zinc. Put the wire gauze on the ring stand, and set the beaker on the gauze. Light the bunsen burner, and put it under the beaker.

When the solution in the beaker begins to boil, use the tongs to place two of the copper cents on top of the zinc. Leave them there until they turn silver colored, about 45 seconds. Use the tongs to remove the cents from the beaker, and dip them into the beaker of water to rinse off the NaOH solution. Dry both cents. They both have a silvery appearance.

With the tongs, hold one of the silvered cents in the blue part of the Bunsen burner flame, and turn the coin to heat it evenly until it changes to a golden color. Do not overheat the coin, or the golden color will disappear. Dip the hot coin into the beaker of water to cool it. Dry the coin. It will retain its golden appearance.

PROCEDURE B

Preparation

Put about 150 mL of tap water in the 250-mL beaker. Set the Bunsen burner under the ring stand, and adjust the position of the ring so it is about 5 cm above the top of the burner. Clean the three cents by buffing them with fine steel wool until they are shiny. Pour 100 mL of 1M HCl into one of the 150-mL beakers.

Presentation

Pour 25 g of granular tin into the empty 150-mL beaker. Pour 25 mL of 3M NaOH onto the tin. Put the wire gauze on the ring stand, and set the beaker of tin and NaOH on the gauze. Light the Bunsen burner, and put it under the beaker.

When the solution in the beaker begins to boil, use the tongs to place two of the copper cents on top of the tin. Leave them there until they turn gray, in 3–5 minutes. During this time, continue to heat the mixture to keep the liquid boiling gently. Use the tongs to remove the gray cents from the beaker, and dip them into the beaker of water to rinse off the NaOH solution. Dry both coins. They both have turned gray.

With the tongs, hold one of the gray cents in the blue part of the Bunsen burner flame and turn the coin to heat it evenly until it just turns golden and begins to tarnish. Do not overheat the coin, or the golden color will disappear. Dip the hot coin into the beaker of water to cool it. Then dip the coin momentarily in the 1M HCl to remove the tarnish. Rinse the coin in the water and dry it. It will have a golden appearance.

DISPOSAL

Allow the sodium hydroxide solution to cool to room temperature and decant it from the zinc or tin. The sodium hydroxide solution may be saved for repeated presentations of the demonstration, or flushed down the drain with water.

The granular zinc or tin should be thoroughly rinsed with distilled water and allowed to dry in the air. It can be saved for repeated presentations of the demonstration.

Flush the 1M hydrochloric acid down the drain with water.

HAZARDS

The hot sodium hydroxide solution is extremely caustic to skin and can damage clothing. Do not heat this solution so strongly that it boils vigorously and spatters.

DISCUSSION

Although this demonstration appears to show the transmutation of copper to silver to gold, this is not the answer to the alchemists' dream. The silvery coating on the copper coins is zinc in Procedure A and tin in Procedure B. The golden coatings formed when the silvery coins are heated are brass in Procedure A and bronze in Procedure B. When both procedures are presented, the zinc-, tin-, brass-, and bronze-coated coins can be compared. Their colors are discernibly different. The tin coating is slightly more yellow than the zinc coating, and the bronze coating is redder than the brass coating.

When the zinc-coated and tin-coated copper coins are heated in the burner flame, the coatings merge with the underlying copper, forming alloys. The melting point of zinc is 420°C and that of tin is 232°C [*1*]. The Bunsen burner flame is hot enough to melt the zinc and tin quickly (see Figure 1). When zinc or tin melts on the surface of the copper coin, it dissolves some of the underlying copper, forming an alloy. The copper cent itself, which is actually an alloy containing 95% copper and 5% zinc having a melting point of about 1050°C, must be heated much longer to melt it in the flame. Cents minted after 1983 are made of a pure zinc core with a thin copper coating. These cents melt easily in a Bunsen burner flame, making it difficult to use them in this demonstration. When a hot copper coin is removed from the flame it will turn black as a result of the formation of a coating of copper(II) oxide. If the blackened coin is held in the blue cone of the Bunsen burner flame, the portion of the coin in the cone will become coppery again. This occurs because the inner cone of the flame contains gases, such as hot methane, that can reduce copper oxide.

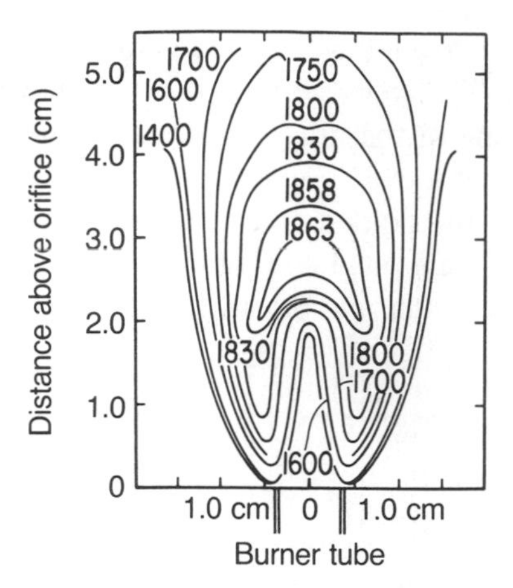

Figure 1. Temperatures in a Bunsen burner flame, in degrees Celsius. (Source: Figure 32 [2].)

The alloy of zinc and copper is brass. The color of the alloy varies with its composition, as the chart below shows [*3*].

Mole ratio Cu:Zn	Color of alloy
10:1 to 7:1	reddish yellow
6:1 to 4:1	yellowish red
3:3	pale yellow
2:1 to 1:2	deep yellow
8:17 to 1:2	silver-white
8:19 to 8:23	light gray
1:3 to 1:5	dark gray

The alloy generally called brass has a composition of 60–82% copper and 18–40% zinc, and is bright yellow. Brass is valued for its color, similar to that of gold. Like gold, brass is a rather soft metal.

Bronze is a reddish yellow metal containing 70–95% copper and 5–30% tin. Commercial bronze frequently contains small fractions of other materials, including zinc and phosphorus. Bronze looks less like gold than does brass, but it can be much harder and more resistant to corrosion than brass. Therefore, it is frequently used for decorative metal work out of doors. The color of bronze depends on its composition, as the chart below indicates [*4*].

Mass percent Cu	Color of bronze
82.8–84.3	reddish yellow
72.8–81.1	yellowish red
82.8–76.3	pale red
68.2	ash gray
61.7	dark gray
51.75	grayish white
34.9	white

An extensive discussion of the properties and uses of both brass and bronze can be found in references 2 and 3.

The coins produced in this demonstration can be conveniently saved in commer-

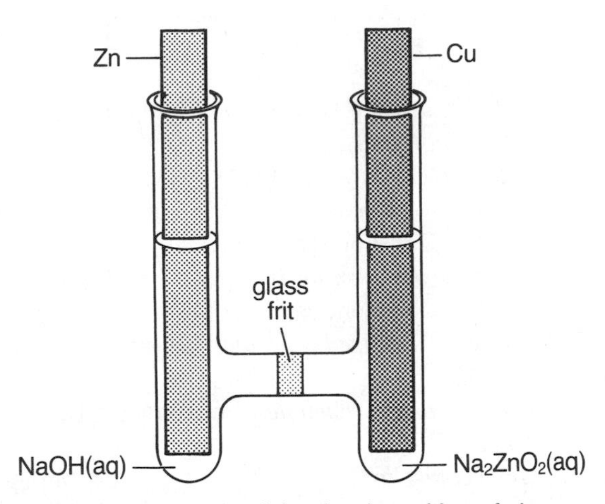

Figure 2. Electrochemical cell for the deposition of zinc on copper.

cial coin holders available at coin shops. The tin-, brass-, and bronze-coated coins will keep their appearance indefinitely. The silvery appearance of the zinc-coated coin will fade to copper color over a period of several weeks.

When zinc and tin are placed in a strongly basic solution, they slowly dissolve. In aqueous sodium hydroxide, zinc is oxidized to the zincate ion and hydroxide ions are reduced to hydrogen.

$$\mathrm{Zn(s) + 2\,OH^-(aq) \longrightarrow ZnO_2{}^{2-}(aq) + H_2(g)}$$

In aqueous solution, the zincate ions are complexed to hydroxide ions and water in the form of $[\mathrm{Zn(OH)_3(H_2O)}]^-$ [*5*]. Tin dissolves in strong bases in a similar fashion, forming the stannate ions and hydrogen gas [*6*].

$$\mathrm{Sn(s) + 2\,OH^-(aq) + 4\,H_2O(l) \longrightarrow [Sn(OH)_6]^{2-}(aq) + 2\,H_2(g)}$$

The bubbles of hydrogen gas can be observed in the solutions before they are heated. Heating the solutions speeds the reactions considerably.

In the demonstration, the zinc and tin coatings will form only if the copper coin is in contact with the zinc or tin metal. This suggests that the process which forms the coating is an electrochemical process. The exact nature of the electrochemical process is not known. The potentials of zinc and tin in 1.0M sodium hydroxide are reported as −1.24 volts and −0.94 volt, respectively [*7, 8*]. Both of these potentials are consistent with the metals' spontaneous reactions with the solution and with the reduction potential of water in basic solution, namely −0.828 volt [*9*]. We have found that zinc can be deposited on a copper electrode in an electrochemical cell containing materials similar to those in this demonstration (see Figure 2). This cell was constructed from two test tubes connected by a tube having a fritted glass plug at the center. One tube was filled with a 3M sodium hydroxide solution and the other with a sodium zincate solution. A zinc electrode was immersed in the sodium hydroxide solution, and a copper electrode was immersed in the sodium zincate solution. When the two electrodes were connected with a wire, a coating of zinc deposited on the copper electrode. In this cell, zincate ions are reduced to zinc at the cathode, and zinc is oxidized to zincate ions at the anode. These two electrode reactions are simply the reverse of each other, and this cell is a concentration cell, driven by the difference in concentration of zincate ions in the two half cells. Perhaps a similar driving force produces the zinc and tin coatings in this demonstration.

REFERENCES

1. R. C. Weast, Ed., *CRC Handbook of Chemistry and Physics,* 66th ed., pp. B-153, B-158, CRC Press: Boca Raton, Florida (1985).
2. B. Lewis and G. Van Elbe, *J. Chem. Phys.* 11:75 (1943).
3. J. W. Mellor, *A Comprehensive Treatise on Inorganic and Theoretical Chemistry,* Vol. 4, p. 673, Longmans, Green, and Co.: London (1923).
4. J. W. Mellor, *A Comprehensive Treatise on Inorganic and Theoretical Chemistry,* Vol. 8, p. 351, Longmans, Green, and Co.: London (1927).
5. F. A. Cotton and G. Wilkinson, *Advanced Inorganic Chemistry,* 5th ed., p. 605, John Wiley and Sons: New York (1988).
6. *Gmelins Handbuch der anorganische Chemie,* No. 46, Part B, p. 364, Verlag Chemie: Weinheim, Germany (1971).
7. *Gmelins Handbuch der anorganische Chemie,* No. 32, p. 412, Verlag Chemie: Weinheim, Germany (1956).
8. *Gmelins Handbuch der anorganische Chemie,* No. 46, Part B, p. 263, Verlag Chemie: Weinheim, Germany (1971).
9. R. C. Weast, Ed., *CRC Handbook of Chemistry and Physics,* 66th ed., p. D-152, CRC Press: Boca Raton, Florida (1985).

Index to Volumes 1–4

Numbers in bold type are volume numbers; numbers in roman type are page numbers.

Demonstrations in Volume 1

1 THERMOCHEMISTRY

2 CHEMILUMINESCENCE

3 POLYMERS

4 COLOR AND EQUILIBRIA OF METAL ION PRECIPITATES AND COMPLEXES

Demonstrations in Volume 2

5 PHYSICAL BEHAVIOR OF GASES

6 CHEMICAL BEHAVIOR OF GASES

7 OSCILLATING CHEMICAL REACTIONS

Demonstrations in Volume 3

8 ACIDS AND BASES

9 LIQUIDS, SOLUTIONS, AND COLLOIDS

Demonstrations in Future Volumes

The following demonstration topics will be included in future volumes:

atomic structure
chemical periodicity
chromatography
corridor demonstrations and exhibits
cryogenics
fluorescence and phosphorescence
kinetics and catalysis
lasers in chemistry
organic chemistry
photochemistry
radioactivity
solid state and materials chemistry
spectroscopy and color

DESIGNED BY HERBERT JOHNSON
COMPOSED BY G&S TYPESETTERS, INC., AUSTIN, TEXAS
MANUFACTURED BY KINGSPORT PRESS, KINGSPORT, TENNESSEE
TEXT IS SET IN TIMES ROMAN, DISPLAY LINES IN HELVETICA

Library of Congress Cataloging-in-Publication Data
(Revised for volume 4)
Shakhashiri, Bassam Z.
Chemical demonstrations.
Includes bibliographies and index.
I. Chemistry—Experiments. I. Title.
QD43.S5 1983 540'.7'8 81-70016
ISBN 0-299-08890-1 (v.1)
ISBN 0-299-10130-4 (v.2)
ISBN 0-299-11950-5 (v.3)
ISBN 0-299-12860-1 (v.4)